W9-CSP-133

THEORY OF FULLY IONIZED PLASMAS

THEORY OF FULLY IONIZED PLASMAS

Günter Ecker

PROFESSOR OF PHYSICS
INSTITUT FÜR THEORETISCHE PHYSIK
RUHR-UNIVERSITÄT
BOCHUM, GERMANY

1972

ACADEMIC PRESS New York and London

COPYRIGHT © 1972, BY ACADEMIC PRESS, INC.
ALL RIGHTS RESERVED
NO PART OF THIS BOOK MAY BE REPRODUCED IN ANY FORM,
BY PHOTOSTAT, MICROFILM, RETRIEVAL SYSTEM, OR ANY
OTHER MEANS, WITHOUT WRITTEN PERMISSION FROM
THE PUBLISHERS.

ACADEMIC PRESS, INC.
111 Fifth Avenue, New York, New York 10003

United Kingdom Edition published by
ACADEMIC PRESS, INC. (LONDON) LTD.
24/28 Oval Road, London NW1 7DD

LIBRARY OF CONGRESS CATALOG CARD NUMBER: 77-154385

PRINTED IN THE UNITED STATES OF AMERICA

QC
718
E24
1972
PHYSICS

Contents

1941

Chapter IV Nonequilibrium States of the Coulomb System with Individual Particle Correlations

Part 2 **The Fully Ionized System in the General Electromagnetic Field**

Chapter V Single-Particle Radiation

Chapter VI Many-Particle Interactions with Electromagnetic Fields

Preface

Due to early contacts with some of the pioneers in gaseous electronics, the exotic complexity of the experimental work in this field has always fascinated me. My interest turned to the analysis of these phenomena which, I soon realized, required a sound insight into the general features of plasma physics. From this background a course of lectures originated which I taught for many years, mainly at the Universities of Bonn and Bochum in Germany, and occasionally at American Universities that I visited.

In the development of this course I found myself more and more intrigued by the mathematical purity and transparency of the basic concepts of the fully ionized system. Therefore the emphasis in the course shifted gradually from gaseous electronics to the treatment of fully ionized systems.

Texts were available which treated many aspects of the fully ionized system under specific conditions. But there seemed to be hardly any text that developed—in an organized, graded, and comprehensive form—the basic theoretical features of the description of equilibrium and non-equilibrium states from first principles. The aim of this book is to provide such a presentation.

Of course, if one wants to develop this general background with some rigor the model must be a simple one. Therefore neither effects caused by external fields nor applications were considered.

The text addresses itself to graduate students and to research workers in the field of plasma physics, astrophysics, or related topics. The reader should have command of a basic understanding of classical theoretical physics in general.

Credit is due to many. There are all those who performed the original work which somehow is abstracted and summarized here. I wish to mention in particular some of the outstanding people in the field of plasma physics—M. Rosenbluth, A. Kaufmann, W. Kunkel, and P. Sturrock—from whom this text has profited through lectures and discussions. I am also grateful for innumerable discourses with my collaborators. Special thanks are given for the criticism, the persistant help, and the contributions by K. G. Fischer, G. Frömling, K. U. Riemann, and K. H. Spatschek.

Looking back over the pages I myself feel that in many places there is room for improvement and extension. But then, as have so many others, I see the truth in Goethe's statement: "So eine Arbeit wird eigentlich nie fertig, man muß sie für fertig erklären, wenn man nach Zeit und Umständen das Möglichste getan hat."

List of Symbols

General Remarks

(a) A great many functions occur in this text. The number of symbols available is not sufficient to distinguish them. Therefore the same function symbol with different argument may characterize a different functional dependence.
(b) Symbols used for abbreviation in a very limited range of the text are not contained in the following list.
(c) Indices at the upper left distinguish μ-space variables from Γ-space variables which are indexed at the lower right.
(d) In a few cases double use of symbols was conceded if the meaning of the symbol was clear from the context.

Symbols of General Importance

$\tilde{y}$	Fourier transform of y
$\hat{y}$	Laplace transform of y
$\mathbf{y}$	vector
$\lvert\mathbf{y}\rvert = y$	modulus of the vector $\mathbf{y}$
$\hat{\mathbf{y}}$	unit vector
$\underset{n}{\mathsf{y}}$	tensor of n-th rank ($n = 2,\ldots$)
$\mathbf{y})(\mathbf{y}$	dyade
$\mathscr{Y}$	operator
$\{x, y\}$	Poisson brackets
$\bar{y}$	set symbol
$\langle y \rangle$	average of y

Latin Symbols

$\mathbf{a}$	thermodynamic external parameters
A	intensive work coefficient
$\mathbf{A}$	vector potential

b	collision parameter
b_l	cluster integral of order l
$b_{\nu\mu}$	Grad coefficient
$\mathbf{B}$	magnetic induction
B_ω	black body radiation
$(1 - N) B_N$	Nth virial coefficient
B_{jk}	coefficient of direct heat transfer of two components
c	velocity of light
C	Euler's constant
C_v	specific heat
$\mathbf{D}$	electric displacement
$\mathbf{D}_\mathrm{e}$	electric dipole moment
$\mathbf{D}_\mathrm{m}$	magnetic dipole moment
D_s^σ	domain of collisions
$D(k, p)$	dielectric function
e	elementary charge
ze, e_+	charge of ions
$-e$, e_-	charge of electrons
E	total energy
$\mathbf{E}$	electric field
f_N	density distribution in Γ-space
$f_N^{(0)}$	equilibrium density distribution in Γ-space
f_s	specific distribution function of order s
$f^{(s)}$	general distribution function of order s
$f_\nu^{(s)}$	general distribution function of the νth component
$f_\nu^{(1)} = f_\nu$	general one-particle distribution function
$\bar{f}_s$	normalized distribution function of order s
f_0	zero-order approximation of the distribution function
f	perturbation of the distribution function
f_{ij}	cluster function
F	Helmholtz free energy, respectively, force
$F_i(\mathbf{r}, \mathbf{p})$	Klimontovich density of the particle i
$F^{(k)}$	Klimontovich density of all particles of component k
$F(u)$	distribution function for one velocity component
$\mathbf{g}$	relative velocity
g_{ij}	pair correlation function of particles i, j
g_{ijk}	triplet correlation function
$\mathfrak{g}_{ij}$	Mayer cluster function
$G, \bar{G}, \bar{\bar{G}}$	Gaunt factors
$\hbar$	Planck's constant
h_l	spherical Hankel function
$\mathbf{h}_\mu$	heat current
$\mathsf{h}_{3\mu}$	heat current tensor
H	Boltzmann's H-function
H	Hamiltonian
$\mathbf{H}$	magnetic field
H_l	Hankel function

$I({}^{(r)}f \mid {}^{(s)}f)$	collision term of particles with distribution function ${}^{(r)}f$, respectively, ${}^{(s)}f$
I_ω	radiation intensity
$\mathbf{j}$	electric current density
j_l	spherical Bessel function
j_ω	emission coefficient
$\mathbf{J}$	flux in the velocity space
J_l	Bessel function
$\mathbf{k}$	wave number vector
k_{D}	reciprocal Debye length
K_ν	modified Bessel function
l	mean free path
$\mathscr{L}_0$	Liouville operator
m_+	mass of ions
m_-	mass of electrons
m_0	rest mass
$\mathbf{M}$	magnetization
$M_{\mu\nu}$	reduced mass
n	particle density, or respectively, refractive index
N	total particle number
p	pressure
$\mathbf{p}$	linear momentum
$\mathbf{P}$	polarization
P_N	normalized distribution function in configuration space
P_s	specific molecular distribution function of order s
$P^{(s)}$	general molecular distribution function of order s
$P(s+1 \mid s)$	conditional probability density
$⨍$	Cauchy principal value
P_j	Legendre polynomial
$\mathbf{q}_\mu$	energy current
$\underset{3}{q_\mu}$	energy current tensor
q_{ij}	shielded cluster function
$q^{l/m}$	thermodynamic transport coefficient
Q	heat
$\underset{2}{\mathbf{Q}}$	scattering tensor
$\underset{2}{\tilde{\mathbf{Q}}_{\mathrm{e}}}$	electric quadrupole tensor
$Q(g_{\mu\nu})$	cross section
$\mathbf{r}$	configuration space coordinate
r	distance
r_0	average particle distance
r_{w}	classical interaction radius
R	reflection coefficient
S	entropy
$\mathbf{S}$	Poynting vector
$\mathscr{S}_s$	s-particle streaming operator

t	time
t_c	two particle interaction time
T	thermodynamic temperature
u	velocity component
u_p	phase velocity
u_g	group velocity
U	internal energy
v	specific volume
$\mathbf{v}$	velocity
v_{th}	mean thermal velocity
V	volume
$w(\Omega)$	energy radiated per solid angle and second
w^T	total energy radiated per second
W	energy
$W(G)$	probability distribution function of the quantity G
$\mathbf{x}$	configuration space coordinate in radiation problems
Z	partition function
Z_g	grand partition function
$\mathbf{Z}$	Hertz vector
$Z(\xi)$	dispersion function

Greek Symbols

α	absorption coefficient
β	autocorrelation coefficient, or respectively, relativistic velocity factor
β_k	irreducible cluster integral of order k
γ_s	configuration space of the sth particle
Γ	phase space
$\Gamma(x)$	Gamma function
δ	number of particles in the Debye sphere
$\delta(x)$	Dirac function
δ_{ij}	Kronecker symbol
Δ	Laplace operator
$\Delta(x, y)$	Dirichlet's function
ϵ	dielectric tensor
ϵ_j	cluster function
$\mathscr{E}$	electron energy
$\mathscr{E}_V$	energy density per unit volume and angular frequency in vacuum
$\mathscr{E}_P$	energy density per unit volume and angular frequency in plasma
η	friction coefficient
η_ω	differential emissivity
η^T	total emissivity

Θ	Gibb's module
κ	heat conduction coefficient
κ_B	Boltzmann's constant
κ_D	Debye constant
κ_ω	spatial absorption coefficient
λ	wavelength
λ	thermal de Broglie wavelength
λ_c	mean-free path
λ_D	Debye length
Λ	plasma parameter
μ	phase space
μ	chemical potential
μ	mobility of the electrons
μ_0	permeability
ν_c	collision frequency
Π_{co}	coupling parameter
Π_d	density parameter
ρ	mass density, or respectively, charge density
ρ	resistivity
$\underset{2}{\boldsymbol{\sigma}}$	conductivity tensor
σ	differential cross section
τ	characteristic time
τ_c	free flight time
τ_p	plasma oscillation time
τ_{in}	interaction time
τ_h	hydrodynamic time scale
ψ	general potential energy
χ_e	dielectric susceptibility
χ_m	magnetic susceptibility
ω_p	plasma frequency
ω_L	Larmor frequency
Ω	solid angle
Φ	potential
ϕ_{ij}	potential energy
ϕ_c	characteristic value of the potential energy

Model of the Fully Ionized Plasma and the Coulomb System

Electrons and nuclei are the resistant groups of our fully ionized plasma. The individuals of the resistant groups are characterized by their mass and their charge alone. All particles are considered to be points.

In principle, we allow the individuals of the resistant groups to form subsystems (atoms and molecules). Particles in such a subsystem are characterized by the fact that their correlation time is much longer than the average fluctuation time in the system, which is given by the ratio of the average particle distance to the average particle velocity. We assume that the number of neutral subsystems is so small that their effect is negligible. This excludes partially ionized systems.

Our **model of the fully ionized plasma** is further characterized by the following conditions:

a. $h = 0$, except within the subsystems and within the semiclassical approximation of the partition function.

There is no general criterion for the applicability of classical mechanics. The neglect of degeneracy effects is justified if the condition

$$\lambda = h/(2\pi m\Theta)^{1/2} \ll r_w = e^2/\Theta \tag{1}$$

is fulfilled. r_w is the classical interaction radius and λ is the thermal de Broglie wavelength. Otherwise the possibility of neglecting quantum-mechanical effects depends on the phenomenon studied. Most quantum-mechanical influences can be omitted if the condition

$$\lambda \ll r_0 = (3/4\pi n)^{1/3} \tag{2}$$

holds, where r_0 is the average particle distance. If condition (2) is violated, then scattering processes with strong deflections may require quantum-mechanical corrections. However, in judging these deviations one should remember that, to first order, the classical result for Coulomb scattering is identical with the quantum-mechanical one.

b. $v^n/c^n = 0$ for $n \geqslant 2$. We do not consider relativistic effects or magnetic interaction.

c. We assume the system to be sufficiently large so that the behavior of its individuals is not affected by the presence of boundaries.

d. $c = \infty$. In that part of the work where this assumption is applied we call our fully ionized plasma a **Coulomb system**.

Part 1 THE FULLY IONIZED SYSTEM IN THE QUASISTATIC ELECTROMAGNETIC FIELD, THE COULOMB SYSTEM

Chapter I Equilibrium States of the Coulomb System

1. Review of the Basic Concepts

Observation of a quality $F(\mathbf{p}, \mathbf{r})$ which is a function of the variables $\mathbf{p}, \mathbf{r}$ of our system produces an **observation average**

$$\langle F\rangle_\tau = (1/\tau) \int_{t'}^{t'+\tau} F(\mathbf{p}, \mathbf{r})\, dt \tag{1.1}$$

where τ designates the duration and t' the beginning of the measurement.

To discuss the observation average $\langle\ \rangle_\tau$ it is practical to consider the evolution of our system in Γ space. The Γ space is a phase space built from the coordinates and the conjugate momenta of all particles of our system. In this space our system is represented by a single point, its time development by a trajectory which is called "the ergode."

The direct—although naïve—approach would be to calculate the ergode by solving the Hamilton equations of our mechanical system for a given set of initial conditions. This procedure is impossible for obvious principal and practical reasons. Instead we use a method which is due to Gibbs. Let us consider a whole ensemble of systems of identical qualities. Any such virtual Gibbs' ensemble is represented by a set of points which can be described by a density distribution $f_N(\mathbf{p}, \mathbf{r}, t)$ in

Γ space. Assuming that the density distribution for such a Gibbs' ensemble in equilibrium is known and given by $f_N^{(0)}(\mathbf{p}, \mathbf{r})$ we postulate—following Gibbs—that the **ensemble average**

$$\langle F \rangle_s = \int F(\mathbf{p}, \mathbf{r}) f_N^{(0)}(\mathbf{p}, \mathbf{r})\, d\mathbf{p}\, d\mathbf{r} \tag{1.2}$$

is identical with the observation average of the measurement. This means

$$\langle F \rangle_s = \langle F \rangle_\tau \tag{1.3}$$

Clearly this postulate implies three assumptions:

1. The observation average is identical with the **time average** defined by

$$\langle F \rangle_t = \lim_{\tau \to \infty} \langle F \rangle_\tau \tag{1.4}$$

2. The time average $\langle F \rangle_t$ is identical with the ensemble average $\langle F \rangle_s$.

3. The ensemble average can be calculated from the "coarsegrained density distribution" in Γ space.

These assumptions have played an important role in the historical development of statistics and are discussed under headlines like "ergodic hypothesis," "irreversibility," or "Liouville paradox."

Accepting Gibbs' postulate we have to find information about the average density distribution $f_N^{(0)}$ of a virtual ensemble of systems in equilibrium.

1.1. Liouville's Law

The members of Gibbs' ensemble neither interact with each other nor are they destroyed or created. Therefore for any density distribution, the simple continuity relation holds

$$\int (f_N \mathbf{v}_\Gamma) \cdot d\boldsymbol{\sigma}_\Gamma = -\frac{\partial}{\partial t} \int f_N\, d\mathbf{p}\, d\mathbf{r} \qquad \text{with} \quad \mathbf{v}_\Gamma = \begin{pmatrix} \dot{\mathbf{r}} \\ \dot{\mathbf{p}} \end{pmatrix} \tag{1.5}$$

or

$$\mathbf{v}_\Gamma \cdot \nabla_\Gamma f_N + f_N \nabla_\Gamma \cdot \mathbf{v}_\Gamma + \partial f_N / \partial t = 0 \tag{1.6}$$

Applying the canonical equations of classical mechanics in the form

$$\nabla_\Gamma \cdot \mathbf{v}_\Gamma = 0 \tag{1.7}$$

it follows from (1.6)

$$df_N/dt = \partial f_N/\partial t + \mathbf{v}_\Gamma \cdot \nabla_\Gamma f_N = \partial f_N/\partial t + \{f_N, H\} = 0 \tag{1.8}$$

where we have used Poisson's brackets { }. Equation (1.8) formulates Liouville's law.

We now claim: A stationary density distribution in the Γ space may always be expressed by the invariants of motion only.

The proof is straightforward for any distribution function satisfying Liouville's equation: From Hamilton's theory we know that it is always possible to find a canonical transform so that the new coordinates and momenta **R, P** are constants

$$d\mathbf{R}(\mathbf{r}, \mathbf{p}, t)/dt = d\mathbf{P}(\mathbf{r}, \mathbf{p}, t)/dt = 0 \tag{1.9}$$

In these new variables, Liouville's equation reads

$$\frac{d}{dt} f_N(\mathbf{R}, \mathbf{P}, t) = \frac{\partial f_N}{\partial \mathbf{R}} \cdot \frac{d\mathbf{R}}{dt} + \frac{\partial f_N}{\partial \mathbf{P}} \cdot \frac{d\mathbf{P}}{dt} + \frac{\partial f_N}{\partial t} = \frac{\partial f_N}{\partial t} = 0 \tag{1.10}$$

This shows that f_N is a function of the motion invariants **R, P** only.

For a conservative system, the Hamilton function H is such an invariant, although, of course, not the only one. Depending on the physical situation, other invariants may be important. However, following Gibbs, we will restrict ourselves to density distributions which are functions of the Hamiltonian.

1.2. The Statistical Ensembles

Using the normalization

$$\int f_N^{(0)}(\mathbf{p}, \mathbf{r})\, d\mathbf{p}\, d\mathbf{r} = 1 \tag{1.11}$$

Gibbs defined the following ensembles:

The **microcanonical ensemble** is a set of conservative systems in Γ space with given values of the total energy E, the particle number N, and external parameters **a**. The corresponding phase space density is

$$f_N^{(0)} = (1/\Omega)\, \delta[H(\mathbf{p}, \mathbf{r}, \mathbf{a}) - E] \tag{1.12}$$

where Ω is a normalization factor.

The **canonical ensemble** is a set of systems in Γ space with given values of the statistical parameter Θ, the particle number N, and external parameters $\mathbf{a}$. The corresponding phase space density is

$$f_N^{(0)} = (1/h^{3N}N!Z)\exp[-H(\mathbf{p}, \mathbf{r}, \mathbf{a})/\Theta] \tag{1.13}$$

where Z is the partition function which, according to the normalization (1.11), reads

$$Z = (1/h^{3N}N!)\int \exp[-H(\mathbf{p}, \mathbf{r}, \mathbf{a})/\Theta]\, d\mathbf{p}\, d\mathbf{r} \tag{1.14}$$

The derivation of the thermodynamical relations from this statistical distribution and comparison with the thermodynamical temperature scale shows that the parameter Θ is related to the thermodynamic temperature T by

$$\Theta = \kappa_B T \tag{1.15}$$

where κ_B is Boltzmann's constant.

The **macrocanonical ensemble** is a set of systems in Γ space with given values of two statistical parameters Θ and m' and external parameters $\mathbf{a}$. The corresponding phase space density is

$$f_N^{(0)} = (1/h^{3N}N!Z_g)\exp\{-[H(\mathbf{p}, \mathbf{r}, \mathbf{a}) - m'N]/\Theta\} \tag{1.16}$$

where the grand partition function Z_g is defined according to a modified normalization condition where the summation over all particle numbers N has to be taken into account:

$$Z_g = \sum_{N=0}^{\infty} (1/h^{3N}N!)\int \exp[-(H - m'N)/\Theta]\, d\mathbf{p}\, d\mathbf{r} \tag{1.17}$$

The parameter Θ is related to the temperature T according to (1.15). By similar reasoning, the parameter m' can be proved to be identical with the chemical potential μ.

Obviously, the relation

$$Z_g = \sum_{N=0}^{\infty} Z(N, \Theta, \mathbf{a})\exp[m'N/\Theta] \tag{1.18}$$

holds.

It is interesting to discuss what type of physical system these ensembles represent.

The **microcanonical ensemble** is, according to its definition, a virtual ensemble of systems with constant energy and particle number. It represents thermodynamically closed systems.

The **canonical ensemble** describes systems with constant particle number but varying energy. The variation of the energy is governed by the statistical parameter Θ. The physical meaning of this energy variation can be understood by studying a subsystem of a microcanonical system with many degrees of freedom. This study shows that the canonical ensemble has a distribution similar to that of such subsystems and the energy exchange corresponds to that of a system in contact with a thermostat, the temperature of which is governed by the statistical parameter Θ.

The **macrocanonical ensemble** has a variation in energy governed by the parameter Θ as well as of particle number governed by the parameter m'. Again treating a subsystem of a microcanonical system with many degrees of freedom, one can show that the macrocanonical ensemble describes systems which are in contact with a large thermostat characterized by the parameter Θ and a large particle reservoir characterized by the parameter m'. It therefore represents thermodynamically open systems.

The density distributions described by the ensembles together with Gibb's postulate enable us to calculate the microscopical and macroscopical qualities of systems in equilibrium. We shall first study the **macroscopic properties**, in particular, the relation between the partition function and the thermodynamic potentials. After that we shall investigate **microscopic qualities**.

2. Macroscopic Qualities

2.1. Relations between the Thermodynamic Potentials and the Canonical Partition Function

The differential of the Helmholtz free energy is defined by

$$dF = dU - T\,dS - S\,dT \tag{2.1}$$

With the first and second law of thermodynamics,

$$\begin{aligned} dQ &= dU - \mathbf{A}\cdot d\mathbf{a} - \mu\,dN \\ dS &= dQ/T \end{aligned} \tag{2.2}$$

(2.1) is transformed into

$$dF = \frac{F - U}{T} dT + \mathbf{A} \cdot d\mathbf{a} + \mu \, dN \tag{2.3}$$

a denotes again the extensive work variables and **A** the conjugated intensive work coefficients.

Now we study the differential of the function $(-\Theta \ln Z)$. This expression is a function of the parameters Θ, $\mathbf{a}$, and the particle number N. In formulating the total differential, we apply the relation

$$\begin{aligned} dZ &= \frac{\partial Z}{\partial N} dN + \frac{\partial Z}{\partial \Theta} d\Theta + \frac{\partial Z}{\partial \mathbf{a}} \cdot d\mathbf{a} \\ &= \frac{\partial Z}{\partial N} dN + \frac{1}{N! h^{N3}} \int \left(\frac{H}{\Theta^2} d\Theta - \frac{1}{\Theta} \frac{\partial H}{\partial \mathbf{a}} \cdot d\mathbf{a} \right) \exp\left(-\frac{H}{\Theta} \right) d\mathbf{p} \, d\mathbf{r} \end{aligned} \tag{2.4}$$

which is a consequence of (1.14). With the relations

$$U = \langle H \rangle = (1/N! h^{3N} Z) \int H \exp(-H/\Theta) \, d\mathbf{p} \, d\mathbf{r} \tag{2.5}$$

and

$$\mathbf{A} = \langle \partial H / \partial \mathbf{a} \rangle = (1/N! h^{3N} Z) \int \partial H / \partial \mathbf{a} \exp(-H/\Theta) \, d\mathbf{p} \, d\mathbf{r} \tag{2.6}$$

which follow from the definition of the average value in (1.2), we find

$$d(-\Theta \ln Z) = (-\Theta \ln Z - U) \, d \ln \Theta + \mathbf{A} \cdot d\mathbf{a} + \frac{\partial}{\partial N} (-\Theta \ln Z) \, dN \tag{2.7}$$

Now anticipating the relation

$$\Theta = \kappa_B T \tag{2.8}$$

which subsequently can be proved by applying the definition of the thermodynamic temperature scale we see that—apart from a constant which is not of interest—the relation

$$F = -\Theta \ln Z \tag{2.9}$$

holds.

This relation links the canonical partition function and the thermodynamic potentials. Having the Helmholtz free energy F we find all other thermodynamical qualities by simple differentiation. This yields, using (2.3) and choosing $a = V$,

$$\begin{aligned}
A &= -p \\
S &= -\left.\frac{\partial F}{\partial T}\right|_{N,V} = \left.\frac{\partial}{\partial T}(\Theta \ln Z)\right|_{N,V} \\
p &= -\left.\frac{\partial F}{\partial V}\right|_{T,N} = \left.\frac{\partial}{\partial V}(\Theta \ln Z)\right|_{T,N} \\
\mu &= \left.\frac{\partial F}{\partial N}\right|_{T,V} = -\left.\frac{\partial}{\partial N}(\Theta \ln Z)\right|_{T,V} \\
U &= F + TS = -\Theta \ln Z + T\left.\frac{\partial}{\partial T}(\Theta \ln Z)\right|_{N,V} = \left.\frac{\Theta^2}{\kappa_B}\frac{\partial}{\partial T}\ln Z\right|_{N,V} \\
C_V &= \left.\frac{\partial U}{\partial T}\right|_{N,V} = \left.\frac{\partial}{\partial T}\left(\frac{\Theta^2}{\kappa_B}\frac{\partial}{\partial T}\ln Z\right)\right|_{N,V}
\end{aligned} \tag{2.10}$$

2.2. Relations between the Thermodynamic Potentials and the Macrocanonical Partition Function

For our aim, it is advised to study the quality pV which is the thermodynamic potential to the variables T, $V = a$, and μ. For the differential we have

$$d(pV) = (dT/T)(pV + U - \mu N) + N\,d\mu + p\,dV \tag{2.11}$$

which follows readily from the relation

$$F = \left.\frac{\partial F}{\partial V}\right|_{N} V + \left.\frac{\partial F}{\partial N}\right|_{V} N = -pV + N\mu = U - TS \tag{2.12}$$

The left-hand side of (2.12) is a consequence of Euler's theorem, provided that the Helmholtz energy is a linear homogeneous function in V and N. This assumption is correct within the frame of Van Hove's theorem.

We now turn our interest to the differential of $(\Theta \ln Z_g)$ as a function of the parameters Θ, m', and $\mathbf{a}$. The average energy and the work coefficient of the system are, according to (1.2),

$$U = \langle H \rangle = \sum_{N=0}^{\infty} (1/Z_g N! h^{3N}) \int H \exp[-(H - m'N)/\Theta)\, d\mathbf{p}\, d\mathbf{r} \tag{2.13}$$

and

$$\mathbf{A} = \langle \partial H/\partial \mathbf{a} \rangle = \sum_{N=0}^{\infty} (1/Z_g N! h^{3N}) \int \frac{\partial H}{\partial \mathbf{a}} \exp[-(H - m'N)/\Theta] \, d\mathbf{p} \, d\mathbf{r} \tag{2.14}$$

Further—since N is a variable—we have, in addition, the average particle number

$$\langle N \rangle = \sum_{N=0}^{\infty} (1/Z_g N! h^{3N}) \int N \exp[-(H - m'N)/\Theta] \, d\mathbf{p} \, d\mathbf{r} \tag{2.15}$$

Using these qualities and the differential

$$\begin{aligned} dZ_g &= \frac{\partial Z_g}{\partial \Theta} d\Theta + \frac{\partial Z_g}{\partial m'} dm' + \frac{\partial Z_g}{\partial \mathbf{a}} \cdot d\mathbf{a} \\ &= \sum_{N=0}^{\infty} \frac{1}{N! h^{3N}} \int \left[\left(\frac{H - m'N}{\Theta^2} \right) d\Theta + \frac{N}{\Theta} dm' - \frac{1}{\Theta} \frac{\partial H}{\partial \mathbf{a}} \cdot d\mathbf{a} \right] \\ &\quad \times \exp\left(- \frac{H - m'N}{\Theta} \right) d\mathbf{p} \, d\mathbf{r} \end{aligned} \tag{2.16}$$

we find

$$d(\Theta \ln Z_g) = [\Theta \ln Z_g + U - m'\langle N \rangle] \, d \ln \Theta + \langle N \rangle \, dm' - \mathbf{A} \cdot d\mathbf{a} \tag{2.17}$$

Anticipating again the relations

$$\Theta = \kappa_B T, \qquad m' = \mu \tag{2.18}$$

which can be proved subsequently from the thermodynamic definitions of these qualities, we find by comparing (2.17) and (2.11)

$$pV = \Theta \ln Z_g \tag{2.19}$$

Other thermodynamic qualities of interest can be found by simple differentiation as (2.11) shows. We have

$$\begin{aligned} S &= \frac{\partial (pV)}{\partial T}\bigg|_{\mu,V} = \frac{\partial}{\partial T} (\Theta \ln Z_g)\bigg|_{\mu,V} \\ p &= \frac{\partial (pV)}{\partial V}\bigg|_{\mu,T} = \frac{\partial}{\partial V} (\Theta \ln Z_g)\bigg|_{\mu,T} \\ N &= \frac{\partial (pV)}{\partial \mu}\bigg|_{T,V} = \frac{\partial}{\partial \mu} (\Theta \ln Z_g)\bigg|_{T,V} \\ U &= \mu N + ST - pV \\ &= \mu \frac{\partial}{\partial \mu} (\Theta \ln Z_g)\bigg|_{T,V} + T \frac{\partial}{\partial T} (\Theta \ln Z_g)\bigg|_{\mu,V} - V \frac{\partial}{\partial V} (\Theta \ln Z_g)\bigg|_{\mu,T} \end{aligned} \tag{2.20}$$

Since all macroscopic thermodynamic qualities of our system can be expressed in terms of the partition function, the problem of the calculation of these macroscopic qualities has been reduced to the evaluation of the partition function.

3. Partition Function of the Coulomb System

3.1. The Problem

In the preceding section we demonstrated that the partition function plays a central role in the calculation of the thermodynamic qualities of a system.

Without lack of generality, we may restrict ourselves in this section to the study of the canonical partition function, since the determination of the grand canonical partition function can be related to the canonical one. Also it seems sufficient to consider a Coulomb system consisting of an equal number N of electrons and singly charged ions. Such a system displays all the typical difficulties and the evaluation procedures can readily be generalized to a system with more than one ion component.

Naïvely applying (1.14) to our Coulomb system we should have to evaluate

$$Z = [1/h^{6N}(N!)^2] \int \exp(-H/\Theta)\, d\mathbf{p}\, d\mathbf{r} \tag{3.1}$$

with the Hamilton function

$$\begin{aligned} H = {} & \frac{1}{2}\sum_i \left[\frac{p_{+i}^2}{m_+} + \frac{p_{-i}^2}{m_-}\right] \\ & + \frac{1}{2}\sum_{i,j}{}' \left[\frac{e^2}{|\mathbf{r}_{+i} - \mathbf{r}_{+j}|} + \frac{e^2}{|\mathbf{r}_{-i} - \mathbf{r}_{-j}|}\right] - \sum_{i,j} \frac{e^2}{|\mathbf{r}_{+i} - \mathbf{r}_{-j}|} \end{aligned} \tag{3.2}$$

The impossibility of this approach is obvious. It accounts in no way for the formation of bound hydrogen states. Even more apparently, the small distance interaction of oppositely charged particles causes the partition function to diverge.

The reason is that (1.14) gives a semiclassical approximation of the true partition function. To account for the contributions of the bound states which are not open to a semiclassical treatment, we have to use the quantum-mechanical representation of the sum over all states

$$Z = \operatorname{tr} \exp(-\mathscr{H}/\Theta) \tag{3.3}$$

where $\mathscr{H}$ is the Hamilton operator of our system.

If the matrix elements of $\mathscr{H}$ could be determined in all generality, we would have an exact result for the partition function. We are, however, still far away from such a general solution.

Instead, it is common use in the practical calculation of the partition function to apply—explicitly or tacitly—the **free–bound model** (Ecker and Kröll, 1966). This free–bound model assumes that the stationary state of the system can be considered as composed of two independent groups of particle states of electrons and nuclei only: weakly interacting classical free particle states on the one hand, and strongly interacting quantum-mechanical bound states of two oppositely charged individuals on the other.

Of course, this approximation goes far beyond the naïve picture described by (3.1) and (3.2). On the other hand, it is still a crude approximation to the true behavior. In each stationary state there is a group of particle states which due to free–bound interactions cannot really be attributed to either of the two above groups. This makes the limit between "free" and "bound" an uncertain quantity. Moreover, whether a particle is to be termed "free" or "bound" can depend on the effect studied and different values of the free–bound limit may be expected for different phenomena.

The details of these involved relations cannot be given here. We present only the results of the corresponding considerations of Ecker and Kröll.

These authors show on the basis of a quantum-mechanical perturbation theory that the "free–bound limit" may roughly be characterized by the relation

$$\varepsilon_b = -e^2/r_0 \tag{3.4}$$

where ε_b is the pair energy of a particle with its nearest oppositely charged neighbor in their center of mass system, neglecting all other interactions. We remark that the energy has positive or negative values for states above or below, respectively, the series limit of the unperturbed atom. The introduction of the energy limit ε_b remains necessarily artificial and afflicted with inherent uncertainty.

Neglecting the interaction between the above defined groups of free and bound particles, the Hamilton operator $\mathscr{H}$ is separable into two independent parts $\mathscr{H}_{\text{f}}$ and $\mathscr{H}_0$ related to the free and bound particle groups, respectively. Under these circumstances, the partition function (3.3) can be factored

$$Z = Z_f Z_0 = \operatorname{tr} \exp(-\mathscr{H}_f/\Theta) \operatorname{tr} \exp(-\mathscr{H}_0/\Theta) \tag{3.5}$$

and applying the eigenfunctions of $\mathscr{H}_0$ in the calculation of Z_0 we arrive at

$$Z_0 = (N_0!)^{-2} \left[(2\pi(m_+ + m_-)\Theta/h^2)^{3/2} \sum_n \exp(-\varepsilon_n/\Theta)\right]^{N_0} \tag{3.6}$$

where N_0 is the number of bound particles, ε_n is the nth energy eigenvalue of the bound states, and the sum includes all bound energy levels. It should be noted that the well-known divergence of the atomic partition function due to the accumulation of terms at the limit $\varepsilon_n \to 0$ does not occur, on account of the limitation stated in (3.4).

To calculate Z_f one starts with the momentum eigenfunctions and applies a perturbation procedure developed by Kirkwood. If r_w designates the classical interaction radius and λ the de Broglie wavelength, then it is shown that the classical canonical partition function without energy limitation and without the contributions from interparticle distances $r^2 < \lambda r_w$ is a sound approximation for Z_f. The exclusion of these short-range contributions removes the divergence difficulty which we mentioned in the beginning as one of the basic obstacles. It does not affect the numerical results of the cluster-evaluation procedures for Z_f outlined in the following, since these calculations implicitly exclude short-range interactions.

Within the frame of the free–bound approximation, the total partition function of a Coulomb system is therefore represented as the product of two contributions, one, Z_0, related to the bound states and the other, Z_f, related to the free states.

The contribution Z_0 is described by the relation (3.6) and presents no difficulties since the energy levels ε_n can be calculated from the quantum-mechanical solution of the two particle problem. Since we are only interested in the contributions of the free particles, we denote in the following the number of free ions or electrons by N.

Therefore the factor Z_f is given by

$$Z_f = [1/h^{6N}(N!)^2] \int \exp(-H/\Theta)\, d\mathbf{p}\, d\mathbf{r}, \qquad |\mathbf{r}_+ - \mathbf{r}_-|^2 > r_w \lambda \tag{3.7}$$

with the Hamilton function (3.2). The evaluation of the momentum contributions in (3.7) is trivial. The configuration part of the integral (3.7) constitutes a problem because of the long-range interaction of the Coulomb forces.

We therefore study in the following the evaluation of

$$Z' = \int \exp\left\{-\frac{1}{2\Theta}\sum_{i,j}{}' \left(\frac{e^2}{|\mathbf{r}_{+i} - \mathbf{r}_{+j}|} + \frac{e^2}{|\mathbf{r}_{-i} - \mathbf{r}_{-j}|}\right) + \frac{1}{\Theta}\sum_{i,j}\frac{e^2}{|\mathbf{r}_{-i} - \mathbf{r}_{+j}|}\right\} d\mathbf{r} \tag{3.8}$$

on the basis of cluster expansions. To prepare this we first recall the basic features of cluster calculus.

3.2. Elements of Cluster Expansions

For the sake of simplicity, we formally rewrite (3.8) as

$$Z' = \int \exp\left(-(1/\Theta)\sum_{i,j}\phi_{ij}\right) d\mathbf{r} \tag{3.9}$$

where the sign of the potential energy ϕ_{ij} depends on the indices in accord with (3.8).

Now we may expand Z':

$$Z' = \int \prod_{i,j}(1 + f_{ij})\, d\mathbf{r}$$

$$= \int \left(1 + \sum f_{ij} + \sum\sum f_{ij}f_{kl} + \sum\sum\sum f_{ij}f_{kl}f_{mn} + \cdots\right) d\mathbf{r} \tag{3.10}$$

where we have used

$$f_{ij} = \exp(-\phi_{ij}/\Theta) - 1 \tag{3.11}$$

To distinguish the various products of f-functions in (3.10), we use the following scheme. The indices are represented by a labeled point pattern and the f_{ij}-functions by lines connecting the corresponding points. The example given in Figure I.1 represents the term

$$f_{2,3}f_{4,5}f_{4,10}f_{13,14}f_{16,17}f_{20,21}f_{16,23}f_{17,23}f_{18,26}f_{20,27}f_{20,28}f_{21,28}f_{27,28} \tag{3.12}$$

We define a **special cluster** as a scheme of connections between labeled diagram points which all are directly or indirectly connected with each other. A special cluster is consequently characterized by the designation of all its members and the scheme of connections between them.

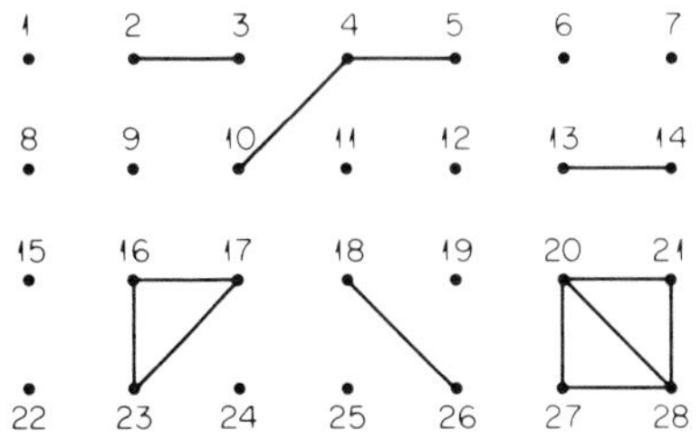

FIGURE I.1. Diagrammatic representation of products of f-functions.

If we drop the distinction of the individuals in the special cluster, we come to a **general cluster**. A general cluster is characterized by the number of particles of the various kinds and the connection scheme between them.

If a given cluster comprises a total number of l individuals composed of l_k individuals of kind k, then the number of selections of sets $\bar{l} = \{l_1, \ldots, l_k, \ldots, l_\sigma\}$ from the set $\bar{N} = \{N_1, \ldots, N_k, \ldots, N_\sigma\}$ is given by the corresponding number of combinations

$$\prod_k N_k!/(N_k - l_k)!\,l_k! \tag{3.13}$$

Each of these sets allows the realization of many special clusters since the permutation of the individuals within the connection scheme produces new special clusters. In the following we only consider clusters of one particle kind. For the purpose of demonstration, we show the group of general four-clusters and indicate the number of special clusters which it contains.

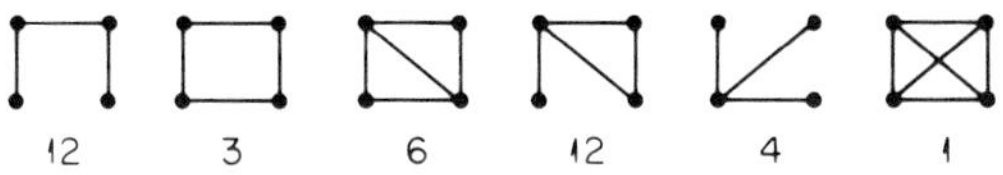

(3.14)

Since any diagram of the type shown in Figure I.1 may be represented as a set of special clusters and on the other hand any term of (3.10) is representable by such a diagram, it is clear that any term of (3.10) can be given as a set of special clusters. This is a unique characterization of such a term.

We will see that for many uses such a detailed specification is not required. It may be sufficient to give the set $\bar{m}_l$ of the number of general clusters of the type l contained in the diagram. For instance, for our diagram of Figure I.1 the corresponding set of numbers is

$$m_1 = 12, \qquad m_2 = 3, \qquad m_3 = 2, \qquad m_4 = 1, \qquad m_5 = 0 \tag{3.15}$$

We now define the **cluster integral** of order l as

$$b_l = (1/l!V) \int \sum \prod f_{ij}\, d\mathbf{r}_1 \cdots d\mathbf{r}_l \tag{3.16}$$

which will prove to be useful in the evaluation of (3.10). In the integrand of (3.16) we sum over all f-bond products which correspond to special clusters under the general l-type cluster.

The dimension of the cluster integral (3.16) is obviously V^{l-1}. As an example we give the first four cluster integrals

$$b_1 = (1/V) \int d\mathbf{r}_1 = 1$$

$$b_2 = (1/2V) \int \bullet\!\!-\!\!\bullet \; d\mathbf{r}_1\, d\mathbf{r}_2 = (1/2V) \int f_{12}\, d\mathbf{r}_1\, d\mathbf{r}_2 = \tfrac{1}{2} \int_0^\infty f(r)\, 4\pi r^2\, dr$$

$b_3 = (1/6V) \int$ 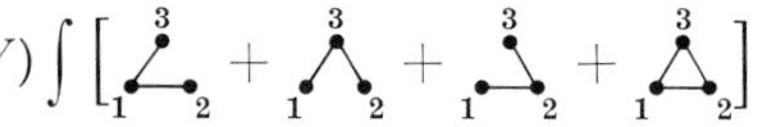 $d\mathbf{r}_1\, d\mathbf{r}_2\, d\mathbf{r}_3$

$$= (1/6V) \int [f_{31}f_{21} + f_{32}f_{31} + f_{32}f_{21} + f_{32}f_{31}f_{21}]\, d\mathbf{r}_1 d\mathbf{r}_2 d\mathbf{r}_3 \tag{3.17}$$

$b_4 = (1/24V) \int$ 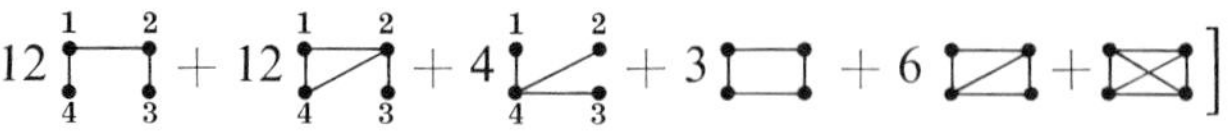

$$\times\, d\mathbf{r}_1\, d\mathbf{r}_2\, d\mathbf{r}_3\, d\mathbf{r}_4$$

$$= (1/24V) \int [12 f_{41}f_{32}f_{21} + 12 f_{42}f_{41}f_{32}f_{21} + 4 f_{43}f_{42}f_{41} + 3 f_{43}f_{41}f_{32}f_{21}$$

$$+ 6 f_{43}f_{42}f_{41}f_{32}f_{21} + f_{43}f_{42}f_{41}f_{32}f_{31}f_{21}]\, d\mathbf{r}_1\, d\mathbf{r}_2\, d\mathbf{r}_3\, d\mathbf{r}_4$$

If V_w designates the volume characterizing the range of pair interactions, then the cluster integral is independent of the volume V if the relation

$$V/l \gg V_w \tag{3.18}$$

holds.

We now aim to express the partition function (3.9) by cluster integrals. To this end, we first select all those diagrams from the sum which group the same specific points to clusters. We allow that the scheme of bonds for these clusters may be different. Then we select from this group those diagrams which all contain one definite special cluster. Let us characterize this cluster by $C(1,\ldots,s)$, where C designates the f-bond scheme and $1,\ldots,s$ the particles contained. In the contribution of this subgroup to (3.9) we may then factor the f-bonds of the specific cluster in the form

$$\sum \prod f_{ij} = \Big(\prod_{C} f_{ij}\Big) R(f_{ij}) \tag{3.19}$$

where R stands for the contributions of the rest of the diagram. We may now proceed in the same way with all special clusters connecting the diagram points 1,..., s. Each of them will produce a term of the type given in (3.19). Summing over all these subgroups and carrying out the integrations in (3.9) we find for the selected group of diagrams

$$(b_l l! V)\, R(f_{ij}) \tag{3.20}$$

Successive application of this procedure yields for the selected set of diagrams

$$\prod_{l} (b_l l! V)^{m_l} \tag{3.21}$$

Therewith we have calculated the contribution of all diagrams connecting the same points to clusters. The partition function Z' is consequently given as the sum of all contributions (3.21).

We recognize from this result that the contributions of those groups are identical which have the same set $\bar{m}_l$ of general clusters. The summation over such subgroups is therefore simply given by the multiplication with the number of diagrams having the same set $\bar{m}_l$. This number is

$$N!/\prod_{l} m_l!(l!)^{m_l} \tag{3.22}$$

for the following reason: $N!$ is the number of permutations of all particles. Within each cluster there are $l!$ uneffective permutations since they are already contained in b_l. Since the diagram contains the l-cluster altogether m_l times, we have $(l!)^{m_l}$ uneffective permutations. Finally, the exchange of whole groups of special clusters does not produce a new realization.

Multiplying the factors (3.22) and (3.21) we find the contribution of all diagrams with the same general cluster subdivision $\bar{m}_l$ in the form

$$N! \prod_{l} (Vb_l)^{m_l}/m_l! \tag{3.23}$$

The partition function is therefore given by

$$Z' = N! \sum_{\bar{m}_l} \prod_{l} (Vb_l)^{m_l}/m_l! \tag{3.24}$$

under the condition

$$\sum_{l=1}^{N} m_l \, l = N \tag{3.25}$$

following from the conservation of the particle number.

The evaluation of (3.24) is very difficult. Fortunately one can replace the sum by its maximum term T_m for large values of N.

This follows from the relation

$$T_m < Z'/N! < T_m \nu_p \tag{3.26}$$

where ν_p is the total number of terms in the sum. ν_p—the *partitio numerorum*—is identical with the number of possibilities to present N as a sum of numbers l with repetition but without consideration of the sequence. It is given by

$$\ln \nu_p \simeq \pi(\tfrac{2}{3}N)^{1/2} \tag{3.27}$$

Considering the logarithm of (3.26), using (3.27), and the fact that the logarithm of the maximum term $\ln T_m$ is proportional to N—as will be shown—it is readily verified that T_m is identical with $Z'/N!$ in the limit $N \to \infty$.

Applying the method of Lagrange's multiplier F and Stirling's formula, we get the following relation for m_l:

$$m_l = N v b_l F^l \qquad \text{with} \quad v = V/N \tag{3.28}$$

where the multiplier F is provided by the condition (3.25) through

$$\sum_{l=1}^{N} l v b_l F^l = 1 \tag{3.29}$$

Consequently, the maximum term T_m, and with that the partition function Z', is determined by

$$\ln T_m = \ln (Z'/N!) = N \left(\sum_{l=1}^{N} v b_l F^l - \ln F \right) \tag{3.30}$$

together with condition (3.29).

In this formulation, the evaluation of the partition function reduces to the calculation of cluster integrals. This problem may be facilitated by introducing irreducible cluster integrals. To show this we need some additional general definitions.

We define:

Two representative points in a diagram are termed **directly connected** if there is no further point on the bond between them; connected by a **chain of order** n if the bond between them is occupied by n points each of them having no other connections; **multiply connected** if there are at least two completely independent paths of bonds and points from one to the other point.

A representative point is termed **independently connected** with n other points if there are n independent paths to the n other points; **knot of order** n if it is directly connected with n points; **junction** if it is a knot of order three or higher; **branch point** of a cluster if its removal together with its bonds leaves at least two independent clusters.

A cluster is termed **irreducible** if there are no branch points in it. Thus in an irreducible cluster each point is at least doubly connected with each other point.

3.3. Irreducible Clusters

To reduce the ordinary cluster integrals to irreducible cluster integrals, we first indicate all branch points in the ordinary cluster under concern. For the purpose of demonstration, consider Figure I.2, where the points 2, 3, 4, and 6 are branch points.

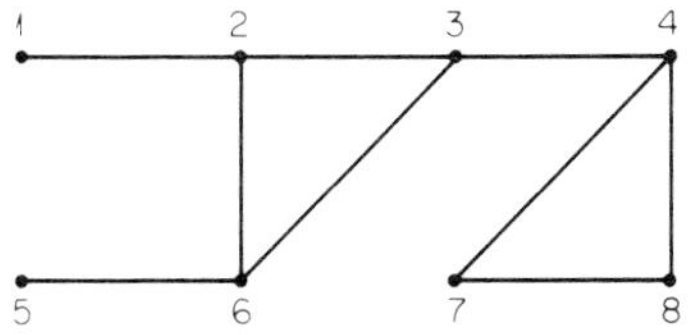

FIGURE I.2. Diagrammatic representation of a reducible cluster.

We now arbitrarily choose one of these branch points and introduce into (3.16) relative coordinates for all diagram points connected with it. Then the coordinate of the branch point does not occur in the f-bonds and contributes only a volume factor. The cluster integral represents itself as a product of two independent factors, each of them involved only with one of the cluster parts which we find by separating the original cluster at the branch point. Successive application of this procedure to all branch points of the cluster obviously yields a product of integrals over irreducible clusters.

We now define the **irreducible cluster integral** of order k as

$$\beta_k = (1/k!V) \int \sum \prod f_{ij} \, d\mathbf{r}_1 \cdots d\mathbf{r}_{k+1} \tag{3.31}$$

Here the sum extends over all products f_{ij} belonging to an irreducible cluster formation of $(k+1)$ diagram points.

The first four irreducible cluster integrals are

$\beta_1 = (1/V) \int$ $d\mathbf{r}_1 \, d\mathbf{r}_2$

$\beta_2 = (1/2V) \int$  $d\mathbf{r}_1 \, d\mathbf{r}_2 \, d\mathbf{r}_3$

$\beta_3 = (1/6V) \int \Big(3$  $+ 6$  $+$  $\Big) \, d\mathbf{r}_1 \cdots d\mathbf{r}_4$ (3.32)

$\beta_4 = (1/24V) \int \Big(12$ 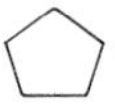 $+ 60$ 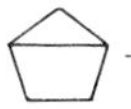 $+ 10$  $+ 60$ 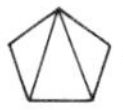 $+ 30$ 

$+ 10$  $+ 15$ 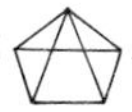 $+ 30$  $+ 10$  $+$ 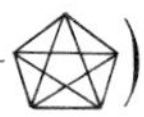 $\Big) \, d\mathbf{r}_1 \cdots d\mathbf{r}_5$

In analytical formulation β_1, β_2, and β_3 may be written as

$$\begin{aligned}
\beta_1 &= (1/V) \int f_{12} \, d\mathbf{r}_1 \, d\mathbf{r}_2 \\
\beta_2 &= (1/2V) \int f_{32} f_{21} f_{13} \, d\mathbf{r}_1 \, d\mathbf{r}_2 \, d\mathbf{r}_3 \\
\beta_3 &= (1/6V) \int [3 f_{43} f_{32} f_{21} f_{14} + 6 f_{43} f_{32} f_{21} f_{14} f_{31} + f_{43} f_{32} f_{21} f_{14} f_{31} f_{42}] \\
&\quad \times d\mathbf{r}_1 \, d\mathbf{r}_2 \, d\mathbf{r}_3 \, d\mathbf{r}_4
\end{aligned} \tag{3.33}$$

For obvious reasons we refrain from presenting β_4.

Applying the above procedure it is easy to show that the following relations hold between the first four ordinary cluster integrals and the irreducible ones:

$$b_1 = 1, \qquad b_2 = \tfrac{1}{2}\beta_1, \qquad b_3 = \tfrac{1}{2}\beta_1^2 + \tfrac{1}{3}\beta_2, \qquad b_4 = \tfrac{2}{3}\beta_1^3 + \beta_1\beta_2 + \tfrac{1}{4}\beta_3 \tag{3.34}$$

In general, the relation

$$b_l = (1/l^2) \sum_{\bar{n}} \prod_k (l\beta_k)^{n_k} / n_k! \tag{3.35}$$

holds, where the sum is to be extended over all products satisfying the condition

$$\sum_{k=1}^{l-1} kn_k = l - 1 \tag{3.36}$$

Since our attention is focused on the application of the cluster calculus to Coulomb systems we will not elaborate the proof of (3.35). The details can be found in any book on cluster calculus (for example, Mayer and Mayer, 1940).

We are now ready to express Z' with the help of irreducible cluster integrals by introducing (3.35) into (3.30) and (3.29). As may be verified by insertion, the relation

$$F = (1/v) \exp\Big(-\sum_k \beta_k v^{-k}\Big) \tag{3.37}$$

is a solution of (3.29). Using (3.37) and (3.35) in (3.30) and developing into a series of falling powers of the specific volume $v = V/N$, we finally find

$$\begin{aligned} \ln (Z'/N!) &= N\Big[1 - \sum_k [k/(k+1)]\, \beta_k v^{-k} + \sum_k \beta_k v^{-k} + \ln v\Big] \\ &= N\Big[1 + \sum_{k \geqslant 1} [1/(k+1)]\, \beta_k v^{-k} + \ln v\Big] \end{aligned} \tag{3.38}$$

3.4. The Prototype Cluster Expansions

In the preceding section, we expressed the contribution Z' of the Coulomb interaction to the partition function in terms of irreducible cluster integrals β_k. This relation shown in (3.38) may be rewritten in the form

$$\ln (Z'/N!) = N(1 + \ln v + S_1) \tag{3.39}$$

where the decisive term is the quantity

$$S_1 = \sum_{k \geqslant 1} \frac{1}{k+1} \beta_k v^{-k} \tag{3.40}$$

The bearing of this term S_1 on the physical phenomena may be judged by formulating the equation of state according to (2.10)

$$p/\Theta = v^{-1} + \partial S_1/\partial v \tag{3.41}$$

Obviously S_1 determines the deviations due to Coulomb interactions. Consequently, S_1 will also determine the coefficients of the virial expansion.

It is customary in the literature to use densities instead of specific volumes and virial coefficients instead of irreducible cluster integrals. This transition may readily be achieved by applying the following transformations to (3.39)–(3.41):

$$n = 1/v; \qquad k = N - 1; \qquad S = S_1/v$$
$$B_N = \beta_{N-1}/N = (1/N!) \int \sum \prod f_{ij}\, d\mathbf{r}_1 \cdots d\mathbf{r}_{N-1} \tag{3.42}$$

It follows

$$S = S_1/v = \sum_{k \geqslant 1} [1/(k+1)]\, \beta_k\, v^{-(k+1)} = \sum_{N \geqslant 2} B_N n^N \tag{3.43}$$

and

$$p/\Theta = n + S - n\, \partial S/\partial n \tag{3.44}$$

and with that

$$p/\Theta = n - \sum_{N \geqslant 2} (N-1)\, B_N n^N \tag{3.45}$$

As this relation demonstrates $(1 - N)\, B_N$ is the Nth virial coefficient.

For the sake of simplicity, our consideration so far was restricted to a single component system. Single component Coulomb systems of appreciable density are difficult to realize due to the corresponding field and energy problems. Normally we are confronted with the two component electron–ion system, to which we will apply our studies in the following. The generalization necessary to this end does not bring up principally new problems. However, it is necessary to introduce an appropriate generalization of the nomenclature.

Without going into the details of its justification, we present here the results of the studies of Mayer (1950), and Meeron (1958), which will appear "more or less evident" from comparison with (3.39)–(3.45). (The critical reader is referred to the cited literature.)

Let our system have σ components distinguished by the indices $1, \ldots, s, \ldots, r, \ldots, \sigma$. Let

$$\bar{N} = \{N_1, \ldots, N_s, \ldots, N_r, \ldots, N_\sigma\} \tag{3.46}$$

be a set grouping the particle numbers of the various components, and

$$\bar{n} = \{n_1, \ldots, n_s, \ldots, n_r, \ldots, n_\sigma\} \tag{3.47}$$

a set characterizing the densities of the components. Further, let us use

$$\bar{n}^{\bar{N}} = n_1^{N_1} n_2^{N_2} \cdots n_\sigma^{N_\sigma}, \qquad \bar{N}! = N_1!\, N_2! \cdots N_\sigma! \tag{3.48}$$

$$N = \sum_i N_i\,, \qquad n = \sum_i n_i \tag{3.49}$$

$$\sum_{\bar{N}(N \geqslant 2)} = \sum_{N \geqslant 2} \sum_{N_1} \sum_{N_2} \cdots \sum_{N_\sigma} \quad \text{with} \quad N = \sum_{i=1}^{\sigma} N_i \tag{3.50}$$

and

$$B_{\bar{N}} = B_{N_1,\ldots,N_\sigma} = (1/\bar{N}!) \int \sum \prod f_{ij}\, d\mathbf{r}_1 \cdots d\mathbf{r}_{N-1} \tag{3.51}$$

where in accordance with the definition of the irreducible clusters we have under the integral in (3.51) all products corresponding to graphs in which $N = \sum N_s$ diagram points are multiply connected.

Applying this nomenclature to the formulas of the multicomponent system we find in analogy to the one-component system

$$S = \sum_{\bar{N}(N \geqslant 2)} B_{\bar{N}}\, \bar{n}^{\bar{N}} \tag{3.52}$$

and for the equation of state of the system

$$p/\Theta = n + S - \sum_{i=1}^{\sigma} n_i\, \partial S/\partial n_i \tag{3.53}$$

respectively in analogy to (3.45)

$$p/\Theta = n - \sum_{\bar{N}(N \geqslant 2)} (N-1)\, B_{\bar{N}}\, \bar{n}^{\bar{N}} \tag{3.54}$$

Evaluating the coefficients $B_{\bar{N}}$ for our Coulomb system we encounter a characteristic difficulty which occurs in the calculation of cluster integrals.

The problem is best demonstrated with the simple example of two particles of kind s and i, respectively. The second virial coefficient B_{si} giving the effect of their interaction is

$$B_{si} = [1/(1 + \delta_{si})] \int_0^\infty 4\pi r_{si}^2\, f_{si}(r_{si})\, dr_{si} \tag{3.55}$$

where δ_{si} is the Kronecker symbol and r_{si} the distance of the two particles.

For large distances, we may develop the function f_{si} in the form

$$f_{si} = \exp(-z_i z_s/\Theta r_{si}) - 1 = \sum_{\nu=1}^{\infty} (1/\nu!)(-z_i z_s/\Theta r_{si})^{\nu} \tag{3.56}$$

where z_i, z_s designate the charges of the two particles. Introducing this into (3.55) we find that the first term of the development (3.56) results in a diverging contribution. This problem is not critical, since in a neutral system summation over all particle components before integration removes the divergence. On the other hand, the next term of the development (3.56) diverges too, and here the neutrality does not prove helpful, since the sign of the particle charges does not affect the contribution. Also the third term diverges logarithmically.

If we had chosen the usual procedure for the evaluation of S—integrating over the product of f in $B_{\bar{N}}$, summing over all the graphs belonging to one of these sets $\bar{N}$, and later over all possible sets—then due to the divergence the integrations could not have been performed.

To overcome this long-range divergence, Mayer suggests the following procedure:

First, we introduce a convergence factor in the form

$$\phi_{ij} = z_i z_j \, e^{-\alpha r_{ij}}/r_{ij} = z_i z_j \, g(r_{ij}) \tag{3.57}$$

In the final results we perform the transition $\alpha \to 0$.

Secondly, considering the sum over all graphs and over all particle sets Mayer suggests to reverse the sequence of the summations by first summing over all particle sets.

Finally we introduce a new system of graphs emerging from the development

$$f_{ij} = \exp[-z_i z_j \, g(r_{ij})/\Theta] - 1 = \sum_{\nu=1}^{\infty} [(-z_i z_j/\Theta)^{\nu}/\nu!](g_{ij})^{\nu} \tag{3.58}$$

if we relate each g-factor to a g-bond. At last we have to sum over the different graphs.

It is necessary to classify the g-clusters in (3.58) observing the restrictions imposed on the f-clusters in (3.51) and the properties of irreducible clusters. Since in the irreducible f-clusters all particles are multiply connected, this is true even more for the corresponding g-clusters, which in contrast to the f-clusters have multiple direct bonds.

We distinguish three essentially different groups of g-clusters characterized by specific qualities of the associated graphs:

1. Single graphs are represented by

(3.59)

2. Cycles are characterized by the fact that each point of the graph is a knot of order two. Some examples are

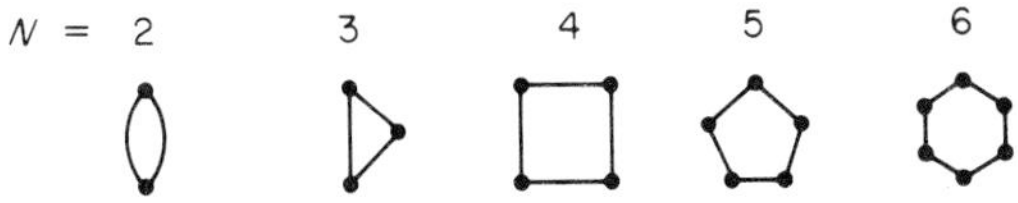

(3.60)

3. All other graphs fall into this group. It is characteristic that in these graphs always two or more points are more than doubly connected. The examples

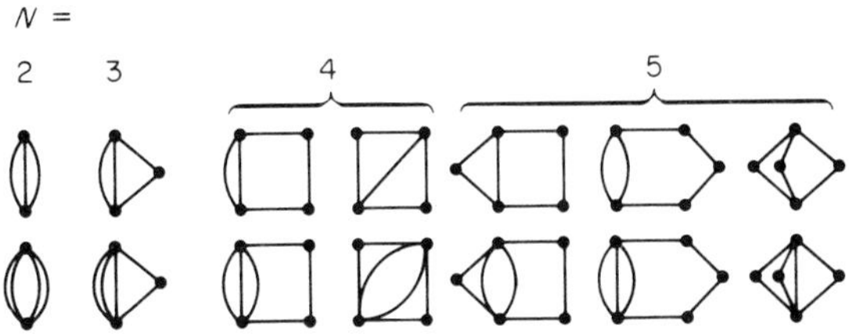

(3.61)

show graphs where only two points are more than doubly connected, whereas in the following graphs more than two points show higher-order connections:

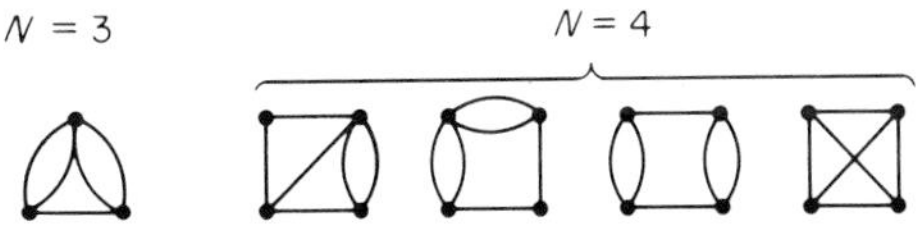

(3.62)

Within Group 3 **prototype graphs** are distinguished by the fact that all their points are knots of higher than second order. We characterize the prototype graph by the symbols $\overline{m}$ and ν. $\overline{m}$ designates the particle set and ν the scheme of connections.

All graphs of Group 3 can be coordinated with prototype graphs by grouping to each prototype graph $(\overline{m}, \nu)$ all graphs which originate from it by substitution of direct connections by chains. Within each chain arbitrary sequences of different particle kinds may occur.

Using the above classification we claim:

I. The single graphs do not contribute in a neutral system.
II. The contributions of the cycles result in the Debye–Hückel law.
III. The total contribution of the graphs of Group 3 can be represented by the prototype graphs only—provided that we replace the g-bonds by q-bonds given by

$$q_{ij} = \exp\left(-\kappa_D r_{ij}\right)/r_{ij} \tag{3.63}$$

where

$$\kappa_D^2 = (4\pi/\Theta) \sum_i z_i^2\, n_i \tag{3.64}$$

is the Debye constant.

Ad I. The first statement is trivial.
Ad II. The proof of the second claim is most easily deduced from the proof of the third statement. It will be derived from there.
Ad III. To prove the third claim we consider an arbitrary g-graph and first introduce a number of symbols.

We recall that $\bar{N}$ is the set of all particles in our g-graph. N_s is the number of particles of the kind s. N designates the total number of all particles in the set. Our g-graph is coordinated to a certain prototype graph which contains a subset $\bar{m}$ of the particles of set $\bar{N}$. m_s designates the number of particles of kind s and m the total number of particles in the prototype graph.

$\bar{\mu}$ is the complementary set of $\bar{m}$ in $\bar{N}$. That means the set $\bar{\mu}$ comprises all particles situated within chains. Again μ_s designates the number of particles of kind s, μ the total number of particles within $\bar{\mu}$.

We further subdivide the set $\bar{\mu}$ into subsets $\bar{\mu}_i$ referring to the particles in the ith chain. Again μ_{is} and μ_i give the number of particles of kind s and the total number of particles in this chain, respectively.

To characterize the chain structure we use three quantities. ν gives the total number of bonds in the prototype graph or the number of chains in the g-graph. ν_s determines the number of chain ends connected to particles of the kind s of the prototype set $\bar{m}$. ν_{si} designates the number of direct bonds or zero chains connecting the particles i and s in the prototype graph.

The following trivial relations hold for the above defined quantities:

$$\bar{m} + \bar{\mu} = \bar{N}, \qquad \bar{\mu} = \sum_i \bar{\mu}_i\,, \qquad \nu = \sum_s \tfrac{1}{2}\nu_s$$
$$N = \sum_s N_s = \sum_s (m_s + \mu_s), \qquad m = \sum_s m_s\,, \qquad \mu_i = \sum_s \mu_{si}\,, \qquad \mu = \sum_i \mu_i \tag{3.65}$$

The contribution to $B_{\bar{N}}$ of the graph under consideration is designated by $B^{\bar{\mu}}_{(\bar{m},\nu)}$. The lower index characterizes the structure of the prototype, the upper index the particle sets in the chains. The corresponding contribution follows from (3.51) introducing the development (3.58) and using the above definitions

$$B^{\bar{\mu}}_{(\bar{m},\nu)} = \frac{1}{(\overline{m+\mu})!} \cdot \frac{(-1/\Theta)^{\nu}}{\bar{\nu}_{sr}!} \prod_s z_s^{\nu_s} \int \prod_i \Big[\prod_s \Big(-\frac{4\pi}{\Theta} z_s^{\,2} \Big)^{\mu_{is}} \Big] \chi_i^{(\mu_i)}(r_{kj}) \, d\mathbf{r}_{\bar{m}} \qquad (3.66)$$

where we have used the abbreviation

$$\chi_i^{(\mu_i)} = \int (g r_{k1})\, g(r_{12}) \cdots g(r_{(\mu_i - 1)\mu_i})\, g(r_{j\mu_i})\, r_{k1}^{\,2} \cdots r_{j\mu_i}^{\,2}\, dr_{k1} \cdots dr_{j\mu_i} \qquad (3.67)$$

We draw particular attention to the fact that within the χ-function we integrate only with respect to the particles within the chains, whereas the integration with respect to particles in the set $\bar{m}$ is represented by the symbol $\int d\mathbf{r}_{\bar{m}}$ in (3.66).

Following Mayer's suggestion (see p. 26) we try to sum over appropriate particle sets $\bar{N}$, as required in (3.52). To this end it is, of course, necessary to multiply first $B^{\bar{\mu}}_{(\bar{m},\nu)}$ with the multipliers $\bar{n}^{\bar{N}}$.

We now consider given sets of $\bar{N}$, $\bar{m}$, $\bar{\mu}$, and $\bar{\mu}_i$. Keeping these constant we may produce new graphs belonging to the same value of $B^{\bar{\mu}}_{(\bar{m},\nu)}$ by

a. interchange of the particles within μ_{is},
b. interchange of the particles within m_s,
c. interchange of the particles between m_s and μ_s.

a. The interchange within μ_{is} corresponds to a permutation within the chains. The number of graphs produced by this procedure is given by

$$\prod_i \bar{\mu}_i! \qquad (3.68)$$

b. The permutation of the individuals within the m_s will not be considered here since the corresponding new graphs are taken into account in the summation of (3.51) as new prototype graphs.

c. The number of new graphs emerging from the exchange of individuals between the groups m_s and μ_s is given by

$$(\overline{m+\mu})!/\bar{m}! \prod_i \bar{\mu}_i! \qquad (3.69)$$

The contribution of all graphs produced by the particle interchange (a, c) is described by multiplication of $\bar{n}^{\bar{N}} B^{\bar{\mu}}_{(\bar{m},\nu)}$ of (3.66) with the factors (3.68), (3.69). We find

$$[\bar{n}^{\bar{N}} B^{\bar{\mu}}_{(\bar{m},\nu)}]' = \frac{\bar{n}^{\bar{m}}}{\bar{m}!} \frac{(-1/\Theta)^{\nu}}{\bar{\nu}_{sr}!} \prod_s z_s^{\nu_s} \int \prod_i \bar{\mu}_i! \prod_s \frac{[-(4\pi/\Theta)\, z_s n_s]^{\mu_{is}}}{\mu_{is}!} \chi_i^{(\mu_i)}\, d\mathbf{r}_{\bar{m}} \tag{3.70}$$

To derive the contribution presented by all graphs having the same chain length μ_i but different construction of the chains, we sum the result (3.70) over all sets $\bar{\mu}_i$ keeping the numbers μ_i constant. In doing so we use the relation

$$\sum_{\substack{\Sigma_s \mu_{si} = \mu_i \\ =\text{const.}}} \bar{\mu}_i! \prod_s \frac{[-(4\pi/\Theta) z_s^2 n_s]^{\mu_{si}}}{\mu_{si}!} = [\sum_s (-(4\pi/\Theta)\, z_s^2\, n_s)]^{\Sigma \mu_{si}} = (-\kappa_{\mathrm{D}}^2)^{\mu_i} \tag{3.71}$$

and find the result

$$[\bar{n}^{\bar{N}} B^{\bar{\mu}}_{(\bar{m},\nu)}]'' = \frac{\bar{n}^{\bar{m}}}{\bar{m}!} \frac{(-1/\Theta)^{\nu}}{\bar{\nu}_{sr}!} \prod_s z_s^{\nu_s} \int \prod_i (-\kappa_{\mathrm{D}}^2)^{\mu_i} \chi_i^{(\mu_i)}\, d\mathbf{r}_{\bar{m}} \tag{3.72}$$

where κ_{D}^2 is given by

$$\kappa_{\mathrm{D}}^2 = \sum_s (4\pi/\Theta)\, z_s^2\, n_s \tag{3.73}$$

The result in (3.72) summarizes the contributions of all graphs of a certain prototype graph with the same chain lengths. To find the total contribution of the whole prototype graph group, we sum the result (3.72) over all possible chain lengths. This problem is complicated due to the factor $\bar{\nu}_{sr}!$ in the denominator of (3.72) because of the fact that independent summation for each chain produces too many states. We would count two graphs different in which only two chains joining the same diagram points are exchanged. These states, however, are not distinguished. Without going through the details of the arguments (see, for instance, Friedman, 1962) we may be allowed to quote the result that for all chains the summation from $\mu_i = 0$ to $\mu_i = \infty$ may be carried out independently provided that we use in the denominator of (3.72) the quantity $\bar{\nu}^0_{sr}!$ instead of $\bar{\nu}_{rs}!$, where ν^0_{sr} is the number of direct bonds between the particles r and s of the prototype cluster.

With that we have for the contribution of a certain prototype graph group

$$[\bar{n}^{\bar{N}} B^{\bar{\mu}}_{(\bar{m},\nu)}]''' = \frac{\bar{n}^{\bar{m}}}{\bar{m}!} \frac{(-1/\Theta)^{\nu}}{\bar{\nu}^0_{sr}!} \prod_s z_s^{\nu_s} \int \prod_i \chi_i\, d\mathbf{r}_{\bar{m}} = S^{(\text{Prototype})} \tag{3.74}$$

where χ_i is given by

$$\chi_i(r_{kj}) = \sum_{\mu_i} (-\kappa_D{}^2)^{\mu_i} \chi_i^{(\mu_i)}(r_{kj}) \tag{3.75}$$

We now calculate the effective interaction law (3.63). Applying the convolution theorem of Fourier analysis to (3.67) we know that we have the relation

$$\tilde{\chi}_i^{(\mu_i)}(\xi) = [\tilde{g}(\xi)]^{\mu_i+1} \tag{3.76}$$

between the spectral function $\tilde{\chi}_i^{(\mu_i)}(\xi)$ and the spectral function $\tilde{g}(\xi)$. On the other hand, it follows from (3.75)

$$\tilde{\chi}_i = \sum_{\mu_i} (-\kappa_D{}^2)^{\mu_i} \tilde{\chi}_i^{(\mu_i)} \tag{3.77}$$

Introducing the result (3.76) into (3.77) the summation may be readily performed, and we arrive at

$$\tilde{\chi}_i = \frac{\tilde{g}}{1 + \tilde{g}\,\kappa_D{}^2} \tag{3.78}$$

The spectral function $\tilde{g}$ (including the convergence factor α) is

$$g(\xi) = \int_0^\infty r^2 g(r) \sin(\xi r)/\xi r \, dr = (\alpha^2 + \xi^2)^{-1} \tag{3.79}$$

from which follows with $\alpha \to 0$

$$\tilde{\chi}_i = (\kappa_D{}^2 + \xi^2)^{-1} \tag{3.80}$$

and the inverse transformation finally yields

$$\chi_i(r) = \exp(-\kappa_D\, r)/r \tag{3.81}$$

With this result, claim III is proved.

Ad II: The proof of claim II is readily related to the above conclusions for III. We consider a cycle consisting of particle groups a_s. We may regard this cycle as a closed chain. The numbers a and a_s correspond to μ_i and μ_{is} in the preceding. In strict analogy to (3.70) we find for the contribution of this cycle to the function S

$$S^{(\bar{a},c)} = \prod_s \left\{ \frac{[-(4\pi/\Theta)\, z_s{}^2 n_s]^{a_s}}{a_s!} \right\} (1/4\pi)\, \chi^{(a-1)}(r = 0) \tag{3.82}$$

As in the preceding we now first interchange all particles keeping the set $\bar{a}$ constant. Then we allow for all possible sets keeping only the number a constant. Next we sum over all possible values of a from 0 to ∞. The result is

$$S^{(c)} = \sum_{a\geqslant 2} (1/2a)(-\kappa_D^2)^a (1/4\pi)\, \chi^{(a-1)}(r = 0) \tag{3.83}$$

The sum in (3.83) cannot be summed up directly to the desired function $S^{(c)}$ because the factors have different indices, and also the index a occurs explicitly in the denominator. If, however, we differentiate $S^{(c)}$ with respect to κ_D^2 we find

$$\frac{dS^{(c)}}{d(\kappa_D^2)} = -(1/8\pi) \sum_{a\geqslant 1} (-\kappa_D^2)^a\, \chi^{(a)}(r = 0) = -(1/8\pi)[\chi(r = 0) - \chi^0(r = 0)] \tag{3.84}$$

and have the direct transition to the function $\chi(r)$. Remembering that χ^0 is the function g, we find in the limit $r \to 0$ and $\alpha \to 0$

$$\frac{dS^{(c)}}{d(\kappa_D^2)} = \kappa_D/8\pi \tag{3.85}$$

and with that

$$S^{(c)} = \kappa_D^3/12\pi \tag{3.86}$$

Introducing this result into the equation of state we have the well-known Debye–Hückel law

$$p/\Theta = n - \kappa_D^3/24\pi \tag{3.87}$$

The cycles contribute the Debye–Hückel part to the virial expansion.

3.5. The Giant Cluster Expansions

Mayer's g-graph expansion removes the long-range divergence. Unfortunately, it introduces at the same time a new divergence for the small range $r \to 0$. This is no surprise, since the power expansion of $\exp(-1/r)$ breaks down for $r \to 0$.

This divergence does not affect Mayer's original investigations since he is interested in ions in an electrolyte where he considers, in addition, short-range interactions which remove the divergence. In our Coulomb system of point charges, however, it remains a problem. We emphasize that this problem is purely a consequence of the special mathematical approach we have used.

To overcome this short-range divergence, Abé suggested the **giant cluster expansion** (Abé, 1959).

It is sufficient to consider one component only ($z_1^2 = e^2$). The previous results for S may then be written in the form

$$\begin{aligned} S &= S^{(c)} + S^{(P)} \\ &= S^{(c)} + \sum_{m=2}^{\infty} (n^m/m!) \int \sum_{\nu} \prod (-e^2\chi(r_{ik})/\Theta)^{\nu_{rs}}/\nu_{rs}!\, d\mathbf{r}_1 \cdots d\mathbf{r}_{m-1} \end{aligned} \tag{3.88}$$

where $S^{(c)}$ is the Debye contribution of the cycles and $S^{(P)}$ that of the prototypes. The first summation in (3.88) covers all prototypes with different particle numbers m, the second extends over all prototype graphs (ν) belonging to a given particle set $\bar{m}$. Since the integrand is assumed not to depend on the position of the clusters in space, we have carried out the integration with respect to $\mathbf{r}_m$ and the corresponding contribution cancelled against a volume factor in the denominator.

The simplest prototype graph and its contribution to S is

$$\begin{aligned} : S^{(m,\nu)} = S^{(2,3)} &= (n^2/2!) \int_0^{\infty} [(-\chi e^2/\Theta)^3/3!]\, 4\pi r^2\, dr \\ &= -(e^6 n^2 \Theta^{-3}\pi/3) \int_0^{\infty} \exp(-3\kappa_D r)/r\, dr \end{aligned} \tag{3.89}$$

The integral diverges.

Since we discussed that the short-range divergence was due to Mayer's expansion into a g-power series, one can hope to remove the problem by summing up the critical expansion.

We demonstrate this with the example of the **watermelon graphs** shown in the following:

(3.90)

The terms in this summation correspond to the powers of Mayer's g-expansion with the difference that we have here χ instead of g. For small values of r, however, these functions are identical. We can therefore hope that the summation (3.90) eliminates the divergence.

The contribution of the watermelon graphs of (3.90) may be written in the form

$$\begin{aligned} S^{(2)} &= (n^2/2) \int_0^{\infty} \sum_{\nu=3}^{\infty} [(-e^2\chi/\Theta)^{\nu}/\nu!]\, 4\pi r^2\, dr \\ &= 2\pi n^2 \int_0^{\infty} [e^{-e^2\chi(r)/\Theta} - 1 + e^2\chi(r)/\Theta - \tfrac{1}{2}[e^2\chi(r)/\Theta]^2]\, r^2\, dr \end{aligned} \tag{3.91}$$

or with the definition

$$w_0 = -(1/\Theta)e^2\chi, \qquad w_l = \sum_{\nu=l}^{\infty} (-e^2\chi/\Theta)^\nu/\nu! \qquad (l > 0) \tag{3.92}$$

in the form

$$S^{(2)} = (n^2/2) \int_0^\infty w_3(r)\, 4\pi r^2\, dr \tag{3.93}$$

$S^{(2)}$ is not divergent anymore.

Let us now study prototype graphs with three junctions. We consider the basic types

(3.94)

To each of them, we associate all other prototypes which originate from the basic ones through replacement of the double bonds by multiple bonds. We sum over all these attached graphs and the four types in (3.94), which results in

$$S^{(3)} = (3n^3/3!) \int w_0(12)\, w_2(23)\, w_2(31)\, d\mathbf{r}_1\, d\mathbf{r}_2$$
$$+ (n^3/3!) \int w_2(12)\, w_2(23)\, w_2(31)\, d\mathbf{r}_1\, d\mathbf{r}_2 \tag{3.95}$$

In this relation the first term represents the contribution of the first three graphs.

The extension of the above procedure to prototype graphs with more than three junctions is obvious.

We see that the contributions $S^{(P)}$ present themselves again in the form of integrals over products of w_i-functions. This suggests that we introduce w-graphs as an ordering scheme. In contrast to the f- and g-graphs, we have to distinguish different types of bonds w_i. The symbolism we use is

$w_0 \rightarrow$ - - - -; $\quad w_1 \rightarrow$ ———; $\quad w_2 \rightarrow$ ═══; $\quad w_3 \rightarrow$ ≡≡≡; $\quad \cdots$ (3.96)

With that we may write $S^{(2)}$ and $S^{(3)}$ in the form

$S^{(2)} = (n^2/2) \int$  $4\pi r^2\, dr$

$S^{(3)} = (n^3/3!) \int \Big($  $\Big)\, d\mathbf{r}_1\, d\mathbf{r}_2$

(3.97)

Again the key function S is given as a sum over certain irreducible graphs determined through cluster integrals of w_i-bonds. In complete analogy to Mayer's presentation (3.43), we write S in the form

$$S = S^{(c)} + \sum_{k=1}^{\infty} [\gamma_k/(k+1)]\, n^{k+1} \tag{3.98}$$

where we define as an irreducible cluster integral of the w_i-bonds

$$\gamma_k = (1/k!) \int \sum \prod_{k+1 \geqslant i > j \geqslant 1} w_i(i, j) \prod d\mathbf{r}_k \tag{3.99}$$

In the sum of (3.99) only the "irreducible w_i-graphs" should occur.

The relation (3.99) yields for $k = 1$ and $k = 2$ the results (3.97). For $k = 3$, we have

$S^{(4)} = (n^4/4!) \int \Big($ 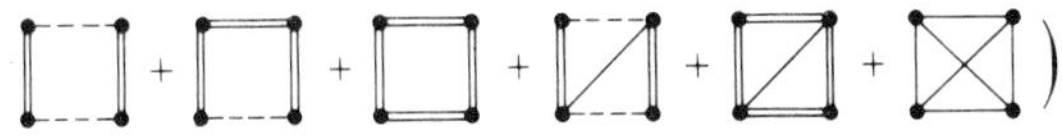 $\Big)\, d\mathbf{r}_1\, d\mathbf{r}_2\, d\mathbf{r}_3$

(3.100)

Abé applied his idea to an electron system with a smeared out ion background. He also evaluated the term $S^{(2)}$ to estimate the contribution of higher-order terms.

The straightforward calculation yields

$$S^{(2)} = (\kappa_D{}^4\, e^2/4\pi\Theta)\{A + \tfrac{1}{12}\ln(\kappa_D e^2/\Theta)\} \tag{3.101}$$

where the constant A is

$$A = \tfrac{1}{6}C + \tfrac{1}{12}\ln 3 - \tfrac{11}{72} \tag{3.102}$$

and C is Euler's constant.

The order of magnitude of the third-order terms is

$$S^{(3)} = O((\kappa_D r_w)^5) \tag{3.103}$$

If we remember that the contribution of the cycles was given by

$$S^{(c)} = \kappa_D{}^3/12\pi \tag{3.104}$$

we see that the contribution of the second-order prototype graphs is small by one order of $\kappa_D r_w$ and that of the third-order prototype graphs by two orders of $\kappa_D r_w$ in comparison to the Debye–Hückel value $S^{(c)}$.

Figure I.3 demonstrates the ratio of the prototype cluster and the

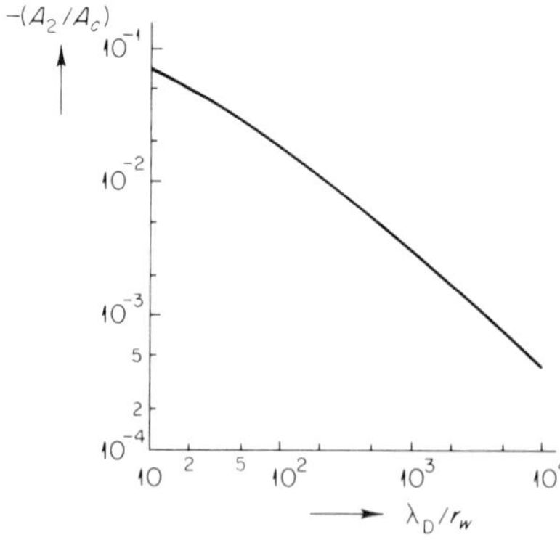

FIGURE I.3. Ratio of the prototype- to the cycle-cluster contribution as a function of the ratio of the Debye length and the classical interaction radius.

cycle contributions to the equation of state

$$\frac{A_2}{A_c} = \frac{S^{(2)} - n(\partial S^{(2)}/\partial n)}{S^{(c)} - n(\partial S^{(c)}/\partial n)} = \frac{0.92 - \ln(\lambda_D/r_w)}{\lambda_D/r_w} \tag{3.105}$$

which depends only on the ratio λ_D/r_w of the Debye length $\lambda_D = 1/\kappa_D$ and the classical interaction radius.

We want to recall that the short-range divergence treated by Abé and others (Morita, 1959) is a purely mathematical effect in Mayer's expansion. It should be distinguished from the principal physical divergence encountered by two oppositely charged particles at small distances. The latter has its origin in the fact that for point charges the quantum-mechanical effects cannot be omitted. In his giant cluster treatment, Abé does not encounter this phenomenon, since he considers only a one-component system with smeared out oppositely charged background.

4. Microscopic Qualities of the Coulomb System

In this section we intend to investigate distribution functions in the Coulomb system. We distinguish two groups. In the first group we consider the distribution of phase variables like the geometric coordinates, the momenta, and the internal states in the phase space. For instance, we shall investigate the distributions of selected coordinate sets and with that correlation properties. The second group comprises distributions of properties which are dependent functions of a given set of phase variables. We shall investigate, for example, the microfield distribution.

4.1. Distribution of the Individuals in Phase Space

The statistical behavior of our system is described by the ensemble distribution function in the Γ phase space. With reference to the free–bound model presented in Section I.3, this phase space distribution can be written as a product of two factors, the one containing the contributions of the bound states which in general have to be treated quantum-mechanically, the other representing the contribution of the free particles which may be described classically.

We concentrate on the investigation of the classical factor since the problems of the quantum-mechanical contribution of the bound states are not of interest here.

According to (1.13) the distribution of the "free particles" is given by

$$f_N^{(0)} = e^{-H/\Theta}/h^{3N}N!Z \tag{4.1}$$

where the Hamiltonian H is given by

$$H = \sum_{i=1}^{N} p_i^2/2m_i + \phi(\mathbf{r}_1, ..., \mathbf{r}_N) \tag{4.2}$$

and the potential energy of the interaction is

$$\phi = \tfrac{1}{2} \sum_{i,j}{}' \phi_{ij} = \tfrac{1}{2} \sum_{i,j}{}' e_i e_j / | \mathbf{r}_i - \mathbf{r}_j | \tag{4.3}$$

Introducing (4.2) into (4.1), we find the representation

$$f_N^{(0)} \propto \left\{ \prod_i \exp[-(1/\Theta)(p_i^2/2m_i)] \right\} e^{-\phi/\Theta} \tag{4.4}$$

In thermal equilibrium there is no correlation between configurations in the configuration space and in the momentum space. This result is not trivial and only true because the potential energy does not depend on the momentum coordinates.

Due to the lack of correlations, we may treat the distributions in the configuration and momentum spaces independently.

Equation (4.4) indicates that for the momentum space no new phenomena are to be expected. We shall have a Maxwell distribution in the momentum space where the statistical parameter Θ is correlated to the mean-square of the momentum by

$$\langle p^2/2m \rangle = \tfrac{3}{2}\Theta \tag{4.5}$$

Since the Maxwell distribution is well known, we turn to the distributions in the configuration space.

According to (4.4) the normalized distribution function in the configuration space is

$$P_N(\mathbf{r}_1, ..., \mathbf{r}_N) = \frac{\exp[-(1/2\Theta) \sum'_{i,j} \phi_{ij}]}{\int \exp[-(1/2\Theta) \sum'_{i,j} \phi_{ij}] \, d\mathbf{r}_1 \cdots d\mathbf{r}_N} \tag{4.6}$$

The direct information contained in this distribution function is of little practical value, as it is experimentally impossible to gain such a detailed knowledge of the system.

Our theoretical investigations will mainly be concerned with the reduced **specific molecular distribution functions**

$$P_s(\mathbf{r}_1, ..., \mathbf{r}_s) = \int P_N(\mathbf{r}_1, ..., \mathbf{r}_N) \, d\mathbf{r}_{s+1} \cdots d\mathbf{r}_N \tag{4.7}$$

of low order, that is, $s = 1, 2, 3$.

Comparing (3.9) and (4.6), it is logical to apply the cluster procedure to the evaluation of the molecular distribution functions. This has been done by several authors (e.g., Meeron, 1960, and De Witt, 1965). We shall not follow their path since we already studied the basic features of cluster expansions in the section on the partition function. Rather we intend to make ourselves familiar with the hierarchy approach to the problem.

4.2. The Hierarchy

It is advisable to introduce a number of definitions first:

The s-**configuration** is a certain arrangement of s individuals in the configuration space. The number of particles of kind m is given by ν_m.

The **specific molecular distribution function of order** s determines in accord with its definition in (4.7) the probability density to find s specified individuals in the positions $\mathbf{r}_1, ..., \mathbf{r}_s$.

The **general molecular distribution function of order** s—which is of particular practical interest—determines the probability density to find a set of altogether s particles in a configuration characterized by $({}^{\alpha}\mathbf{r}_1, \ldots, {}^{\alpha}\mathbf{r}_{\nu_\alpha}, \ldots, {}^{\sigma}\mathbf{r}_1, \ldots, {}^{\sigma}\mathbf{r}_{\nu_\sigma})$. ν_m designates the number of particles of kind m in the configuration. One should note carefully that dealing with the specific molecular distribution function P_s we have attached to the νth of the distinguishable particles the coordinate vector $\mathbf{r}_\nu$. Dealing with the general molecular distribution function $P^{(s)}$ the position of the ν_m particles of kind m in the s-configuration is described by the set of coordinate vectors ${}^{m}\mathbf{r}_1, \ldots, {}^{m}\mathbf{r}_{\nu_m}$, where the distinction of the particles within the group ν_m has been dropped. Obviously $\sum_{\mu=1}^{\sigma} \nu_\mu = s$ holds.

If $P^{(s)}$ designates the general molecular distribution function, N_m the total number of particles of kind m in our whole system, and ${}^{m}\mathbf{r}_i$ the coordinate of the ith particle of kind m in the configuration, then we have the following relation between the general molecular distribution function and the specific one:

$$P^{(s)}({}^{\alpha}\mathbf{r}_1, \ldots, {}^{\alpha}\mathbf{r}_{\nu_\alpha}, \ldots, {}^{\sigma}\mathbf{r}_1, \ldots, {}^{\sigma}\mathbf{r}_{\nu_\sigma}) = \prod_m [N_m!/(N_m - \nu_m)!]\, P_s(\mathbf{r}_1, \ldots, \mathbf{r}_s) \tag{4.8}$$

The factor in front of P_s is the number of realizations of a group s with subgroups ν_m out of a given set of N distinguishable individuals in subgroups N_m $(m = 1, \ldots, \sigma)$.

We derive the hierarchy by applying the gradient with respect to the coordinate i to the specific distribution function P_s resulting in

$$\nabla_i P_s = -(1/\Theta h^{3N} N! Z) \int e^{-\Phi/\Theta} \sum_{j=1}^{N}{}' \nabla_i \phi_{ij}\, d\mathbf{r}_{s+1} \cdots d\mathbf{r}_N \tag{4.9}$$

The sum under the integral is subdivided into two parts

$$\sum_{j=1}^{N}{}' = \sum_{j=1}^{s}{}' + \sum_{j=s+1}^{N} \tag{4.10}$$

where the first summation includes all the particles within the s-configuration and the rest all other particles. This yields

$$\nabla_i P_s = -(1/\Theta) \sum_{j=1}^{s}{}' (\nabla_i \phi_{ij})\, P_s - (1/\Theta) \sum_{j=s+1}^{N} \int (\nabla_i \phi_{ij})\, P_{s+1}(\mathbf{r}_1, \ldots, \mathbf{r}_s, \mathbf{r}_j)\, d\mathbf{r}_j \tag{4.11}$$

respectively, after division through P_s

$$\Theta \nabla_i \ln P_s = -\sum_{j=1}^{s}{}' (\nabla_i \phi_{ij}) - \sum_{j=s+1}^{N} \int (\nabla_i \phi_{ij}) \frac{P_{s+1}(\mathbf{r}_1, \ldots, \mathbf{r}_s, \mathbf{r}_j)}{P_s(\mathbf{r}_1, \ldots, \mathbf{r}_s)}\, d\mathbf{r}_j \tag{4.12}$$

This is already the **hierarchy of the specific molecular distribution functions.**

To gain the corresponding hierarchy for the general molecular distribution functions, we introduce (4.8) into (4.12). In doing so, let us observe that the sum in (4.12) covers like and unlike particles.

Within one particle group, however, all terms yield the same contribution to the sum, and consequently we may rewrite (4.12) in the form

$$\Theta \nabla_i \ln P_s(\mathbf{r}_1, \dots, \mathbf{r}_s)$$
$$= - \sum_{j=1}^{s}{}' (\nabla_i \phi_{ij}) - \sum_k (N_k - \nu_k) \int (\nabla_i \phi_{ij}) \frac{P_{s+1}(\mathbf{r}_1, \dots, \mathbf{r}_s, {}^k\mathbf{r}_j)}{P_s(\mathbf{r}_1, \dots, \mathbf{r}_s)} d^k\mathbf{r}_j \tag{4.13}$$

The summation index k extends over all particle kinds. It follows with

$$\frac{P_{s+1}(\mathbf{r}_1, \dots, \mathbf{r}_s, {}^k\mathbf{r}_j)}{P_s(\mathbf{r}_1, \dots, \mathbf{r}_s)} = \frac{(N_k - \nu_k - 1)!}{(N_k - \nu_k)!} \frac{P^{(s+1)}}{P^{(s)}} = \frac{1}{(N_k - \nu_k)} \frac{P^{(s+1)}}{P^{(s)}} \tag{4.14}$$

the relation

$$\Theta \nabla_i \ln P^{(s)} = - \sum_{j=1}^{s}{}' (\nabla_i \phi_{ij}) - \sum_k \int (\nabla_i \phi_{ij}) \frac{P^{(s+1)}({}^\alpha\mathbf{r}_1, \dots, {}^\sigma\mathbf{r}_{\nu_\sigma}, {}^k\mathbf{r}_j)}{P^{(s)}({}^\alpha\mathbf{r}_1, \dots, {}^\sigma\mathbf{r}_{\nu_\sigma})} d^k\mathbf{r}_j \tag{4.15}$$

for the **hierarchy of the general distribution functions.**

Before we attempt an approximate solution of the above hierarchy, we give a physical interpretation of the terms in this hierarchy. This, in particular, since we will encounter concepts like the **average force,** the **potential of the average force,** and the **average potential energy** which will be of importance in the following treatment.

4.3. Physical Interpretation of the Hierarchy

Let us first consider the average force which the particle i in the s-configuration experiences. The force which the particle i experiences in a N-configuration is

$$\mathbf{F}_i = - \sum_{j=1}^{N}{}' (\nabla_i \phi_{ij}) \tag{4.16}$$

The average force on the particle in the s-configuration is therefore defined by

$$\langle \mathbf{F}_i \rangle_s = - \frac{\int e^{-\phi/\Theta} \sum_{j=1}^{N}{}' (\nabla_i \phi_{ij}) \, d\mathbf{r}_{s+1} \cdots d\mathbf{r}_N}{\int e^{-\phi/\Theta} \, d\mathbf{r}_{s+1} \cdots d\mathbf{r}_N} \tag{4.17}$$

Applying the definition (4.7), this means

$$\langle \mathbf{F}_i \rangle_s = \Theta \nabla_i \ln P_s(\mathbf{r}_1, \ldots, \mathbf{r}_s) \tag{4.18}$$

If we further define the potential $\langle W_i \rangle_s$ of the average force of the ith particle in the s-configuration through

$$\langle \mathbf{F}_i \rangle_s = - \nabla_i \langle W_i \rangle_s \tag{4.19}$$

we find the general relation

$$P_s = \frac{\exp\{-(1/\Theta)\langle W_i \rangle_s\}}{\int \exp\{-(1/\Theta)\langle W_i \rangle_s\} \, d\mathbf{r}_1 \cdots d\mathbf{r}_s} \tag{4.20}$$

Observe that relation (4.20) holds quite generally and no assumptions have been introduced into its derivation. Observe further that the probability density to find the ith particle at the point r_i in the s-configuration is given by a **Boltzmann distribution** with the change that the **Coulomb potential energy** has been replaced by the **potential of the average force.**

The **average potential energy of the** ith **particle in the** s**-configuration** should be clearly distinguished from the **potential of the average force** defined in (4.19). The potential energy which the ith particle experiences in the N-configuration is

$$\phi_i = \sum_{j=1}^{N}{}' \phi_{ij} \tag{4.21}$$

and with that the significant definition of the **average potential energy** of the ith particle in the s-configuration is

$$\langle \phi_i \rangle_s = \frac{\int e^{-\phi/\Theta} \sum_{j=1}^{N}{}' \phi_{ij} \, d\mathbf{r}_{s+1} \cdots d\mathbf{r}_N}{\int e^{-\phi/\Theta} \, d\mathbf{r}_{s+1} \cdots d\mathbf{r}_N} \tag{4.22}$$

Again we split the sum into the two contributions of (4.10) and find

$$\langle \phi_i \rangle_s = \sum_{j=1}^{s}{}' \phi_{ij} + \sum_{j=s+1}^{N} \int \phi_{ij} \frac{P_{s+1}(\mathbf{r}_1, \ldots, \mathbf{r}_s, \mathbf{r}_j)}{P_s(\mathbf{r}_1, \ldots, \mathbf{r}_s)} \, d\mathbf{r}_j \tag{4.23}$$

or with (4.20)

$$\langle\phi_i\rangle_s = \sum_{j=1}^{s}{}'\phi_{ij} + \sum_{j=s+1}^{N} \frac{\int \phi_{ij}\, e^{-(1/\Theta)[\langle W_i\rangle_{s+1} - \langle W_i\rangle_s]}\, d\mathbf{r}_j \int e^{-(1/\Theta)\langle W_i\rangle_s}\, d\mathbf{r}_1 \cdots d\mathbf{r}_s}{\int e^{-(1/\Theta)\langle W_i\rangle_{s+1}}\, d\mathbf{r}_1 \cdots d\mathbf{r}_{s+1}} \tag{4.24}$$

This result reflects the complicated relation between the average potential energy of the ith particle and its potential of the average force. We draw attention to the fact that (4.24) includes also terms of the $(s + 1)$-configuration.

We are now in the position to discuss the physical meaning of the hierarchy. It follows from (4.18) that the left-hand side of the system (4.12) is the average force on the ith particle in the s-configuration. The first term on the right-hand side of (4.12) describes the forces exerted by the other particles in the s-configuration. To discuss the second term on the right-hand side, we consider the quantity P_{s+1}/P_s. Elementary knowledge of probability calculus says that this quantity is the conditional probability density $P(s + 1 \mid s)$ to find the particle $s + 1$ in the volume element $d\mathbf{r}_{s+1}$ under the condition that we know that the particles $1, \ldots, s$ are in the volume elements $d\mathbf{r}_1 \cdots d\mathbf{r}_s$. That is

$$P(s + 1 \mid s) = P_{s+1}/P_s \tag{4.25}$$

Accordingly, the second term on the right-hand side of (4.12) gives the average force of the particles $s + 1, \ldots, N$ on the ith particle of the s-configuration.

We see that the hierarchy formulates the trivial fact that the force on the ith particle in the s-configuration is composed of the contribution of the $(s - 1)$ remaining particles in the s-configuration and the average action of all other particles of the system.

Since the hierarchy of the general distribution functions follows from that of the special distribution functions by a simple substitution, it is clear that the physical meaning of the terms in the general hierarchy is identically the same.

As a supplement and in the interest of the sequel, we will demonstrate in addition the physical meaning of the general molecular distribution function. We know that the average density which ν_m particles of kind m in the s-configuration contribute to the total density of the system is given by

$$\begin{aligned} \langle n\rangle_{\nu_m} &= \int P_s \sum_{j=1}^{\nu_m} \delta(\mathbf{r} - \mathbf{r}_j)\, d\mathbf{r}_1 \cdots d\mathbf{r}_s \\ &= \sum_{j=1}^{\nu_m} \int {}^mP_1(\mathbf{r}_j)\, \delta(\mathbf{r} - \mathbf{r}_j)\, d\mathbf{r}_j = \nu_m\, {}^mP_1(\mathbf{r}) \end{aligned} \tag{4.26}$$

where the prefix m denotes the particle kind under consideration. Of course, this result is independent of the rank s of the configuration.

We see that the first-order specific distribution function of kind m represents the average contribution of a single particle of kind m to the density.

Employing relation (4.8), it follows that the general distribution function of first order ${}^{m}P^{(1)}(\mathbf{r})$ describes the average density of all particles of kind m in the system.

Let us now try to calculate the average density contribution of the $(N - s)$ particles not in the s-configuration. Recalling the conditional probability density

$$P(N \mid s) = P_N/P_s \tag{4.27}$$

we formulate this density contribution as

$$\langle n\rangle_{N-s} = \int \frac{P_N}{P_s} \sum_{j=s+1}^{N} \delta(\mathbf{r} - \mathbf{r}_j)\, d\mathbf{r}_{s+1} \cdots d\mathbf{r}_N = \sum_{s+1}^{N} \frac{P_{s+1}(\mathbf{r}_1, \ldots, \mathbf{r}_s, \mathbf{r})}{P_s(\mathbf{r}_1, \ldots, \mathbf{r}_s)} \tag{4.28}$$

Considering (4.14) we have

$$\langle n\rangle_{N-s} = \sum_k (N_k - \nu_k) \frac{P_{s+1}}{P_s} = \sum_k \frac{{}^{k}P^{(s+1)}({}^{\alpha}\mathbf{r}_1, \ldots, {}^{\sigma}\mathbf{r}_{\nu_\sigma}, \mathbf{r})}{P^{(s)}({}^{\alpha}\mathbf{r}_1, \ldots, {}^{\sigma}\mathbf{r}_{\nu_\sigma})} \tag{4.29}$$

where the prefix k denotes again the particle kind at the point of observation $\mathbf{r}$. This relation shows that for a given s-configuration the density contribution of all rest particles of kind k is given by the ratio ${}^{k}P^{(s+1)}/P^{(s)}$.

4.4. Approximate Solution of the Hierarchy

The hierarchy confronts us with an infinite system of coupled integro–differential equations. This problem is in general unsolvable. To make it tractable, one usually truncates the hierarchy at a certain stage $s = s'$ by expressing $P^{(s'+1)}$ through a more or less justified function or a functional of $P^{(s)}$ with $s = 1, \ldots, s'$.

We consider the two most important cases:

1. **Single-particle approximation.** In this case, we only study the first equation of the hierarchy replacing the pair distribution by a product of one particle distributions.

2. **Pair approximation.** In this approximation we study the first two equations of the hierarchy. We neglect all correlations of higher than second order and assume the correlations of second order to be small.

4.5. Single-Particle Approach

We start from the first equation of the hierarchy (4.15):

$$\Theta \nabla_i \ln {}^mP^{(1)}({}^m\mathbf{r}_i) = -\sum_k \int (\nabla_i \phi_{ij}) \frac{{}^{mk}P^{(2)}({}^m\mathbf{r}_i, {}^k\mathbf{r}_j)}{{}^mP^{(1)}({}^m\mathbf{r}_i)} \, d^k\mathbf{r}_j \tag{4.30}$$

We define the pair correlation function g_{ij} of the special molecular distribution functions by[†]

$${}^{mk}P_2({}^m\mathbf{r}_i, {}^k\mathbf{r}_j) = {}^mP_1({}^m\mathbf{r}_i){}^kP_1({}^k\mathbf{r}_j)[1 + g_{mk}({}^m\mathbf{r}_i, {}^k\mathbf{r}_j)] \tag{4.31}$$

For the case $m \neq k$ it follows with relation (4.8) for the general molecular distribution functions that

$${}^{mk}P^{(2)}({}^m\mathbf{r}_i, {}^k\mathbf{r}_j) = {}^mP^{(1)}({}^m\mathbf{r}_i)\, {}^kP^{(1)}({}^k\mathbf{r}_j)(1 + g_{mk}) \tag{4.32}$$

In the case $m = k$, we have

$${}^{kk}P^{(2)}({}^k\mathbf{r}_i, {}^k\mathbf{r}_j) = {}^kP^{(1)}({}^k\mathbf{r}_i)\, {}^kP^{(1)}({}^k\mathbf{r}_j)(1 + g_{kk})(N_k - 1)/N_k \tag{4.33}$$

In the limit $N_k \to \infty$, (4.33) reduces to the form of (4.32).

The approximation which characterizes the single-particle approach is

$$g_{mk} \equiv 0 \tag{4.34}$$

Combination of (4.30), (4.32), and (4.34) yields

$$\nabla_i \ln {}^mP^{(1)}({}^m\mathbf{r}_i) = -(1/\Theta) \sum_k \nabla_i \int {}^kP^{(1)}({}^k\mathbf{r}_j)\phi_{ij} \, d^k\mathbf{r}_j \tag{4.35}$$

We recognize that the decisive effect of the assumption (4.34) is the reduction of the hierarchy to a closed system of only as many equations as we have particle components.

Equation (4.23) together with (4.8), (4.32), and (4.34) yields

$${}^m\langle\phi_i\rangle^{(1)} = \langle\phi_i\rangle_1 = \sum_k \int \phi_{ij}\, {}^kP^{(1)}({}^k\mathbf{r}_j) \, d^k\mathbf{r}_j \tag{4.36}$$

Introducing this into (4.35) and comparing with (4.18) and (4.19) we find

$${}^m\langle\phi_i\rangle^{(1)} = {}^m\langle W_i\rangle_1 = {}^m\langle W_i\rangle^{(1)} \tag{4.37}$$

[†] It should be noted that this is not the only possibility to define pair correlation functions (see, e.g., p. 92).

Within the single-particle approach, the average potential energy is identical with the potential of the average force. This result is important since it is the background for the validity of the Poisson–Boltzmann equation which we want to derive in the following. Integration of (4.35) using the relation (4.36) and the normalization condition produces

$$ {}^{m}P^{(1)}({}^{m}\mathbf{r}_i) = \frac{\exp(-{}^{m}\langle\phi_i\rangle^{(1)}/\Theta)\, N_m}{\int \exp(-{}^{m}\langle\phi_i\rangle^{(1)}/\Theta)\, d^{m}\mathbf{r}_i} \tag{4.38} $$

In the following, we change from the potential energies to the potentials by the relation

$$ \Phi_j^{(1)} = {}^{k}\langle\phi_j\rangle^{(1)}/e_k \tag{4.39} $$

Doing so in (4.36) and applying at the same time the Laplace operator to this equation, we find "Poisson's equation"

$$ \Delta_i\, \Phi_i^{(1)} = -4\pi \sum_k e_k {}^{k}P^{(1)} \tag{4.40} $$

We now introduce the result (4.38) into (4.40) and find for the average potential $\Phi_i^{(1)}$ the differential equation

$$ \Delta_i\, \Phi_i^{(1)} = -4\pi \sum_k \frac{e_k N_k \exp(-e_k \Phi_i^{(1)}/\Theta)}{\int \exp(-\, e_k \Phi_j^{(1)}/\Theta)\, d^{k}\mathbf{r}_j} \tag{4.41} $$

This is the basic Poisson–Boltzmann equation which here has been justified within the single-particle approximation.

4.6. The Poisson–Boltzmann Equation

The Poisson–Boltzmann equation is quasilinear since its highest derivative is linear. The mathematically interesting features are identical with those of the Poisson equation. In general, it would be necessary to find numerical solutions of the Poisson–Boltzmann equation. We will, however, show that the neglect of the pair correlation g_{ij} in the present single-particle approach is identical with the requirement

$$ | e_k \Phi_j^{(1)}/\Theta | \ll 1 \tag{4.42} $$

Therefore it is consequent to linearize the Poisson–Boltzmann equation using (4.42).

Before doing this we would like to stress that so far we have always considered an arbitrary multicomponent system. Without loss of generality we consider in the following a two-component system with positive (e_+) and negative point charges (e_-).† Equation (4.41) then reads

$$\Delta\Phi^{(1)} = -4\pi\left[\frac{e_+N_+\exp(-e_+\Phi^{(1)}/\Theta)}{\int\exp(-e_+\Phi^{(1)}/\Theta)\,d\mathbf{r}} + \frac{e_-N_-\exp(-e_-\Phi^{(1)}/\Theta)}{\int\exp(-e_-\Phi^{(1)}/\Theta)\,d\mathbf{r}}\right] \tag{4.43}$$

where we can safely drop the index i.

To prepare the solution, we may take advantage of the fact that the Poisson–Boltzmann equation is independent of the choice of the zero point of the potential. We may therefore dispose of this quantity through the relation

$$\frac{e_+N_+}{\int\exp(-e_+\Phi^{(1)}/\Theta)\,d\mathbf{r}} = \frac{-e_-N_-}{\int\exp(-e_-\Phi^{(1)}/\Theta)\,d\mathbf{r}} = \rho \tag{4.44}$$

Equation (4.43) is then written

$$\Delta\Phi^{(1)} = -4\pi\rho[\exp(-e_+\Phi^{(1)}/\Theta) - \exp(-e_-\Phi^{(1)}/\Theta)] \tag{4.45}$$

Equation (4.45) shows that through definition (4.44) the zero point of the potential is identical with the point where the plasma is neutral. This does not imply that such a point has to exist within the range of the plasma. For the mathematical evaluation the fact that the neutral point may be outside of the plasma is without significance.

Linearizing (4.45) yields

$$\Delta\Phi^{(1)} = +(1/D_\rho^{\,2})\,\Phi^{(1)} \tag{4.46}$$

with the abbreviation

$$D_\rho^{\,2} = \Theta/4\pi\rho(e_+ - e_-) \tag{4.47}$$

The differential equation (4.46) comprises many solutions. The adapted one will be selected by the boundary conditions. The boundary conditions are determined by the geometrical structure and the physical requirements. In the present discussion we restrict ourselves to a spherically symmetric geometry. Therefore we may present the solution of the differential equation (4.46) as

$$\Phi^{(1)} = (1/r)(c_1\exp(-r/D_\rho) + c_2\exp(+r/D_\rho)) \tag{4.48}$$

This solution contains two arbitrary constants. One of them is used to satisfy the regularity requirement for $\Phi^{(1)}$ at $r = 0$. The other one is prescribed by the boundary condition at R.

Due to equilibrium, particles can neither be absorbed nor emitted

† Note that (e_-) is a negative quantity.

at the spherical surface. Since we have only one parameter at our disposal we can prescribe only one quantity. This may be the potential value at the surface or the net charge of our whole system which then would yield a certain potential value at the surface of our volume.

With the two conditions, i.e., prescribed potential value at the surface $\Phi_R^{(1)}$ and regularity at $r = 0$, the solution is written as

$$\Phi_{\text{in}}^{(1)} = \Phi_R^{(1)} \frac{R}{r} \frac{\sinh(r/D_\rho)}{\sinh(R/D_\rho)} \tag{4.49}$$

where D_ρ is determined by (4.44) and (4.47). Here we attach the index in to designate that this is the solution within $0 < r < R$ in contrast to $\Phi_{\text{ex}}^{(1)}$ in the range $R < r < \infty$ which we will discuss in the following. At this point a moment of recollection is advised.

The Poisson–Boltzmann equation is a second-order differential equation. Consequently, for a given charge density distribution, the potential distribution is completely defined by the regularity condition in the center and the choice of the point of zero potential. In our case, the particle density distributions of the two components follow from (4.49) if the total numbers N_+ and N_- are given. The zero point of the potential is already disposed of by equation (4.44). Consequently, there should be no possibility to choose arbitrarily the value of the potential at $r = R$, provided that we dispose of N_+ and N_-.

This suspicion is correct. As we can see from the first part of (4.44), N_+ or N_- is prescribed if we have chosen $\Phi_R^{(1)}$. In other words, given one of the quantities N_+ or N_- and the potential $\Phi_R^{(1)}$ the total space charge of the system is determined.

If we designate the total charge by

$$Q = e_+ N_+ + e_- N_- \tag{4.50}$$

a simple calculation using relations (4.42) and (4.44) results in

$$\begin{aligned} Q &= e_+ N_+ \left\{ 1 - \frac{\int \exp(-e_- \Phi^{(1)}/\Theta)\, d\mathbf{r}}{\int \exp(-e_+ \Phi^{(1)}/\Theta)\, d\mathbf{r}} \right\} \\ &= e_+ N_+ \left\{ 1 - \frac{V - (1/\Theta)\, e_- \int \Phi^{(1)}\, d\mathbf{r}}{V - (1/\Theta)\, e_+ \int \Phi^{(1)}\, d\mathbf{r}} \right\} \\ &= -(\rho/\Theta)(e_+ - e_-) \int \Phi^{(1)}\, d\mathbf{r} = -\Phi_R^{(1)}\, R \int_0^R \frac{\sinh(r/D_\rho)}{\sinh(R/D_\rho)} \frac{r}{D_\rho^2}\, dr \end{aligned} \tag{4.51}$$

Knowledge of Q allows us to determine the potential in the range

$r > R$, too. Applying the solutions of the charge-free Poisson equation and matching the potential at the surface we get

$$\Phi_{\text{ex}}^{(1)} = Q\left[\frac{1}{r} - \frac{1}{R}\right] + \Phi_R^{(1)} \tag{4.52}$$

Equations (4.49) and (4.52) represent the potential distribution within the whole configuration space for N_+ or N_- given and prescribed surface potential $\Phi_R^{(1)}$.

The parameter D_ρ given in (4.47) can be simplified. It follows for ρ from (4.44) using condition (4.42)

$$\rho = \frac{N_+e_+}{V - (1/\Theta)\, e_+ \int \Phi^{(1)}\, d\mathbf{r}} = \frac{-N_-e_-}{V - (1/\Theta)\, e_- \int \Phi^{(1)}\, d\mathbf{r}} \tag{4.53}$$

and therefore

$$\rho = \frac{e_+e_-(N_+ + N_-)}{(e_- - e_+)V} \tag{4.54}$$

and finally with (4.47) the result

$$1/D_\rho^2 = -4\pi e_+e_-(N_+ + N_-)/\Theta V \tag{4.55}$$

We can, of course, find all possible potential distributions by varying the parameters N_+ (or N_-) and $\Phi_R^{(1)}$. Unfortunately, the grouping achieved in this way is not very transparent to an interpretation and—what is even worse—practically all curves have different zero points of potential. We remark that the approach in the preceding section was chosen with respect to the mathematical appropriateness. To give a suitable graphical representation of our results, we renormalize the distributions so that all have zero potential at infinite distance. Moreover, we normalize the potential at R to 1. We then find

$$\frac{\Phi_{\text{ex}}^{(1)} - \Phi_\infty^{(1)}}{|\Phi_R^{(1)} - \Phi_\infty^{(1)}|} = \frac{R}{r}\,\frac{\Phi_R^{(1)}[1 - (R/D_\rho)\coth(R/D_\rho)]}{|\Phi_R^{(1)}[1 - (R/D_\rho)\coth(R/D_\rho)]|} = -\frac{R}{r} \tag{4.56}$$

and

$$\frac{\Phi_{\text{in}}^{(1)} - \Phi_\infty^{(1)}}{|\Phi_R^{(1)} - \Phi_\infty^{(1)}|} = \frac{\Phi_R^{(1)}\left(\dfrac{R}{r}\,\dfrac{\sinh[(r/R)(R/D_\rho)]}{\sinh(R/D_\rho)} - (R/D_\rho)\coth(R/D_\rho)\right)}{|\Phi_R^{(1)}[1 - (R/D_\rho)\coth(R/D_\rho)]|} \tag{4.57}$$

with

$$\Phi_R^{(1)} - \Phi_\infty^{(1)} = \Phi_R^{(1)}[1 - (R/D_\rho)\coth(R/D_\rho)] = Q/R \tag{4.58}$$

These results are shown in Figure I.4.

Two essential features are recognizable: First, the depth of the potential trough only depends on $(N_+ - N_-)$. Secondly, and this is more surprising, the potential distribution in the interior of our Coulomb system is only dependent on the average total particle number density $(N_+ + N_-)/V$. Naïvely one would have expected here an influence of the net space charge.

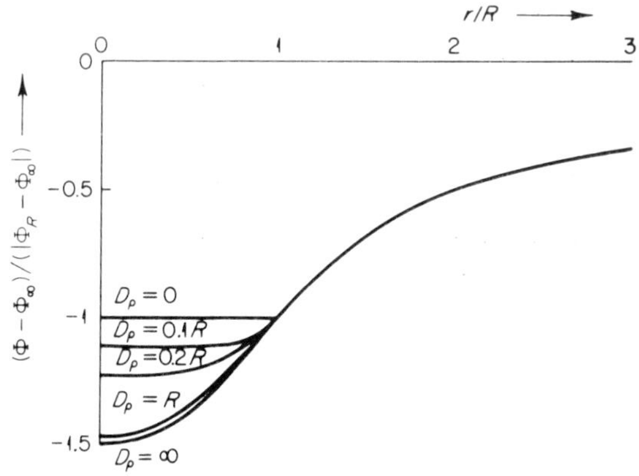

FIGURE I.4. Potential distribution of a plasma sphere in relative coordinates for various parameter values D_ρ.

It is interesting to apply our preceding results to an example of general interest.

The most widely used **definition of a plasma** is: A plasma is a system of charged particles which has field effects practically only in a small environment near its boundaries—in the sheaths.

Let us study what conditions we may derive for such a system from our above results. It is clear from (4.49) that the above definition requires

$$D_\rho \ll R \tag{4.59}$$

With this in mind an approximate solution of (4.51) results in

$$Q/e_+N_+ \simeq -(3/\Theta)(e_+ - e_-)\,\Phi_R^{(1)} D_\rho/R \tag{4.60}$$

Remembering the conditions (4.59) and (4.42) we see that the right-hand side is a second-order term negligible within the frame of our linear theory. Consequently the described definition of the plasma requires necessarily quasineutrality

$$N_+e_+ \simeq -N_-e_- \tag{4.61}$$

Inserting this into relation (4.55), we arrive at

$$\frac{1}{D_\rho^{\ 2}} = \frac{4\pi}{\Theta}\left(\frac{N_+e_+^{\ 2}}{V} + \frac{N_-e_-^{\ 2}}{V}\right) = \frac{1}{D^2} = \kappa_\mathrm{D}^{\ 2} \tag{4.62}$$

and find that our "sheath parameter" D in a plasma is identical with the well-known Debye length λ_D .

A word of caution: In considering the results, we should always bear in mind that we have used a linearized theory. We also should recall the assumption of equilibrium which excludes particle loss to the surface as well as particle emission from the surface. The extension of the surface sheaths may be different from the Debye length given in (4.62) if one of our assumptions is not fulfilled.

4.7. The Pair Approximation

In accordance with the **pair approximation** defined above we neglect all correlations of higher than second order and consider the second-order correlations as small

$$g_{ij} \ll 1 \tag{4.63}$$

Within the frame of this pair approximation we aim to prove the following three claims:

1. In analogy to (4.36) a simple relation holds between the distribution function $P^{(s)}$ and the average potential energy of the s-configuration $\langle\phi_i\rangle^{(s)}$.
2. In analogy to (4.41) a generalized Poisson–Boltzmann equation holds for the average potential energy of the s-configuration.
3. The average potential energy of the s-configuration may be written as the sum of effective pair potential energies between the individuals of the configuration. This effective potential energy depends as well on the kind of the two partners selected as on the set of the rest particles not in the s-configuration. The simplest special case of this law is the wellknown Debye–Hückel interaction within a 2-configuration.

Ad I: We start with (4.15)

$$\Theta \nabla_i \ln P^{(s)} = - \sum_{j=1}^{s}{}' \nabla_i \phi_{ij} - \sum_k \int (\nabla_i \phi_{ij}) \frac{P^{(s+1)}({}^\alpha\mathbf{r}_1, \ldots, {}^\sigma\mathbf{r}_{\nu\sigma}, {}^k\mathbf{r}_j)}{P^{(s)}({}^\alpha\mathbf{r}_1, \ldots, {}^\sigma\mathbf{r}_{\nu\sigma})} d^k\mathbf{r}_j \tag{4.64}$$

Applying the relation (4.14) to (4.23), we find for the average potential energy of the ith particle in the s-configuration

$$\langle\phi_i\rangle^{(s)} = \langle\phi_i\rangle_s = \sum_{j=1}^{s}{}' \phi_{ij} + \sum_k \int \phi_{ij} \frac{P^{(s+1)}({}^\alpha\mathbf{r}_1, \ldots, {}^\sigma\mathbf{r}_{\nu\sigma}, {}^k\mathbf{r}_j)}{P^{(s)}({}^\alpha\mathbf{r}_1, \ldots, {}^\sigma\mathbf{r}_{\nu\sigma})} d^k\mathbf{r}_j \tag{4.65}$$

Further, after application of the operator ∇_i to the relation (4.65) and introduction of the result into (4.64), we have

$$\Theta \nabla_i \ln P^{(s)} = -\nabla_i \langle\phi_i\rangle^{(s)} + \sum_k \int \left[\nabla_i \frac{P^{(s+1)}({}^\alpha\mathbf{r}_1, \ldots, {}^\sigma\mathbf{r}_{\nu\sigma}, {}^k\mathbf{r}_j)}{P^{(s)}({}^\alpha\mathbf{r}_1, \ldots, {}^\sigma\mathbf{r}_{\nu\sigma})}\right] \phi_{ij}\, d^k\mathbf{r}_j \tag{4.66}$$

To advance the evaluation of the second term of the right-hand side we aim to express the ratio $P^{(s+1)}/P^{(s)}$ within the pair approximation.

We consider an expansion of the specific molecular distribution function of order s in the form†

$$P_s = \prod_i P_1(\mathbf{r}_i) \prod_{k(2)} (1 + {}^s g_{ij}) \prod_{k(3)} (1 + {}^s g_{ijk}) \prod_{k(4)} \cdots \tag{4.67}$$

In this equation, the quantities ${}^s g_{i\cdots k}$ are the correlation functions, attributed to the set of s particles of P_s. For the calculations in the rest of this chapter we use the assumption that the correlation functions ${}^s g_{i\cdots k}$ are the same for all s, which is a good approximation as long as $s \ll N$ holds.

The order of the correlation function is characterized by the number of indices which also designate the particles engaged in this correlation. The symbol $k(q)$ indicates that the product should be taken over all combinations of order q.

Within the frame of the pair approximation we neglect all correlations

† See footnote on p. 44.

of higher than second order. Equation (4.67) may be rewritten in the form

$$P_s = \prod_i P_1(\mathbf{r}_i) \prod_{k(2)} (1 + g_{ij}) = \frac{\prod_{k(2)} P_2(\mathbf{r}_i, \mathbf{r}_j)}{\prod_i (P_1(\mathbf{r}_i))^{(s-2)}} \tag{4.68}$$

It follows that

$$\frac{P_{s+1}(..., \mathbf{r}_j)}{P_s(\cdots)} = \frac{\prod_i P_2(\mathbf{r}_i, \mathbf{r}_j)}{(P_1(\mathbf{r}_j))^{(s-1)} \prod_i P_1(\mathbf{r}_i)} \tag{4.69}$$

or

$$\frac{P_{s+1}(..., \mathbf{r}_j)}{P_s(\cdots)} = P_1(\mathbf{r}_j) \prod_i [1 + g_{ij}(\mathbf{r}_i, \mathbf{r}_j)] \tag{4.70}$$

According to the relation (4.14) we have

$$\begin{aligned} \frac{P^{(s+1)}(..., {}^k\mathbf{r}_j)}{P^{(s)}(\cdots)} &= (N_k - \nu_k) \frac{P_{s+1}(..., \mathbf{r}_j)}{P_s(\cdots)} \\ P^{(2)}({}^k\mathbf{r}_j, {}^q\mathbf{r}_i) &= P_2(\mathbf{r}_j, \mathbf{r}_i) N_k N_q \qquad \text{for} \quad k \neq q \\ P^{(2)}({}^k\mathbf{r}_j, {}^q\mathbf{r}_i) &= P_2(\mathbf{r}_j, \mathbf{r}_i) N_k(N_k - 1) \qquad \text{for} \quad k = q \end{aligned} \tag{4.71}$$

and it follows from (4.67) for $k \neq q$

$$P^{(2)}({}^k\mathbf{r}_j, {}^q\mathbf{r}_i) = P^{(1)}({}^k\mathbf{r}_j) P^{(1)}({}^q\mathbf{r}_i)[1 + g_{kq}({}^k\mathbf{r}_j, {}^q\mathbf{r}_i)] \tag{4.72}$$

and with that from (4.70)

$$\frac{P^{(s+1)}(..., {}^k\mathbf{r}_j)}{P^{(s)}(...)} = \frac{N_k - \nu_k}{N_k} P^{(1)}({}^k\mathbf{r}_j) \prod_i [1 + g_{kq}({}^k\mathbf{r}_j, {}^q\mathbf{r}_i)] \tag{4.73}$$

The second term on the right-hand side of (4.66) is now written as

$$\sum_k \int \left(\nabla_i \frac{P^{(s+1)}(..., {}^k\mathbf{r}_j)}{P^{(s)}}\right) \phi_{ij} \, d^k\mathbf{r}_j = \sum_k \frac{N_k - \nu_k}{N_k} \int P^{(1)}({}^k\mathbf{r}_j)[\nabla_i \, g_{kq}] \phi_{ij} \, d^k\mathbf{r}_j \tag{4.74}$$

In the derivation of this equation we neglected in accordance with the pair approximation all products of g-functions. We should observe that the operator ∇_i acts on the ith particle of the s-configuration and has therefore no effect on the first-order distribution functions in (4.73), which refer to the group of the remaining particles.

Assuming homogeneity and isotropy, the correlation function g_{kq}

depends only on the absolute value of $\mathbf{r} = {}^k\mathbf{r}_j - {}^q\mathbf{r}_i$ of the particles (i) and (j). It follows that

$$\nabla_i g_{kq} = -\nabla_j g_{kq} \tag{4.75}$$

Using this we have

$$\sum_k \int \left(\nabla_i \frac{P^{(s+1)}(\ldots, {}^k\mathbf{r}_j)}{P^{(s)}(\cdots)}\right) \phi_{ij}\, d^k\mathbf{r}_j = -\sum_k \frac{N_k - \nu_k}{N_k} \int P^{(1)}({}^k\mathbf{r}_j)\, \phi_{ij} \times \nabla_j (g_{kq}({}^q\mathbf{r}_i, {}^k\mathbf{r}_j))\, d^k\mathbf{r}_j \tag{4.76}$$

The interaction potential energy ϕ_{ij} according to its definition also depends only on the absolute value of the distance vector $\mathbf{r}$. We may therefore rewrite (4.76) in the form

$$\sum_k \int \left(\nabla_i \frac{P^{(s+1)}}{P^{(s)}}\right) \phi_{ij}\, d^k\mathbf{r}_j = -\sum_k \frac{N_k - \nu_k}{N_k} \int \frac{P^{(1)}(\mathbf{r}) - P^{(1)}(-\mathbf{r})}{2} \times (\phi_{ij}(\mathbf{r})\, \nabla g_{kq}(\mathbf{r}))\, d\mathbf{r} \tag{4.77}$$

simply by shifting the origin of the coordinate system to ${}^q\mathbf{r}_i$.

Due to the assumption of homogeneity, the quantity $P^{(1)}$ is a constant and therefore it follows that

$$\sum_k \int \left(\nabla_i \frac{P^{(s+1)}}{P^{(s)}}\right) \phi_{ij}\, d^k\mathbf{r}_j = 0 \tag{4.78}$$

and consequently

$$\Theta \nabla_i \ln P^{(s)} = -\nabla_i \langle \phi_i \rangle^{(s)} \tag{4.79}$$

Now we sum (4.79) over all particles in the s-configuration and find

$$\Theta \nabla \ln P^{(s)} = -\sum_i \nabla_i \langle \phi_i \rangle^{(s)} \tag{4.80}$$

with

$$\nabla = \sum_{i=1}^{s} \nabla_i \tag{4.81}$$

being the nabla operator in the s-configuration space.

We want to relate the right-hand side of (4.80) to the average energy of the s-configuration

$$\langle \phi \rangle^{(s)} = \tfrac{1}{2} \sum_{i,j}{}' \phi_{ij} + \sum_{j,q} \int \phi_{qj} \frac{P^{(s+1)}(\ldots, {}^k\mathbf{r}_j)}{P^{(s)}(\cdots)}\, d^k\mathbf{r}_j + C^{(N-s)} \tag{4.82}$$

The first term on the right-hand side of this equation represents the interaction energy of the s individuals in the configuration, the second term describes the average interaction energy of the $(N - s)$ rest particles with the s-configuration, and the last term $C^{(N-s)}$ gives the average interaction energy within the rest group. The term $C^{(N-s)}$ does not depend on the coordinates of the s-configuration.

We apply the operator ∇ to (4.82) first and to (4.65) second, and subtract the resulting equations. Then we find

$$\nabla\langle\phi\rangle^{(s)} = \sum_i \nabla_i \langle\phi_i\rangle^{(s)} + \sum_{\substack{j,q,i \\ q\neq i}} \int \phi_{qj} \nabla_i \frac{P^{(s+1)}(\ldots, {}^k\mathbf{r}_j)}{P^{(s)}(\ldots)} d^k\mathbf{r}_j \tag{4.83}$$

or with (4.74)

$$\nabla\langle\phi\rangle^{(s)} = \sum_i \nabla_i \langle\phi_i\rangle^{(s)} + \sum_{\substack{j,q,i \\ q\neq i}} \frac{N_k - \nu_k}{N_k} \int P^{(1)}({}^k\mathbf{r}_j)\, \phi_{qj} \nabla_i\, g_{mk}({}^m\mathbf{r}_i\,, {}^k\mathbf{r}_j)\, d^k\mathbf{r}_j \tag{4.84}$$

Again the factor $P^{(1)}$ is a constant due to the supposition of homogeneity.

Particular attention should be paid to the fact that in the second term on the right-hand-side of (4.84) the coordinates of the factor ϕ_{qj} have the indices (q, j), whereas the coordinates of the factor g_{mk} have the indices (i, j). If these sets of indices were the same we could omit this term with the argument given in (4.76) and (4.77). Nevertheless the term under consideration disappears because the relation

$$\int \phi_{qi}\, \nabla_i\, g_{mk}({}^m\mathbf{r}_i\,, {}^k\mathbf{r}_j)\, d^k\mathbf{r}_j = - \int \phi_{iq}\, \nabla_q\, g_{mk}({}^m\mathbf{r}_q\,, {}^k\mathbf{r}_j)\, d^k\mathbf{r}_j \tag{4.85}$$

holds. To verify this one must remember that g only depends on the distance. Therefore summation over all (i) and (q) in the second term on the right-hand side of (4.84) renders the term zero.

We have shown

$$\sum_i \nabla_i \langle\phi\rangle^{(s)} = \sum_i \nabla_i \langle\phi_i\rangle^{(s)} = \nabla\langle\phi\rangle^{(s)} \tag{4.86}$$

Introducing this result into (4.80) we arrive at the following simple connection between the general molecular distribution function of order s and the average potential energy of the configuration

$$\Theta\nabla \ln P^{(s)} = -\nabla\langle\phi\rangle^{(s)} \tag{4.87}$$

Ad 2: To derive a generalized Poisson–Boltzmann equation we apply the operator ∇_i to (4.65) at the same time making use of our result (4.78). We find

$$\nabla_i \langle\phi_i\rangle^{(s)} = \sum_{j=1}^{s}{}' \nabla_i \phi_{ij} + \sum_{j=s+1}^{N} \int \frac{P^{(s+1)}(\ldots, {}^k\mathbf{r}_j)}{P^{(s)}(\ldots)} (\nabla_i \phi_{ij})\, d^k\mathbf{r}_j \tag{4.88}$$

Scalar multiplication with the operator ∇_i and use of well-known relations of potential theory yield

$$\Delta_i \langle\phi_i\rangle^{(s)} = -4\pi \sum_{j=1}^{s}{}' \delta({}^k\mathbf{r}_j - {}^q\mathbf{r}_i)\, e_k e_q - 4\pi \sum_{j=s+1}^{N} \int \frac{P^{(s+1)}}{P^{(s)}} \delta({}^k\mathbf{r}_j - {}^q\mathbf{r}_i)\, e_k e_q\, d^k\mathbf{r}_j$$

$$+ \sum_{j=s+1}^{N} \int \left(\nabla_i \frac{P^{(s+1)}(\ldots, {}^k\mathbf{r}_j)}{P^{(s)}(\ldots)}\right) (\nabla_i \phi_{ij})\, d^k\mathbf{r}_j \tag{4.89}$$

Attention should be given to the third term on the right-hand side of this equation whose existence arises from the fact that the "space charge density" in the present case is a function of the test particle coordinate (${}^k\mathbf{r}_j$). According to (4.87) we introduce the relation

$$\frac{P^{(s+1)}}{P^{(s)}} = A_k \exp\left\{\frac{\langle\phi\rangle^{(s+1)} - \langle\phi\rangle^{(s)}}{\Theta}\right\} \tag{4.90}$$

where A_k is a normalization constant. With the normalization conditions

$$\int P_s\, d\mathbf{r}_1 \cdots d\mathbf{r}_s = 1$$
$$\int P_1\, d\mathbf{r}_i = 1 \tag{4.91}$$

using (4.7), (4.8), and the relation (4.68) within the frame of the pair approximation, the condition

$$\int P^{(1)}({}^q\mathbf{r}_i) P^{(1)}({}^k\mathbf{r}_j)\, g_{qk}({}^k\mathbf{r}_j, {}^q\mathbf{r}_i)\, d^k\mathbf{r}_j\, d^q\mathbf{r}_i = 0 \tag{4.92}$$

follows.

We now introduce (4.73) into (4.90), remembering that $P^{(1)}$ is a constant, and make use of the relation (4.92) to find

$$A_k = \frac{(N_k - \nu_k)V^s}{\int \exp[-(\langle\phi\rangle^{(s+1)} - \langle\phi\rangle^{(s)})/\Theta]\, d\mathbf{r}_1 \cdots d\mathbf{r}_{s+1}} \tag{4.93}$$

Introducing this result into (4.89), we get

$$\Delta_i \langle\phi_i\rangle^{(s)} = -4\pi \sum_{j=1}^{s}{}' \delta({}^k\mathbf{r}_j - {}^q\mathbf{r}_i)\, e_k e_q$$

$$-4\pi \sum_{j=s+1}^{N} \int \delta({}^k\mathbf{r}_j - {}^q\mathbf{r}_i)\, e_k e_q V^s \frac{\exp\{-[\langle\phi\rangle^{(s+1)} - \langle\phi\rangle^{(s)}]/\Theta\}(N_k - \nu_k)\, d^k\mathbf{r}_j}{\int \exp\{-[\langle\phi\rangle^{(s+1)} - \langle\phi\rangle^{(s)}]/\Theta\}\, d\mathbf{r}_1 \cdots d\mathbf{r}_{s+1}}$$

$$+ \sum_{j=s+1}^{N} \frac{\int (\nabla_i \phi_{ij})(\nabla_i \exp\{-[\langle\phi\rangle^{(s+1)} - \langle\phi\rangle^{(s)}]/\Theta\})\,(N_k - \nu_k)\, V^s\, d^k\mathbf{r}_j}{\int \exp\{-[\langle\phi\rangle^{(s+1)} - \langle\phi\rangle^{(s)}]/\Theta\}\, d\mathbf{r}_1 \cdots d\mathbf{r}_{s+1}} \tag{4.94}$$

Observe that this generalized Poisson–Boltzmann equation contains not only one function but three different functions: $\langle\phi_i\rangle^{(s)}$, $\langle\phi\rangle^{(s)}$, and $\langle\phi\rangle^{(s+1)}$.

This generalized Poisson–Boltzmann equation is rather complicated, and its solution becomes tractable only by using the pair superposition of the average potential energy of the s-configuration which we will discuss in the following.

Ad 3: We now attempt to describe the average potential energy of the s-configuration as a superposition of effective pair potential energies between the individuals of the configuration. We use the ansatz

$$\langle\phi\rangle^{(s)} = \sum_{k(2)} {}^{\bar\nu}\phi_{qk}^{(2)}({}^q\mathbf{r}_i, {}^k\mathbf{r}_j) \tag{4.95}$$

where the symbol $k(2)$ designates all possible pair combinations of particles (i, j) within the s-configuration.

We expect that the effective pair potential energy depends on the set $\bar\nu$ of the particles in the s-configuration and with that in particular on the number s. The index (q, k) shows that the effective pair potential energy will also depend on the kind of the two particles (q, k) which we consider in our s-configuration.

After choosing the kind (q, k), the effect of the system on the effective particle interactions is essentially due to the rest system of the $(N - s)$ particles. With this in mind it follows from the general approximation

$$N_k \simeq N_k - 1 \tag{4.96}$$

that the relation

$$\overline{{}^{\nu+1}}\phi^{(2)} \simeq {}^{\bar\nu}\phi^{(2)} \tag{4.97}$$

holds. Using this, the ansatz (4.95), and linearizing the exponential function—which, as we will show, is consistent within the pair approximation—we find from (4.94)

$$\Delta_i \sum_{j=1}^{s}{}' \, {}^{\bar\nu}\phi_{qk}^{(2)}({}^q\mathbf{r}_i, {}^k\mathbf{r}_j) = -4\pi \sum_{j=1}^{s}{}' \, \delta({}^k\mathbf{r}_j - {}^q\mathbf{r}_i)\, e_q e_k$$

$$- \frac{4\pi \sum_k^{\overline{N-s}} [e_q e_k (N_k - \nu_k) V^s - \sum_{l=1}^{s} (e_q e_k/\Theta)(N_k - \nu_k) V^s \, {}^{\bar\nu}\phi_{qu}^{(2)}({}^q\mathbf{r}_i, {}^u\mathbf{r}_l)]}{\int \exp\{-[\langle\phi\rangle^{(s+1)} - \langle\phi\rangle^{(s)}]/\Theta\}\, d\mathbf{r}_1 \cdots d\mathbf{r}_{s+1}}$$

$$- \sum_{j=s+1}^{N} \frac{\int (\nabla_i \phi_{ij})[(\nabla_i \, {}^{\bar\nu}\phi_{qk}^{(2)}({}^k\mathbf{r}_j, {}^q\mathbf{r}_i))/\Theta](N_k - \nu_k) V^s \, d^k\mathbf{r}_j}{\int \exp\{-[\langle\phi\rangle^{(s+1)} - \langle\phi\rangle^{(s)}]/\Theta\}\, d\mathbf{r}_1 \cdots d\mathbf{r}_{s+1}} \tag{4.98}$$

Note that the first term on the right-hand side has a prime at the summation sign, while the others do not. The last term does not show the summation with respect to l since the operator ∇_i eliminates all terms except $l = i$.

Integration by parts of the fourth term results in

$$\sum_{j=s+1}^{N} \frac{[(N_k - \nu_k) V^s/\Theta] \int {}^{\bar\nu}\phi_{kq}^{(2)} \Delta_i \phi_{ij} \, d^k\mathbf{r}_j}{\int \exp\{-[\langle\phi\rangle^{(s+1)} - \langle\phi\rangle^{(s)}]/\Theta\}\, d\mathbf{r}_1 \cdots d\mathbf{r}_{s+1}}$$

$$= \frac{4\pi}{\Theta} \sum_{j=s+1}^{N} \frac{e_k e_q (N_k - \nu_k) V^s \int {}^{\bar\nu}\phi_{kq}^{(2)}({}^q\mathbf{r}_i, {}^k\mathbf{r}_j)\, \delta({}^q\mathbf{r}_i - {}^k\mathbf{r}_j)\, d^k\mathbf{r}_j}{\int \exp\{-[\langle\phi\rangle^{(s+1)} - \langle\phi\rangle^{(s)}]/\Theta\}\, d\mathbf{r}_1 \cdots d\mathbf{r}_{s+1}} \tag{4.99}$$

and we see that this term cancels with the term $j = l$, $u = k$, of the third term on the right-hand side of (4.98). We have

$$\Delta_i \sum_{j=1}^{s}{}' \, {}^{\bar\nu}\phi_{qk}^{(2)}({}^q\mathbf{r}_i, {}^k\mathbf{r}_j) = -4\pi \sum_{j=1}^{s} \delta({}^k\mathbf{r}_j - {}^q\mathbf{r}_i)\, e_q e_k$$

$$- \frac{4\pi \sum_k^{\overline{N-s}} e_q e_k (N_k - \nu_k) V^s}{\int \exp\{-[\langle\phi\rangle^{(s+1)} - \langle\phi\rangle^{(s)}]/\Theta\}\, d\mathbf{r}_1 \cdots d\mathbf{r}_{s+1}}$$

$$+ 4\pi \sum_k^{\overline{N-s}} \sum_{\substack{l=1 \\ l \neq i}}^{s} \frac{e_q e_k (N_k - \nu_k) V^s \, {}^{\bar\nu}\phi_{qu}^{(2)}({}^q\mathbf{r}_i, {}^u\mathbf{r}_l)/\Theta}{\int \exp\{-[\langle\phi\rangle^{(s+1)} - \langle\phi\rangle^{(s)}]/\Theta\}\, d\mathbf{r}_1 \cdots d\mathbf{r}_{s+1}} \tag{4.100}$$

The pair interactions satisfy the trivial relations

$${}^{\bar\nu}\phi_{ku}^{(2)} = \frac{e_k}{e_q} \, {}^{\bar\nu}\phi_{qu}^{(2)} \tag{4.101}$$

Further the normalization constant (4.93) is arbitrary and at our disposal. We demand

$$A_k = \frac{(N_k - \nu_k)}{V} \tag{4.102}$$

We regroup the summation over l, too, so that together with (4.102) and (4.101) relation (4.100) yields

$$\Delta_i \sum_{j=1}^{s}{}' \, {}^{\bar{\nu}}\phi_{qk}^{(2)}({}^q\mathbf{r}_i, {}^k\mathbf{r}_j) = -4\pi \sum_{j=1}^{s}{}' \, \delta({}^q\mathbf{r}_i - {}^k\mathbf{r}_j)\, e_q e_k - 4\pi \sum_{k}^{\overline{N-s}} e_q e_k \frac{N_k - \nu_k}{V}$$

$$+4\pi \sum_{\substack{l=1 \\ l \neq i}}^{s} \left(\sum_{k}^{\overline{N-s}} e_k^2 \frac{N_k - \nu_k}{\Theta V} \right) {}^{\bar{\nu}}\phi_{qu}^{(2)}({}^q\mathbf{r}_i, {}^u\mathbf{r}_l) \tag{4.103}$$

It is clear that the sequence of the summation in the third term on the right-hand side of (4.103) is irrelevant. Therefore instead of summing over l with the corresponding particle kind (u), we sum over j with the corresponding particle kind (k).

Further, a homogeneous and isotropic plasma must be a neutral one which requires

$$\sum_{k}^{\bar{N}} e_k N_k = 0 \tag{4.104}$$

With that, the second term of the right-hand side of (4.103) may be written as

$$-4\pi \sum_{k}^{\overline{N-s}} e_q e_k \frac{(N_k - \nu_k)}{V} = + \frac{4\pi e_q}{V} \sum_{k}^{\bar{s}} \nu_k e_k \simeq \frac{4\pi e_q}{V} \sum_{l=1}^{s}{}' \, e_l \tag{4.105}$$

where the prime at the summation sign excludes the charge of the ith particle and $\nu_q - 1 \simeq \nu_q$ must be valid.

Applying (4.105) to (4.103), this relation trivially separates into a set of equations of the same type

$$\Delta_i \, {}^{\bar{\nu}}\phi_{qk}^{(2)}({}^q\mathbf{r}_i, {}^k\mathbf{r}_j) = -4\pi e_q e_k \delta({}^q\mathbf{r}_i - {}^k\mathbf{r}_j) + 4\pi e_q e_k / V$$

$$+4\pi \left(\sum_{k}^{\overline{N-s}} e_k^2 \frac{N_k - \nu_k}{\Theta V} \right) {}^{\bar{\nu}}\phi_{qk}^{(2)}({}^q\mathbf{r}_i, {}^k\mathbf{r}_j) \tag{4.106}$$

We will show that this equation (for the effective potential energy) has solutions of the Debye–Hückel type. Then we shall have proved Claim 3:

The average potential energy of a s-configuration may be presented as a superposition of pair potentials. These pair potential energies can be found from the generalized Poisson–Boltzmann equation (4.106). They depend on the kind of the two particles selected and on the set $\bar{\nu}$ of the s-configuration.

The above result is interesting in many respects. For instance, we see that not only the average energy of a s-configuration may be represented as a sum of effective pair potential energies, but also the interaction of a certain test particle with a s-configuration may be given as a sum of effective pair potential energies. This is readily seen if we consider the s-configuration plus the test particle as one $(s+1)$-system and the s-configuration itself as another one. Subtracting the average energy of the s-configuration from that of the (s + test particle)-configuration, we see that the difference is a given as sum over effective pair potential energies. This is only exact if we identify the quantities ${}^{\bar{\nu}}\phi^{(2)}$ respectively ${}^{\overline{\nu+1}}\phi^{(2)}$. The neglect of the one particle in these quantities is consistent within the general approximation of this section (see, e.g., 4.97).

This result allows us to calculate the field action of such a s-configuration on a test particle. Let the test particle have charge $+1$. We find the field exerted by the s-configuration on the test particle as a superposition of effective fields of each individual of the s-configuration. The effective field is proportional to the gradient of our ${}^{\bar{\nu}}\phi_{qk}^{(2)}$-interaction, where q stands for the test particle.

These examples will prove to be useful in the discussion of the field produced by the total ion component of an ion–electron system.

4.8. Solution of the Generalized Poisson–Boltzmann Equation

Equation (4.106) can be simplified by remembering that ${}^{\bar{\nu}}\phi_{qk}^{(2)}$ is a function only of the absolute value of $\mathbf{r} = ({}^{q}\mathbf{r}_i - {}^{k}\mathbf{r}_j)$.

The **generalized Poisson-Boltzmann equation** is therefore after separation

$$\frac{1}{r^2}\frac{d}{dr}r^2\frac{d}{dr}\,{}^{\bar{\nu}}\phi_{qk}^{(2)} = -4\pi e_q e_k \delta(r) + \frac{4\pi e_q e_k}{V} + \frac{1}{({}^{\bar{\nu}}\lambda_D)^2}\,{}^{\bar{\nu}}\phi_{qk}^{(2)} \tag{4.107}$$

with the abbreviation

$$1/({}^{\bar{\nu}}\lambda_D)^2 = \sum_k 4\pi e_k{}^2(N_k - \nu_k)/\Theta V \tag{4.108}$$

Note that ${}^{\bar{\nu}}\lambda_D$ is the Debye length for a system with particle numbers $(N_k - \nu_k)$.

The general solution of (4.107) is

$$^{\bar{\nu}}\phi_{qk}^{(2)} = \frac{1}{r}\left[C_1 \exp\left(-\frac{r}{^{\bar{\nu}}\lambda_D}\right) + C_2 \exp\left(+\frac{r}{^{\bar{\nu}}\lambda_D}\right)\right] \tag{4.109}$$

Let us consider the limit $V \to \infty$. For $r \to \infty$, the potential energy $^{\bar{\nu}}\phi_{qk}^{(2)}$ should be finite and therefore $C_2 = 0$.

For $r \to 0$, the singularity of the Dirac function in (4.107) must be accounted for, and therefore we have $C_1 = e_s e_k$. Consequently our solution is

$$^{\bar{\nu}}\phi_{qk}^{(2)} = (e_q e_k/r) \exp(-r/^{\bar{\nu}}\lambda_D) \tag{4.110}$$

Thus for the special case $s = 2$, this result is identical with the well-known Debye formula.

Before closing this section, two additional comments are in place.

First, we have claimed above that the linearization of the exponential functions of the pair potential energies is consistent within the pair approximation. To investigate this, we refer to (4.72) and (4.87) and find within the general frame of the pair approximation

$$\begin{aligned} {}^{qk}P^{(2)}/{}^{q}P^{(1)} &= {}^{k}P^{(1)}[1 + g_{qk}({}^{q}\mathbf{r}_i, {}^{k}\mathbf{r}_j)] \\ &= (N_k/V)[1 + g_{qk}] = (N_k/V)\exp(g_{qk}) \\ &= (N_k/V)\exp(-{}^{\bar{\nu}}\phi_{qk}/\Theta) \end{aligned} \tag{4.111}$$

which immediately yields

$$g_{qk} = -{}^{\bar{\nu}}\phi_{qk}^{(2)}/\Theta \tag{4.112}$$

Together with the basic condition (4.63) of the pair approximation, this shows that the linearization of the exponential function is not only meaningful but, in fact, consistent. An attempt to improve on the theory by including higher-order terms in the expansion of the exponential is devious and inconsistent.

Secondly, although it is the usual procedure and a tacit agreement in practically all treatments of this problem, we should not overlook the fact that we extended our solution of the Poisson–Boltzmann equation down to $r = 0$, although the linearization is not possible in the range of small r. In fact, our solution satisfies the condition (4.63) only in the range

$$r > {}^{qk}r_w = e_q e_k/\Theta \tag{4.113}$$

where ${}^{qk}r_w$ is the classical interaction radius.

Therefore it is actually not possible to determine the constant C_1 from the transition $r \to 0$. Rather we should have determined the constant C_1 from the matching of the field at $r \simeq {}^{qk}r_w$. Then, however, such continuity condition would require the knowledge of the potential distribution within the range $0 < r < {}^{qk}r_w$. If in this range the third term on the right-hand side of (4.107) contributed only negligibly, we could use the Coulomb potential energy in that area.

In the range $0 < r < {}^{qk}r_w$, the third term represents the average charge contribution of all other particles not in the s-configuration. We have to admit that we do not know the contributions in that region neither from this theory nor from another one. Quantum-mechanical effects and individual aspects become important in this area.

If we introduce the plausible assumption that the average charge contribution of all other particles in this range is negligible, then the Coulomb approach is correct, provided that we have

$$r_w = e^2/\Theta \ll r_0 = [(4\pi/3)(N/V)]^{-1/3} \tag{4.114}$$

This condition may be written in the form

$$n \ll (3/4\pi)[\Theta/e^2]^3 = n_{\mathrm{cr}} \tag{4.115}$$

where n_{cr} designates the critical density. With the assumption mentioned, we therefore expect the potential energy (4.110) to be a good approximation in systems with densities well below the critical one.

On the other hand, as one approaches the critical density the pair correlation is not small any more and higher order correlations become progressively important. Consequently, our requirement for the validity of the solution (4.110) is not an additional one but just consistent with the overall condition (4.63).

4.9. The Born–Green Equation

In the preceding section we developed a procedure to solve for the distribution functions within the pair approximation, using the average potential energies from the Poisson–Boltzmann equation. For the simplest case of the pair correlation in a 2-configuration the result is presented in (4.110) and (4.111). In this section we intend to study a different approach to the calculation of the distribution functions which is of basic interest and does not necessitate the consideration of the average potential energies. For the sake of conciseness, we restrict our investigations here to the most important case of the 2-configuration and at the same time to a neutral Coulomb system of two components of equal numbers of oppositely charged particles. The basic equation is the Born–Green equation which can be derived in the following way:

Using relations (4.72), (4.8), and (4.71) we have in the frame of the general approximation (4.63)

$$\frac{P^{(3)}(\mathbf{r}_1, \mathbf{r}_2, \mathbf{r}_3)}{P^{(2)}(\mathbf{r}_1, \mathbf{r}_2)} = \frac{P^{(2)}(\mathbf{r}_1, \mathbf{r}_3)\, P^{(2)}(\mathbf{r}_2, \mathbf{r}_3)}{P^{(1)}(\mathbf{r}_1)\, P^{(1)}(\mathbf{r}_2)\, P^{(1)}(\mathbf{r}_3)} \tag{4.116}$$

To simplify the notation let us use the following abbreviations in our two component ensemble

$$\begin{aligned} &{}^{k}P^{(1)}; \qquad k = +, - \to P^{(+)}, P^{(-)} \\ &{}^{qk}P^{(2)}; \qquad qk = ++, +-, -- \to P^{(++)}, P^{(+-)}, P^{(--)} \end{aligned} \tag{4.117}$$

With that, we find from (4.11) for the pair distribution $P^{(++)}$ the Born–Green equation

$$\begin{aligned} &\Theta \nabla_1 P^{(++)}(\mathbf{r}_1, \mathbf{r}_2) + P^{(++)}(\mathbf{r}_1, \mathbf{r}_2)\, \nabla_1 \phi_{12}^{(++)} \\ &\quad = -\frac{P^{(++)}(\mathbf{r}_1, \mathbf{r}_2)}{P^{(+)}(\mathbf{r}_1)\, P^{(+)}(\mathbf{r}_2)} \int \frac{P^{(++)}(\mathbf{r}_1, \mathbf{r}_3)\, P^{(++)}(\mathbf{r}_2, \mathbf{r}_3)}{P^{(+)}(\mathbf{r}_3)}\, \nabla_1 \phi^{(++)}(\mathbf{r}_1, \mathbf{r}_3)\, d\mathbf{r}_3 \\ &\quad\quad - \frac{P^{(++)}(\mathbf{r}_1, \mathbf{r}_2)}{P^{(+)}(\mathbf{r}_1)\, P^{(+)}(\mathbf{r}_2)} \int \frac{P^{(+-)}(\mathbf{r}_1, \mathbf{r}_3)\, P^{(+-)}(\mathbf{r}_2, \mathbf{r}_3)}{P^{(-)}(\mathbf{r}_3)}\, \nabla_1 \phi^{(+-)}(\mathbf{r}_1, \mathbf{r}_3)\, d\mathbf{r}_3 \end{aligned} \tag{4.118}$$

Apparently this equation not only contains the quantity $P^{(++)}$ but in addition the function $P^{(+-)}$. Since we may derive equivalent equations for $P^{(+-)}$ and $P^{(--)}$, we see that we actually have a closed system of three coupled Born–Green equations for the quantities $P^{(qk)}$.

In principle, one may solve these three integro-differential equations simultaneously, but instead we prefer here to use a less orthodox but simpler procedure. Suppose that we have

$$g^{(++)} = g^{(--)} = -g^{(+-)} = ge^2 \tag{4.119}$$

for the pair correlations defined in (4.31), with the notation according to (4.117). (Of course we have to show the consistency of this assumption with the final result.)

This supposition allows to decouple the equations. With the relation

$$\left(\frac{P^{(++)}(\mathbf{r}_i, \mathbf{r}_j)}{P^{(+)}(\mathbf{r}_i)\, P^{(+)}(\mathbf{r}_j)} - 1\right) = -\left(\frac{P^{(+-)}(\mathbf{r}_i, \mathbf{r}_j)}{P^{(+)}(\mathbf{r}_i)\, P^{(-)}(\mathbf{r}_j)} - 1\right) \tag{4.120}$$

and (4.32), (4.119), and using the coordinate transformation

$$\mathbf{r}_1 = \mathbf{r}, \qquad \mathbf{r}_2 = 0, \qquad \mathbf{r}_3 = \mathbf{r}' \tag{4.121}$$

we find from (4.118) the following integro-differential equation for g:

$$\begin{aligned}\Theta \nabla g + (1 + e^2 g)\, \nabla\psi(\mathbf{r}) = {} & -2ne^2 \int g(\mathbf{r}')\, \nabla\psi(|\,\mathbf{r} - \mathbf{r}'\,|)\, d\mathbf{r}' \\ & -2ne^2 \int g(|\,\mathbf{r} - \mathbf{r}'\,|)\, \nabla\psi(|\,\mathbf{r} - \mathbf{r}'\,|)\, d\mathbf{r}'\end{aligned} \tag{4.122}$$

where we have used the abbreviations

$$\begin{aligned} \phi^{(++)} &= e^2\psi, \qquad & P^{(+)} &= P^{(-)} = n \\ \phi^{(+-)} &= -e^2\psi, \qquad & \psi(\mathbf{r}) &= |\,\mathbf{r}\,|^{-1} \end{aligned} \tag{4.123}$$

Considering the fact that we are studying an infinitely extended plasma with no external field (that means that the plasma is homogeneous and isotropic), we may conclude that the second term on the right-hand side of (4.122) vanishes identically due to arguments analogous to those given on p. 54.

We restrict our considerations to the range

$$r \gg r_w = e^2/\Theta \tag{4.124}$$

Further, we will prove the relation

$$\nabla g = O\Big(\frac{g}{r} + \frac{1}{\kappa_{\mathrm{D}}}\Big) \tag{4.125}$$

to be consistent with our final result. Using (4.124) and (4.125) in (4.122) we may neglect the third term of the left-hand side of this equation in comparison to the others and find

$$\Theta \nabla g(r) + \nabla\psi(r) = -2ne^2 \int g(r')\, \nabla\psi(|\,\mathbf{r} - \mathbf{r}'\,|)\, d\mathbf{r}' \tag{4.126}$$

or

$$\Theta g(r) + \psi(r) = -2ne^2 \int g(r')\, \psi(|\,\mathbf{r} - \mathbf{r}'\,|)\, d\mathbf{r}' + C \tag{4.127}$$

where the constant C can be shown to be zero studying the limiting process $r \to \infty$.

Now we Fourier transform (4.127) and find, using the convolution theorem,

$$\tilde{g}(\xi) + 1/2\pi^2\xi^2\Theta = -(8\pi ne^2/\Theta\xi^2)\, \tilde{g}(\xi) \tag{4.128}$$

After simple rearrangement we have

$$\tilde{g}(\xi) = -1/2\pi^2(\xi^2 + \kappa_D^2)\Theta \qquad (4.129)$$

where

$$\kappa_D^2 = 8\pi n e^2/\Theta \qquad (4.130)$$

is Debye's constant of the system under consideration. The inverse transformation yields

$$g = -(1/\Theta r)\exp(-\kappa_D r) \qquad (4.131)$$

and with that the pair correlation of the ions

$$g^{(++)} = -(e^2/\Theta r)\exp(-\kappa_D r) \qquad (4.132)$$

The correlation functions $g^{(--)}$ and $g^{(+-)}$ can be derived in full analogy and show that our assumption (4.119) as well as the condition (4.125) are satisfied within the range $r > r_w$. This latter restriction is no additional requirement but rather a necessary condition due to our linearization process anyhow. It is interesting to note that this condition is identical with the condition (4.115) restricting the applicability of these theories to subcritical densities.

4.10. Distributions of Dependent Quantities

So far we have studied how the particles of our system are distributed on the phase variables. The canonical distribution of (4.1) in the Γ space allows us to calculate the probability distribution of any quantity which is a function of the phase variables.

Let $G(\mathbf{r}_1, ..., \mathbf{r}_N, \mathbf{p}_1, ..., \mathbf{p}_N)$ be such a given function. The probability distribution function $W(G)$ can then be expressed with the help of the function $f_N^{(0)}$ through

$$W(G) = \int \delta(G - G(\mathbf{r}_1, ..., \mathbf{p}_N))\, f_N^{(0)}\, d\mathbf{r}\, d\mathbf{p} \qquad (4.133)$$

where the integration has to be carried out over the whole Γ space. Introducing $f_N^{(0)}$ from (4.1) we find

$$W(G) = \frac{\int \delta(G - G(\mathbf{r}_1, ..., \mathbf{p}_N)) \exp[-H(\mathbf{r}_1, ..., \mathbf{p}_N)/\Theta]\, d\mathbf{r}\, d\mathbf{p}}{\int \exp[-H(\mathbf{r}_1, ..., \mathbf{p}_N)/\Theta]\, d\mathbf{r} d\mathbf{p}} \qquad (4.134)$$

with $H(\mathbf{r}_1, ..., \mathbf{p}_N)$ given by (4.2).

To proceed with the evaluation of the probability distribution function

$W(G)$, it is advised to transform (4.134) to the Fourier spectral function

$$\tilde{W}_G(\xi) = \int e^{-i\xi G} W(G)\, dG = \int e^{-i\xi G(\mathbf{r}_1,\ldots,\mathbf{p}_N)} f_N^{(0)}\, d\mathbf{r}\, d\mathbf{p} \tag{4.135}$$

If $\mathbf{G}$ is a vector quantity, we of course have correspondingly

$$\tilde{W}_{\mathbf{G}}(\boldsymbol{\xi}) = \int e^{-i\boldsymbol{\xi}\cdot\mathbf{G}} W(\mathbf{G})\, d\mathbf{G} = \int e^{-i\boldsymbol{\xi}\cdot\mathbf{G}(\mathbf{r}_1,\ldots,\mathbf{p}_N)} f_N^{(0)}\, d\mathbf{r}\, d\mathbf{p} \tag{4.136}$$

Further evaluation of these relations is possible only if we have knowledge about the special type of the function G. We therefore turn to a special case.

4.11. The Microfield Distribution

The microfield distribution is one of the most important and typical distributions within a plasma. The field observed at a given neutral test point can be written as a linear superposition of the contributions of all individuals of the system

$$\mathbf{E} = \sum_j \mathbf{E}_j(\mathbf{r}_j) \tag{4.137}$$

Introducing this into (4.136) we find for the Fourier spectral function $\tilde{W}_{\mathbf{E}}(\boldsymbol{\xi})$ of the microfield distribution

$$\begin{aligned} \tilde{W}_{\mathbf{E}}(\boldsymbol{\xi}) &= \int e^{-i(\boldsymbol{\xi}\cdot\mathbf{E})} W(\mathbf{E})\, d\mathbf{E} \\ &= \int \cdots \int \exp\left[-i\left(\boldsymbol{\xi}\cdot\sum_j \mathbf{E}_j\right)\right] P_N\, d\mathbf{r}_1 \cdots d\mathbf{r}_N \end{aligned} \tag{4.138}$$

Since there is no correlation between the momentum and the configuration space coordinates and since the field contributions do not depend on the momentum coordinates, we have integrated over the momenta, changing from the distribution $f_N^{(0)}$ in Γ space to the distribution function P_N in the configuration space.

The difficulty in the evaluation of this equation is of course presented by the exponential factor of the distribution function P_N.

In the light of the general success of cluster expansions, Baranger and Mozer (1959, 1960) used a cluster approach with the bonds ε_j defined through

$$\varepsilon_j = e^{-i(\boldsymbol{\xi}\cdot\mathbf{E}_j)} - 1 \tag{4.139}$$

Introducing this into (4.138) and carrying out the ε-product, we may write

$$\tilde{W}_{\mathbf{E}}(\xi) = \sum_{s=0}^{N} \tilde{W}_{\mathbf{E}}^{s}(\xi) \tag{4.140}$$

where

$$\tilde{W}_{\mathbf{E}}^{s}(\xi) = \int \cdots \int \sum_{j_1 < \cdots < j_s} \varepsilon_{j_1} \cdots \varepsilon_{j_s} P_N \, d\mathbf{r}_1 \cdots d\mathbf{r}_N \tag{4.141}$$

holds. Introducing our specific distribution function (4.7), we may write†

$$\tilde{W}_{\mathbf{E}}^{s}(\xi) = [N!/(N-s)!s!] \int \cdots \int \varepsilon_1 \cdots \varepsilon_s P_s \, d\mathbf{r}_1 \cdots d\mathbf{r}_s \tag{4.142}$$

where the combinatorial factor in front of the integral reflects the fact that every group of s particles yields the same contribution and we have to count the integral as many times as we may subdivide N particles into subsets of s particles.

We now use for P_s the expansion (see p. 51)

$$P_s = \prod_{k(1)} P_1(\mathbf{r}_j) \prod_{k(2)} (1 + {}^{s}g_{jl}) \prod_{k(3)} (1 + {}^{s}g_{jlm}) \cdots \tag{4.143}$$

where the symbols $k(i)$, as we have indicated already in (4.67), comprise all combinations of i particles out of s given individuals. After we have carried out the product in (4.143), the distribution P_s represents itself as the sum over terms, each of them being the product of ${}^{s}g_h$ and/or P_1 functions, where the index h gives the number of indices $jl \cdots m$ and therefore denotes the rank of the correlation function.

Let us now consider one of these terms. The ${}^{s}g_i$ structure of the term suggests a division of the s individuals into subgroups. Within each subgroup the individuals are directly or indirectly connected. Between different subgroups there is no connection.

We turn to the summation over all terms which exhibit the same subdivision of the s individuals into subgroups. In addition we introduce new correlation functions by the relations

$${}^{s}\bar{g}_i = \sum \prod^{\nu} {}^{s}g_{\nu} \tag{4.144}$$

† It should be noted that we consider one particle component only. Actually, however, we are interested in the situation within a true plasma consisting of an electron and ion component. We will understand in the following how the "one-component results" may be applied to the plasma.

where the products under the summation sign cover all possible connection schemes for the ν particles. In the particular case $\nu = 3$, for instance, we have

(4.145)

The contribution of this subsum to our spectral function is now given as the product over the ${}^s\bar{g}_i$ functions and fully determined by the set of the numbers $\bar{v}$, where v_h gives the number of subgroups which contain h particles each. Trivially the numbers v_h have to fulfill the condition

$$\sum_{h=1}^{s} v_h h = s \tag{4.146}$$

Now let us extend the summation over all groups which have the same substructure $\bar{v}$ but different distribution of the s individuals into these subgroups. Each of them makes the same contribution to $\tilde{W}_E^s$. Therefore we cover this summation by the factor

$$C(s) = s!\Big/\prod_{h=1}^{s} v_h!(h!)^{v_h} \tag{4.147}$$

which contains in the numerator all possible permutations of the s individuals and removes in the denominator by the first factor those permutations which should not be counted since they represent the exchange of whole subgroups and by the second factor those which correspond to permutations within a particular subgroup. Recalling that

$$P_1 = 1/V \tag{4.148}$$

holds, using the abbreviation

$${}^s\mathscr{E}_h(\xi) = \int \cdots \int \varepsilon_1 \cdots \varepsilon_h \, {}^s\bar{g}_h \, d\mathbf{r}_1 \cdots d\mathbf{r}_h \tag{4.149}$$

and the approximation

$$N!/(N-s)! \simeq N^s \tag{4.150}$$

we may write the spectral function $\tilde{W}_\mathrm{E}(\xi)$ in the form

$$\tilde{W}_\mathrm{E}(\xi) = \sum_{s=0}^{N} (n^s/s!) \left(\sum C(s)\, {}^s\mathscr{E}_1^{v_1} \cdots {}^s\mathscr{E}_s^{v_s}\right) \tag{4.151}$$

where the sum in the bracket covers all compositions of s individuals into v_h substructures of h particles each satisfying the condition (4.146).

Introducing the coefficient (4.147) and rearranging using (4.146), we find in the limit $N \to \infty$, $V \to \infty$ with constant density $n = N/V$,

$$\tilde{W}_{\mathbf{E}}(\xi) = \prod_{h=1}^{\infty} \sum_{v_h=0}^{\infty} (1/v_h!)[(n^h/h!)\, \mathscr{E}_h(\xi)]^{v_h} \tag{4.152}$$

Due to the fact that we carry the product in (4.152) from one to infinity, this equation contains a number of terms not included in (4.151). The relative contribution of these additional—incorrect—terms disappears in the limit $N \to \infty$. We have therefore

$$\tilde{W}_{\mathbf{E}}(\xi) = \exp\left\{\sum_{h=1}^{\infty} (n^h/h!)\, \mathscr{E}_h(\xi)\right\} \tag{4.153}$$

where—following Baranger and Mozer—we used the approximation

$$\mathscr{E}_h(\xi) = {}^{s}\mathscr{E}_h(\xi), \qquad s = 1, 2, 3, \ldots \tag{4.154}$$

in the last two equations.

With (4.153) our problem has been reduced to the calculation of the quantities $\mathscr{E}_h$ of (4.149). This presents insurmountable difficulties, since the correlation functions $\bar{g}_h$ are unknown for $h > 2$. We will therefore be limited to an approximation of the type

$$\tilde{W}_{\mathbf{E}}(\xi) = \exp\{n\mathscr{E}_1(\xi) + \tfrac{1}{2}n^2\, \mathscr{E}_2(\xi)\} \tag{4.155}$$

hoping that our expansion converges rapidly enough so that higher-order terms are negligible.

The Holtsmark Theory

The first calculation of the microfield distribution neglected correlations altogether (Holtsmark, 1919). In our formulation, that means that we have to deal only with the term $\mathscr{E}_1(\xi)$, whereas the term $\mathscr{E}_2(\xi)$ is identically zero. In this approximation the spectral function reads

$$\tilde{W}_{\mathbf{E}}(\xi) = \exp\left\{n \int (e^{-i(\xi\cdot\mathbf{E})} - 1)\, d\mathbf{r}\right\} \tag{4.156}$$

Let the field be given by $\mathbf{E} = -e\mathbf{r}/r^3$. After introducing polar coordinates, the integral in (4.156) can be easily evaluated and yields

$$\int_0^{2\pi}\int_0^{\pi}\int_0^{\infty} (\exp[i\xi(e/r^2)\cos\vartheta] - 1)\, r^2\, dr \sin\vartheta\, d\vartheta\, d\varphi = -\tfrac{4}{15}(2\pi e\xi)^{3/2} \tag{4.157}$$

Employing the isotropy of the system, the distribution of the absolute value of the field is given by

$$W(E) = 4\pi E^2 W(\mathbf{E}) = (2E/\pi) \int_0^\infty \sin(\xi E) \exp\{-(4n/15)(2\pi e\xi)^{3/2}\} \xi \, d\xi \quad (4.158)$$

or transforming to the reduced field

$$\beta = E/E_0, \qquad E_0 = e/r_0^2 \quad (4.159)$$

we find for the distribution of the reduced field β using the abbreviation $v = \xi E_0$

$$W(\beta) = (2/\pi\beta) \int_0^\infty v \sin v \exp[-(v/\beta)^{3/2}] dv \quad (4.160)$$

For small or large values of β the following asymptotic expansions hold:

$$W(\beta) \propto (4/3\pi)\, \beta^2 \qquad (\beta \to 0) \quad (4.161)$$

and

$$W(\beta) \propto (3/2)\, \beta^{-5/2} \qquad (\beta \to \infty) \quad (4.162)$$

In general, it is necessary to evaluate the integral by numerical methods.

The Holtsmark theory is a good approximation in the limit of high temperatures and/or small densities. Therefore any theory accounting for the interaction should reproduce the Holtsmark distribution in these limits.

A remark is noteworthy with respect to the divergence of the second moment $\langle E^2 \rangle$ of the distribution (4.158). Contrary to a widespread opinion this divergence cannot be removed by a quantum-mechanical correction but is a principal deficiency of our microfield treatment due to the unphysical assumption of point charges (Engelmann, 1962).

The Dressed Particle Approach

The first attempt to account for the correlation for the low-frequency microfield component of the ions was made by Ecker and Müller (1958) using an uncorrelated dressed ion model. This approach uses effective fields of the form

$$\mathbf{E} = (e/r^3)\, \mathbf{r}(1 + \kappa_D r)\, e^{-\kappa_D r} \quad (4.163)$$

Naturally the more complicated dependence of $\mathbf{E}$ on r in (4.163) renders the evaluation more difficult.

Introducing the field (4.163) into (4.156) and using the abbreviations

$$\bar{E} = E/\kappa_D^2, \qquad y = \xi\kappa_D\,, \qquad x = \kappa_D r \tag{4.164}$$

we find the spectral function

$$\tilde{W}_E(\xi) = \exp\left\{-3\delta \int_{E_R}^{\infty} \left\{\frac{\sin y\bar{E}}{y\bar{E}} - 1\right\} \frac{(1+x)^{5/2}\, e^{-3x/2}}{1+[1+x]^2} \frac{d\bar{E}}{\bar{E}^{5/2}}\right\} \tag{4.165}$$

The lower limit in the integral accounts for the limitations imposed by the breakdown of the pair correlation theory which limits the results to the range

$$\delta = 1/6(\pi\, n)^{1/2}\, r_w^{3/2} \gg 1 \tag{4.166}$$

where δ is the number of particles in the Debye sphere.

Holtsmark's distribution together with the dressed particle results are shown in Figure I.5. As expected, Holtsmark's result corresponds to $\delta = \infty$.

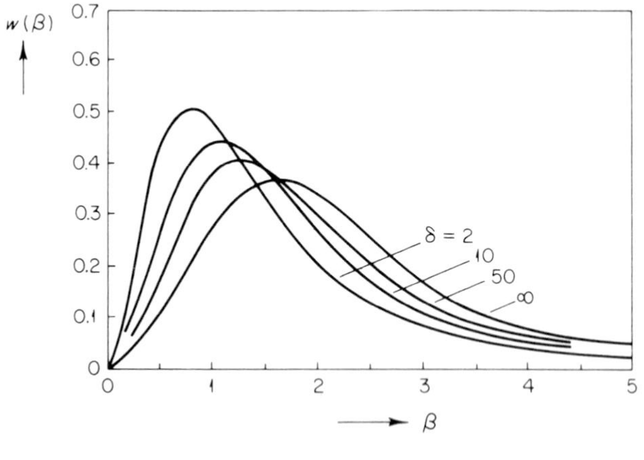

FIGURE I.5. The Holtsmark microfield distribution ($\delta = \infty$) and correlation corrections calculated from the dressed particle approach.

The influence of correlations introduced into this theory via effective field contributions of the ions grows quite remarkably as the system approaches the critical density with decreasing δ.

Studying the low-frequency ion component of the microfield, it seems quite justified to introduce an effective field which accounts for the screening of the electrons. However, it seems not so obvious that effective screening of the ions can account for the correlations of those ions actually represented by the term $\mathscr{E}_2$ of (4.158). For this reason Baranger and Mozer (1959, 1960) used their cluster approach.

Correlated Ions with Debye Fields

Studying a plasma Baranger and Mozer define as the low-frequency component the time average of the field over intervals large in comparison to the lifetime of an electron microstate, but small in comparison to the lifetime of an ion microstate. The difference between the so-defined low-frequency component and the true microfield distribution is, of course, the high-frequency component. It can be found by appropriate deconvolution of the true field component and the low-frequency component.

Without justification, Baranger and Mozer introduce for the effective fields of the low-frequency component screened Coulomb fields of the form

$$\mathbf{E}_i = (Z_i e/r_i^3)\,\mathbf{r}_i(1 + \kappa_{D-} r_i)\exp[-\kappa_{D-} r_i] \tag{4.167}$$

Our discussion of the Poisson–Boltzmann equation on p. 59 provides the basis for this assumption. For the pair correlation functions required in the calculation of $\mathscr{E}_2$, Baranger and Mozer assume the relation

$$\bar{g}_2(\mathbf{r}_1, \mathbf{r}_2) = -(e^2/\Theta)\exp[-\kappa_D r_{12}]/r_{12} \tag{4.168}$$

where κ_D contains both shielding contributions of ions and electrons in the form

$$\kappa_D^2 = \kappa_{D-}^2 + \kappa_{D+}^2 \tag{4.169}$$

Our discussion within the preceding chapters which carefully distinguishes the dependence of the pair correlations on the s-configuration indicates that this assumption is not self-evident.

To evaluate $\mathscr{E}_2$, the authors expand the factors under the integral in terms of spherical harmonics. Introducing polar coordinates and making use of the orthogonality relations and well-known addition theorems of the spherical functions, the integration with respect to the angles can be carried out elementarily. It remains an infinite series of double integrals over the radial parts. An analytical evaluation even of the simplest of these terms is impossible. Fortunately the results of the machine calculation suggest a good convergence of the development, so that one has a good approximation considering only the first three terms.

This is not the place to present the details of the numerical evaluation, and we therefore restrict ourselves to the presentation of the final results in Figure I.6, where we find the Holtsmark result and the results

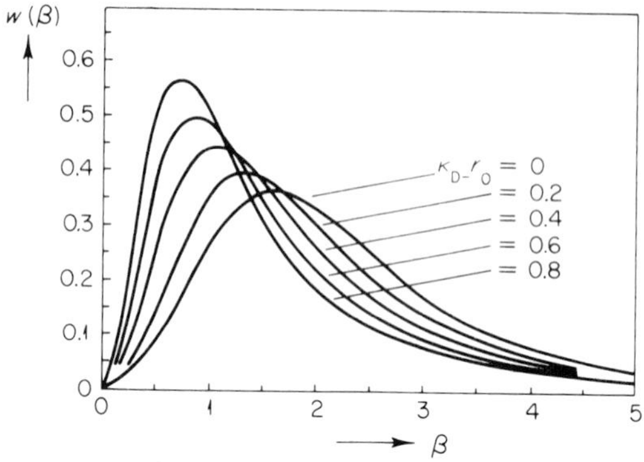

FIGURE I.6. The microfield distribution including correlations according to the approach of Baranger and Mozer.

of the approach of Baranger and Mozer. It should be mentioned that the curves shown there are not actually those evaluated by Baranger and Mozer but are corrected for a numerical error (Pfennig and Trefftz, 1966).

In Figure I.7, we compare for one characteristic value of the number of

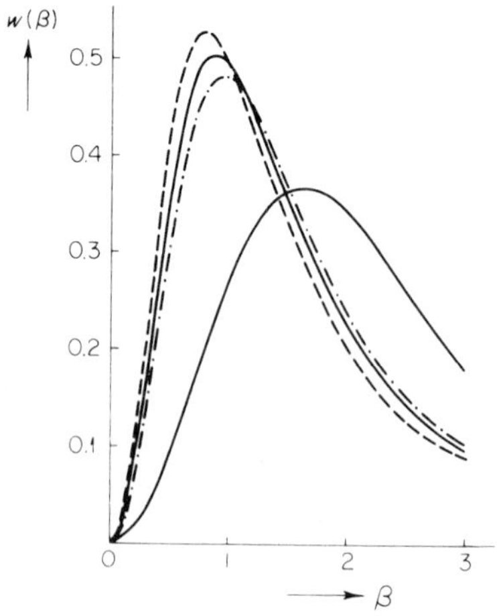

FIGURE I.7. Comparison of the Holtsmark distribution (heavy line), Baranger and Mozer's result (dotted line), the corrected result (narrow line), and Ecker and Müller's result (dashed line).

particles in the Debye-sphere ($\delta \approx 5$) the result of Ecker and Müller's approach and of Baranger and Mozer's calculations. Both theories give remarkable corrections to the Holtsmark distribution. In comparison to this correction their difference is essentially small.

It should be mentioned that both approaches apply a dressed particles model (Ecker and Fischer, 1968) to the calculation of the microfield distribution. A rigorous foundation of these models has not yet been given.

5. The Fluctuation–Dissipation Theorem

5.1. Derivation of the General Theorem

In the preceding section we studied the unrestricted probability distribution of the electric microfield for our system in equilibrium. With respect to the dynamic theory of the Coulomb system close to equilibrium another quantity, namely the conditional probability distribution $W(\mathbf{E}_t \mid \mathbf{E}_{t+\tau})$ is of vital importance. This quantity gives the probability density to find at a given point the field $\mathbf{E}_{t+\tau}$ at the time $t+\tau$ if we know that we had the field $\mathbf{E}_t$ at time t. This quantity is extremely difficult to evaluate (Chandrasekhar, 1943; Kogan and Selidowkin, 1969) and the problem has not yet been solved. Successful attempts, however, have been undertaken to calculate the field autocorrelation coefficient which determines the transport qualities of the plasma (Rostoker, 1960). It is related to the conditional field probability through

$$\langle \mathbf{E}_t)(\mathbf{E}_{t+\tau} \rangle = \int d\mathbf{E} \int d\mathbf{E}' \, \mathbf{E})(\mathbf{E}' \, W(\mathbf{E} \mid \mathbf{E}', \tau) \tag{5.1}$$

In this section we derive the fluctuation–dissipation theorem, which can be used as the basis for the calculation of the autocorrelation coefficient. We study the problem generally considering a system under the influence of a set of generalized external parameters $A_s(t)$ which depend only on time and not on the phase coordinates $\mathbf{r}_1, \ldots, \mathbf{p}_N$ of our system. Then in the approximation of "linear response" the Hamiltonian of our system may be written in the form

$$H(\mathbf{p}_k, \mathbf{r}_k, t) = H_0(\mathbf{p}_k, \mathbf{r}_k) + \sum_s B_s(\mathbf{p}_k, \mathbf{r}_k) \, A_s(t) \tag{5.2}$$

H_0 is the Hamiltonian of the undisturbed system. The coefficients B

are phase functions depending only on $\mathbf{p}_k$, $\mathbf{r}_k$ and not explicitly on the time t.

Let us first study how the dissipation of the system expresses itself in these terms. The momentary dissipation of our system reads

$$\left\langle \frac{\partial H}{\partial t} \right\rangle = \left\langle \frac{dH}{dt} \right\rangle = \left\langle \frac{dH_0}{dt} \right\rangle + \sum_s \left\langle \frac{d}{dt} B_s A_s \right\rangle \tag{5.3}$$

where the brackets $\langle\ \rangle$ denote the ensemble average. Engaging Liouville's theorem and the canonical equations of motion, we find in linear order

$$\begin{aligned} \left\langle \frac{\partial H}{\partial t} \right\rangle &= \sum_s \langle \{H_0, B_s A_s\} \rangle + \frac{d}{dt} \sum_s \langle B_s A_s \rangle \\ &= - \sum_s \langle \dot{B}_s \rangle A_s + \frac{d}{dt} \sum_s \langle B_s A_s \rangle \end{aligned} \tag{5.4}$$

where $\{\ldots,\ldots\}$ denotes the Poisson brackets. With the requirement that the external parameters A_s vanish for $t = \pm\infty$ the total energy dissipation is then given by

$$W = \int_{-\infty}^{+\infty} \langle \partial H / \partial t \rangle \, dt = - \sum_s \int_{-\infty}^{+\infty} \langle \dot{B}_s \rangle A_s \, dt = - \sum_s \int_{-\infty}^{+\infty} \langle \tilde{\dot{B}}_s \rangle_\omega \tilde{A}^*_{s\omega} \, d\omega \tag{5.5}$$

where we made use of Parseval's theorem.

Now we want to relate the dissipation just formulated to the fluctuation spectrum of the system through the application of the Kubo formula. To derive the **Kubo formula** we start from Liouville's equation

$$\partial f_N / \partial t = -\{f_N, H\} = -\{f_N, H_0\} - \sum_s \{f_N, B_s\} A_s \tag{5.6}$$

Introducing the Liouville operator

$$i\mathscr{L}_0 := \{\ldots, H_0\} \tag{5.7}$$

we get

$$\partial f_N / \partial t + i\mathscr{L}_0 f_N = \sum_s \{B_s, f_N\} A_s(t) \tag{5.8}$$

Transforming this linear first-order differential equation into the equivalent integral equation, we have

$$f_N = f_N^{(0)} + \sum_s \int_0^\infty d\tau \, e^{-i\mathscr{L}_0 \tau} \{B_s, f_N\} A_s(t - \tau) \tag{5.9}$$

$f_N(t = -\infty) = f_N^{(0)}$ is assumed to be the equilibrium distribution $f_N^{(0)} = \exp[-H_0/\Theta]/ZN!\,h^{3N}$.

Equation (5.9) can be solved by iteration. For our purpose we restrict ourselves to terms linear in $\{A_s\}$. This means we replace f_N by $f_N^{(0)}$ on the right-hand side. Remembering $df_N^{(0)} = -(1/\Theta) f_N^{(0)}\, dH_0$ and $\{B_s, H_0\} = \dot{B}_s$ we arrive at

$$f_N = f_N^{(0)} - (1/\Theta) \sum_s \int_0^\infty d\tau\, e^{-i\mathscr{L}_0\tau} f_N^{(0)} \dot{B}_s A_s(t-\tau) \tag{5.10}$$

or since $e^{-i\mathscr{L}_0\tau}$ commutes with $f_N^{(0)}$ and shifts all variables $\mathbf{p}_k(t)$, $\mathbf{r}_k(t)$ to $\mathbf{p}_k(t-\tau)$, $\mathbf{r}_k(t-\tau)$, we have

$$f_N(t) = f_N^{(0)} - (1/\Theta) \sum_s \int_0^\infty f_N^{(0)} \dot{B}_s(\mathbf{p}_k(t-\tau), \mathbf{r}_k(t-\tau))\, A_s(t-\tau)\, d\tau \tag{5.11}$$

With that we find for the ensemble average of a coordinate function R

$$\langle R\rangle(t) = \langle R\rangle_0 - (1/\Theta) \sum_s \int_0^\infty d\tau \langle R(\mathbf{p}_k(t), \mathbf{r}_k(t))\, \dot{B}_s(\mathbf{p}_k(t-\tau), \mathbf{r}_k(t-\tau))\rangle_0\, A_s(t-\tau) \tag{5.12}$$

The index 0 refers to the equilibrium situation. Equation (5.12) is the **Kubo formula** which we intended to derive.

To find the fluctuation–dissipation theorem, we apply $R = \dot{B}_s$:

$$\langle \dot{B}_s\rangle(t) = -(1/\Theta) \sum_{s'} \int_0^\infty d\tau \langle \dot{B}_s(\mathbf{p}_k(t), \mathbf{r}_k(t))\, \dot{B}_{s'}(\mathbf{p}_k(t-\tau), \mathbf{r}_k(t-\tau))\rangle_0\, A_{s'}(t-\tau) \tag{5.13}$$

We now consider the parameters s, s' as characterizing certain points $\mathbf{r}$, $\mathbf{r}'$ in space and consequently replace the sum by an integral. Note that these variables $\mathbf{r}$, $\mathbf{r}'$ should not be confused with the particle variables $\mathbf{p}_k$, $\mathbf{r}_k$.

Defining the autocorrelation coefficient in space and time for a stationary and homogeneous system by

$$\beta(\mathbf{r} - \mathbf{r}', \tau) := \langle \dot{B}(\mathbf{r}; \mathbf{p}_k(t), \mathbf{r}_k(t))\, \dot{B}(\mathbf{r}'; \mathbf{p}_k(t-\tau), \mathbf{r}_k(t-\tau))\rangle_0 \tag{5.14}$$

we then get

$$\langle \dot{B}\rangle(\mathbf{r}, t) = -(1/\Theta) \int d\mathbf{r}' \int_0^\infty d\tau\, \beta(\mathbf{r} - \mathbf{r}', \tau)\, A(\mathbf{r}', t-\tau) \tag{5.15}$$

It follows from the convolution theorem that we may introduce a "generalized conductivity" $C(\mathbf{k}, \omega)$ through

$$\langle \tilde{\tilde{\dot{B}}}\rangle(\mathbf{k}, \omega) =: -k^2 C(\mathbf{k}, \omega)\, \tilde{\tilde{A}}(\mathbf{k}, \omega) \tag{5.16}$$

Equation (5.5) and the application of Parseval's theorem to **r**, **k** then yields for the dissipated energy

$$\begin{aligned} W &= \int d\mathbf{k} \int d\omega\, k^2 C(\mathbf{k}, \omega)\, \tilde{\tilde{A}}(\mathbf{k}, \omega)\, \tilde{\tilde{A}}^*(\mathbf{k}, \omega) \\ &= \int d\mathbf{k} \int d\omega\, k^2 \mathrm{Re}\, C(\mathbf{k}, \omega)\, |\tilde{\tilde{A}}(\mathbf{k}, \omega)|^2 \end{aligned} \tag{5.17}$$

So Re $C(\mathbf{k}, \omega)$ is the decisive parameter for energy dissipation. From (5.15) and (5.16), we read

$$\Theta k^2 C(\mathbf{k}, \omega) = \int_0^\infty dt \int d\mathbf{r}\, e^{i\omega t - i\mathbf{k}\cdot\mathbf{r}} \beta(\mathbf{r}, t) \tag{5.18}$$

and using the symmetry relation $\beta(\mathbf{r}, -t) = \beta(\mathbf{r}, t)$ we find

$$\begin{aligned} 2\Theta k^2 \mathrm{Re}[C(\mathbf{k}, \omega)] &= \int_{-\infty}^\infty dt \int d\mathbf{r}\, e^{i\omega t - i\mathbf{k}\cdot\mathbf{r}} \beta(\mathbf{r}, t) \\ &= (2\pi)^2\, \tilde{\tilde{\beta}}(\mathbf{k}, \omega) \end{aligned} \tag{5.19}$$

This is the **fluctuation–dissipation theorem** relating the dissipative parameter Re$[C(\mathbf{k}, \omega)]$ to the equilibrium fluctuation spectrum of the response function $\dot{B}_s$.

5.2. Fluctuation–Dissipation Theorem for the Coulomb System

In this case we consider as the disturbance of our system an external electric potential $\Phi_{\mathrm{ex}}(\mathbf{r}, t)$. We then have

$$H - H_0 = \int d\mathbf{r} \sum_i e_i \delta(\mathbf{r} - \mathbf{r}_i)\, \Phi_{\mathrm{ex}}(\mathbf{r}, t) \tag{5.20}$$

and the general quantities of the preceding paragraph can be specified to

$$A(\mathbf{r}, t) = \Phi_{\mathrm{ex}}(\mathbf{r}, t), \qquad B(\mathbf{r}, \mathbf{r}_k) = \sum_i e_i \delta(\mathbf{r} - \mathbf{r}_i) \tag{5.21}$$

Using the identity $(d/dt)\, \delta(\mathbf{r} - \mathbf{r}_i) = -(d/d\mathbf{r}) \cdot \delta(\mathbf{r} - \mathbf{r}_i)\, \mathbf{v}_i$ we find

$$\langle \tilde{\tilde{\dot{B}}} \rangle(\mathbf{k}, \omega) = -i\mathbf{k} \cdot \tilde{\tilde{\mathbf{j}}}(\mathbf{k}, \omega) \tag{5.22}$$

so that (5.16) reads

$$\tilde{\tilde{\mathbf{j}}}_{\|}(\mathbf{k}, \omega) = C(\mathbf{k}, \omega)(-i\mathbf{k}\tilde{\tilde{\Phi}}_{\mathrm{ex}}(\mathbf{k}, \omega)) = C(\mathbf{k}, \omega)\, \tilde{\tilde{\mathbf{E}}}_{\|\mathrm{ex}}(\mathbf{k}, \omega) \tag{5.23}$$

Here $\mathbf{j}$ denotes the electrical current density related to the *internal* electric field $\mathbf{E}$ through the conductivity σ by the relation $\mathbf{j} = \sigma\mathbf{E}$. Since the internal electric field again is related to the external electric field through the complex dielectric constant by $\varepsilon\mathbf{E} = \mathbf{E}_{ex}$ we find for our generalized conductivity

$$C(\mathbf{k}, \omega) = \sigma(\mathbf{k}, \omega)/\varepsilon(\mathbf{k}, \omega) \tag{5.24}$$

Considering only the electrons which have equal charge $(-e)$ the autocorrelation coefficient is given by

$$\beta(\mathbf{r}, t) = e^2\langle \dot{n}(\mathbf{r}, t)\, \dot{n}(\mathbf{0}, 0)\rangle_0 \tag{5.25}$$

where n is the microscopic density. In analogy to the calculations in Chapter VI, Section 4, we find

$$\begin{aligned} \tilde{\tilde{\beta}}(\mathbf{k}, \omega) &= \lim_{T, V\to\infty} [(2\pi e)^2/TV]\langle |\, \tilde{\tilde{\dot{n}}}(\mathbf{k}, \omega)|^2\rangle_0 \\ &= \lim[(2\pi e)^2/TV]\, \omega^2\langle |\, \tilde{\tilde{n}}(\mathbf{k}, \omega)|^2\rangle_0 \end{aligned} \tag{5.26}$$

So the **fluctuation–dissipation theorem** for our system yields

$$\Theta \frac{k^2}{\omega^2} \operatorname{Re}\left(\frac{\sigma(\mathbf{k}, \omega)}{\varepsilon(\mathbf{k}, \omega)}\right) = \lim \frac{(2\pi)^4 e^2}{TV} \langle |\, \tilde{\tilde{n}}(\mathbf{k}, \omega)|^2\rangle_0 \tag{5.27}$$

With $\varepsilon = 1 + 4\pi i\sigma/\omega$ (cf. VI.2.35) we may also write

$$\lim \frac{(2\pi)^4 e^2}{TV} \langle |\, \tilde{\tilde{n}}(\mathbf{k}, \omega)|^2\rangle_0 = \Theta \frac{\mathbf{k}^2}{\omega^2} \frac{\rho_r}{\rho_r^2 + (\rho_i + 4\pi/\omega)^2} \tag{5.28}$$

where ρ_r and ρ_i designate the real and imaginary part of the resistivity $\rho = 1/\sigma$, respectively. Equation (5.28) relates the Fourier transform of the density fluctuation spectrum to the characteristic quantity of dissipation, the resistivity ρ.

REFERENCES AND SUPPLEMENTARY READING

Section 1

Ehrenfest, P., and Ehrenfest, T. (1911). "The Conceptual Foundations of the Statistical Approach in Mechanics." Leipzig [Engl. Transl.: Cornell Univ. Press, Ithaca, New York, 1959].

Gibbs, J. W. (1902). "Elementary Principles in Statistical Mechanics." Dover, New York.

Münster, A. (1959). Prinzipien der statistischen Mechanik. *In* "Handbuch der Physik" (S. Flügge, ed.), Vol. III/2. Springer, Berlin.

Tolman, R. (1938). "The Principles of Statistical Mechanics." Freeman, San Francisco, California.

Section 2

Hill, T. L. (1960). "An Introduction to Statistical Thermodynamics." Addison–Wesley, Reading, Massachusetts.

Landau, L. D., and Lifschitz, E. M. (1959). "Statistical Physics," Vol. V of "Course in Theoretical Physics," Pergamon, New York and London.

Montroll, E. W. (1960). Topics on statistical mechanics of interacting particles. *In* "La théorie des gaz neutres et ionisés", (C. De Witt and J. F. Detoeuf, eds.) Hermann, Paris.

Section 3

Abé, R. (1959). *Progr. Theor. Phys.* **22**, 213.

Ebeling, W. (1967). *Ann. Phys.* (Leipzig) **19**, 104.

Ebeling, W., Hoffmann, H. J., and Kelbg, G. (1967). *Beitr. Plasmaphys.* **3**, 233.

Ecker, G., and Kröll, W. (1966). *Z. Naturforsch. A* **21**, 2012.

Ecker, G., and Kröll, W. (1966). *Z. Naturforsch. A* **21**, 2033.

Friedman, H. L. (1962). "Ionic Solution Theory," p. 147. Wiley (Interscience), New York.

Kelbg, G. (1964). *Ann. Phys.* (Leipzig) **14**, 394.

Kelbg, G. (1966). Extensions of the Debye–Hückel limiting law; Application to ionic solutions and plasmas. *In* "Chemical Physics of Ionic Solutions" (B. E. Conway and R. G. Barradas, eds.). Wiley, New York.

Mayer, J. E. (1950). *J. Chem. Phys.* **18**, 1426.

Mayer, J. E., and Mayer, M. G. (1940). "Statistical Mechanics," p. 455. Wiley, New York.

Meeron, E. (1958). *J. Chem. Phys.* **28**, 630.

Morita, T. (1958). *Progr. Theor. Phys.* **20**, 920.

Section 4

Born, M., and Green, H. S. (1946). *Proc. Roy. Soc. Ser. A* **188**, 10.

De Witt, H. E. (1965). *Phys. Rev.* **140**, 466.

Frisch, H. L., and Lebowitz, J. L., eds. (1964). "The Equilibrium Theory of Classical Fluids." Benjamin, New York.

Graboske, H. C., Jr., Harwood, D. J., and DeWitt, H. E. (1971). *Phys. Rev.* **A3**, 1419.

Gündel, H. (1970). *Beitr. Plasmaphys.* **10**, 455.

Hirt, C. W. (1965). *Phys. Fluids* **8**, 693.

Hirt, C. W. (1967). *Phys. Fluids* **10**, 565.

Kirkwood, J. G. (1934). *J. Chem. Phys.* **2**, 767.

Kirkwood, J. G. (1935). *J. Chem. Phys.* **3**, 300.

Kirkwood, J. G., and Boggs, E. M. (1942). *J. Chem. Phys.* **10**, 394.

Matsuda, K. (1968). *Phys. Fluids* **11**, 328.

Meeron, E. (1957). *J. Chem. Phys.* **27**, 1238.

Meeron, E. (1958). *Phys. Fluids* **1**, 139.

Meeron, E., and Rodemich, E. R. (1958). *Phys. Fluids* **1**, 246.

Meeron, E. (1960). *J. Math. Phys.* **1**, 192.

O'Neil, T., and Rostoker, N. (1965). *Phys. Fluids* **8**, 1109.

Onsager, L. (1933). *Chem. Rev.* **13**, 73.

Vedenov, A. A. (1965). Thermodynamics of a plasma. *In* "Reviews of Plasma Physics," (M. A. Leontovich, ed.), Vol. 1. Consultants Bureau, New York.

Section 4.11

Alyamovskij, V. N. (1962). *Zh. Eksp. Teor. Fiz.* **42**, 1536; *Sov. Phys. JETP* **15**, 1067.

Baranger, M., and Mozer, B. (1959). *Phys. Rev.* **115**, 521.

Broyles, A. A. (1958). *Z. Phys.* **151**, 187.
Ecker, G. (1957). *Z. Phys.* **148**, 593.
Ecker, G., and Fischer, K. G. (1968). *Forschungsber. NRW No.* 1949.
Ecker, G., and Müller, K. G. (1958). *Z. Phys.* **153**, 317.
Engelmann, F. (1962). *Z. Phys.* **169**, 126.
Holtsmark, J. (1919). *Ann. Phys.* (Leipzig) **58**, 577.
Hooper, C. F. (1966). *Phys. Rev.* **149**, 77.
Hooper, C. F. (1968). *Phys. Rev.* **165**, 215.
Jackson, J. L. (1960). *Phys. Fluids* **3**, 927.
Kelbg, G. (1964). *Ann. Phys.* (Leipzig) **13**, 385.
Mozer, B., and Baranger, M. (1960). *Phys. Rev.* **118**, 626.
Pfennig, H., and Trefftz, E. (1966). *Z. Naturforsch. A* **21**, 697.
Weise, K. (1965). *Z. Phys.* **183**, 36.
Weise, K. (1968). *Z. Phys.* **212**, 458.

SECTION 5

Callen, H. B., and Welton, T. A. (1951). *Phys. Rev.* **83**, 34.
Chandrasekhar, S. (1943). *Rev. Mod. Phys.* **15**, 1.
Kogan, W. I., and Selidowkin, A. D. (1969). *Beitr. Plasmaphys.* **9**, 199.
Kubo, R. (1957). *J. Phys. Soc. Jap.* **20**, 439.
Kubo, R. (1959). Some aspects of the statistical mechanical theory of irreversible processes. *In* "Lectures in Theoretical Physics" (W. E. Brittin and L. G. Dunham, eds.), Vol. I. Wiley (Interscience), New York.
Lax, M. (1960). *Rev. Mod. Phys.* **32**, 25.
Nyquist, H. (1928). *Phys. Rev.* **32**, 110.
Rostoker, N. (1960). *Nucl. Fusion* **1**, 101.
Takahashi, H. (1952). *J. Phys. Soc. Jap.* **20**, 439.
Taylor, J. B. (1960). *Phys. Fluids* **3**, 792.
Thompson, W. B., and Hubbard, J. (1960). *Rev. Mod. Phys.* **32**, 714.

Chapter II Nonequilibrium States of the Coulomb System, General Description

1. The Exact Density Distribution of the Single System

1.1. The Klimontovich Equations

In this section, we study the true density distribution of a single system in μ space. This quantity should be clearly distinguished from the corresponding average density distribution of a Gibbs' ensemble.

We describe the contribution of the ith particle to the total density of our system with the help of Dirac functions in the form

$$F_i(\mathbf{r}, \mathbf{p}) = \delta(\mathbf{r} - \mathbf{r}_i(t))\, \delta(\mathbf{p} - \mathbf{p}_i(t)) \tag{1.1}$$

The total density of all particles of the component k is therefore

$$F^{(k)} = \sum_{i(k)} \delta(\mathbf{r} - \mathbf{r}_i(t))\, \delta(\mathbf{p} - \mathbf{p}_i(t)) \tag{1.2}$$

where $i(k)$ designates all particle indices belonging to the component k.

To formulate the laws of motion for the density distribution F_i it seems advisable to recall first some of the rules governing the Dirac functions. The Dirac function is not a true function but belongs to the group of "Schwarz distributions." The relations given in the following are based partly on the definition of the Dirac functions and partly on the theory of Schwarz distributions. In so far as the relations are free of integration, they are to be understood symbolically. We have

$$\delta(x) = 0 \quad \text{for} \quad x \neq 0, \qquad \int_{-\infty}^{+\infty} \delta(x)\, dx = 1$$

$$\int_{-\infty}^{+\infty} f(x)\, \delta(x)\, dx = f(0), \qquad \int_{-\infty}^{+\infty} f(x)\, \delta(x-a)\, dx = f(a)$$

$$x\delta(x) = 0, \qquad f(x)\, \delta(x-a) = f(a)\, \delta(x-a)$$

$$\delta(-x) = \delta(x), \qquad \delta(ax) = \delta(x)/|\, a\,|, \qquad a \neq 0 \tag{1.3}$$

$$\delta(x^2 - a^2) = \{\delta(x-a) + \delta(x+a)\}/2a, \qquad a > 0$$

$$\int_{-\infty}^{+\infty} \delta(a-x)\, \delta(x-b)\, dx = \delta(a-b)$$

$$\int_{-\infty}^{+\infty} f(x)\, \delta'(x)\, dx = [f(x)\, \delta(x)]_{-\infty}^{+\infty} - \int_{-\infty}^{+\infty} f'(x)\, \delta(x)\, dx = -f'(0)$$

$$\int_{-\infty}^{+\infty} f(x)\, \delta^{(n)}(x)\, dx = (-1)^n f^{(n)}(0)$$

The time derivative of F_i is given by

$$\begin{aligned}(\partial F_i/\partial t)_{\mathbf{r},\mathbf{p}} &= \dot{\mathbf{r}}_i \cdot (\partial F_i/\partial \mathbf{r}_i)_{\mathbf{r},\mathbf{p},\mathbf{p}_i} + \dot{\mathbf{p}}_i \cdot (\partial F_i/\partial \mathbf{p}_i)_{\mathbf{r},\mathbf{p},\mathbf{r}_i} \\ &= -\dot{\mathbf{r}}_i \cdot (\partial F_i/\partial \mathbf{r})_{\mathbf{p},t} - \dot{\mathbf{p}}_i \cdot (\partial F_i/\partial \mathbf{p})_{\mathbf{r},t}\end{aligned} \tag{1.4}$$

If $\mathbf{E}_j{}^i$ designates the field exerted by the particle j on the particle i, we find from the laws of classical mechanics

$$\left(\frac{\partial F_i}{\partial t}\right)_{\mathbf{r},\mathbf{p}} = -\frac{\mathbf{p}_i}{m_i} \cdot \left(\frac{\partial F_i}{\partial \mathbf{r}}\right)_{\mathbf{p},t} - e_i \sum_j{}' \mathbf{E}_j{}^i \cdot \left(\frac{\partial F_i}{\partial \mathbf{p}}\right)_{\mathbf{r},t} \tag{1.5}$$

or

$$\frac{\partial F_i}{\partial t} = -\frac{\mathbf{p}_i}{m_i} \cdot \frac{\partial F_i}{\partial \mathbf{r}} - e_i \frac{\partial}{\partial \mathbf{p}} \cdot \Big\{\sum_j{}' \mathbf{E}_j{}^i F_i\Big\} \tag{1.6}$$

since the $\mathbf{E}_j{}^i$ given by

$$\mathbf{E}_j{}^i = -\frac{\partial}{\partial \mathbf{r}_i}\left(\frac{e_j}{|\, \mathbf{r}_i - \mathbf{r}_j\,|}\right) \tag{1.7}$$

are independent of the particle momenta. Utilizing the rules given in (1.3), we have

$$\mathbf{E}_j{}^i F_i = -e_j \left(\frac{\partial}{\partial \mathbf{r}_i} \frac{1}{|\mathbf{r}_i - \mathbf{r}_j|}\right) \delta(\mathbf{r} - \mathbf{r}_i)\, \delta(\mathbf{p} - \mathbf{p}_i)$$

$$= -e_j \int d\mathbf{r}' \int d\mathbf{p}' \left(\frac{\partial}{\partial \mathbf{r}} \frac{1}{|\mathbf{r} - \mathbf{r}'|}\right) \delta(\mathbf{r} - \mathbf{r}_i)\, \delta(\mathbf{p} - \mathbf{p}_i)\, \delta(\mathbf{r}' - \mathbf{r}_j)\, \delta(\mathbf{p}' - \mathbf{p}_j)$$

$$= -e_j \int d\mathbf{r}' \int d\mathbf{p}' \left(\frac{\partial}{\partial \mathbf{r}} \frac{1}{|\mathbf{r} - \mathbf{r}'|}\right) F_i(\mathbf{r}, \mathbf{p}, t)\, F_j(\mathbf{r}', \mathbf{p}', t) \tag{1.8}$$

and consequently

$$\frac{\partial F_i}{\partial t} + \frac{\mathbf{p}}{m_i} \cdot \frac{\partial F_i}{\partial \mathbf{r}} - e_i \frac{\partial}{\partial \mathbf{p}} \cdot \left\{ \int d\mathbf{r}' \int d\mathbf{p}' \left(\frac{\partial}{\partial \mathbf{r}} \frac{1}{|\mathbf{r} - \mathbf{r}'|}\right) F_i \sum_j{}' F_j e_j \right\} = 0 \tag{1.9}$$

It should be noted that (1.9) incorporates only the Coulomb interactions within the system and neglects external fields. If there is such a field $\mathbf{E}_{\mathrm{ex}}$ it has to be added to the sum on the right-hand side of (1.6) and correspondingly in the following formulas. The index i in (1.9) may denote any particle of our system. Therefore (1.9) actually represents a system of simultaneous differential equations the order of which is given by the number of individuals of our system.

We sum in (1.9) over the density contributions of all particles of the same component k. With the definition (1.2) this results in

$$\frac{\partial F^{(k)}}{\partial t} + \frac{\mathbf{p}}{m_k} \cdot \frac{\partial F^{(k)}}{\partial \mathbf{r}}$$
$$- e_k \frac{\partial}{\partial \mathbf{p}} \cdot \left\{ \int d\mathbf{r}' \int d\mathbf{p}' \left(\frac{\partial}{\partial \mathbf{r}} \frac{1}{|\mathbf{r} - \mathbf{r}'|}\right) F^{(k)} \sum_\nu F^{(\nu)}(\mathbf{r}', \mathbf{p}', t)\, e_\nu \right\} = 0 \tag{1.10}$$

where we now include the term originally excluded by the prime in the summation in (1.9). This is justified because this term contributes only negligibly in comparison to the $N - 1$ other terms. Again (1.10) is a system of simultaneous differential equations. However, the order is now given by the number of different particle components.

We may rewrite (1.10) as

$$\frac{\partial F^{(k)}}{\partial t} + \frac{\mathbf{p}}{m_k} \cdot \frac{\partial F^{(k)}}{\partial \mathbf{r}} + e_k \frac{\partial}{\partial \mathbf{p}} \cdot \{\mathbf{E} F^{(k)}\} = 0 \tag{1.11}$$

with

$$\mathbf{E} = - \int d\mathbf{r}' \int d\mathbf{p}' \sum_\nu F^{(\nu)}(\mathbf{r}', \mathbf{p}', t)\, e_\nu \left(\frac{\partial}{\partial \mathbf{r}} \frac{1}{|\mathbf{r} - \mathbf{r}'|}\right) \tag{1.12}$$

or

$$\frac{\partial}{\partial \mathbf{r}} \cdot \mathbf{E} = 4\pi \sum_{\nu} \int d\mathbf{p}' F^{(\nu)}(\mathbf{r}, \mathbf{p}', t)\, e_{\nu} \tag{1.13}$$

remembering that we have

$$\Delta(1/|\mathbf{r} - \mathbf{r}'|) = -4\pi\delta(\mathbf{r} - \mathbf{r}') \tag{1.14}$$

Equations (1.11) and (1.12) are the **Klimontovich equations** for the true density of our system. They are exact and free of approximations.

It should be noted that these Klimontovich equations are identical with the Vlasov equations of statistical mechanics describing the ensemble average of the density. The latter, however, are not exact and are of limited applicability. This fact might raise the question why we look at all for a statistical solution of the approximate Vlasov equations if we show here that the same equations govern the exact behavior of the true density of our system. The reason, of course, is that the solution of the Klimontovich equations is subject to a very restrictive condition: All solutions must at any time be representable in the form given in (1.2). This incisive condition renders a solution of the Klimontovich equations practically impossible, reflecting the well-known difficulties of the ergodic problem.

2. The Average Distributions of Gibbs' Ensemble

2.1. Derivation of the BBGKY Hierarchy from the Klimontovich Equations

As in the case of equilibrium statistics, we now consider a virtual Gibbs' ensemble of systems in Γ space characterized by the probability density

$$f_N(\mathbf{r}_1, \ldots, \mathbf{r}_N, \mathbf{p}_1, \ldots, \mathbf{p}_N; t) \tag{2.1}$$

We define the probability density for the ith particle at a given point $({}^i\mathbf{r}, {}^i\mathbf{p})$ in the μ space through†

$$f_1({}^i\mathbf{r}, {}^i\mathbf{p}; t) = \int f_N F_i \, d\mathbf{r}_1 \cdots d\mathbf{p}_N \tag{2.2}$$

† Here attention is drawn to the distinction of the indices. The index at the lower right ($\mathbf{r}_i$, $\mathbf{p}_i$, $\mathbf{r}_j$, $\mathbf{p}_j$) characterizes the position of the particles (points) in Γ space. The index at the upper left (${}^i\mathbf{r}$, ${}^i\mathbf{p}$, ${}^j\mathbf{r}$, ${}^j\mathbf{p}$) characterizes the point of observation in μ space at which we study the effect of the particles i and j.

In a more general form, the probability density to find the set of particles $(1,\ldots, s)$ at the coordinates $({}^1\mathbf{r},\ldots, {}^s\mathbf{r}, {}^1\mathbf{p},\ldots, {}^s\mathbf{p})$ is given by the specific distribution function of order s

$$f_s({}^1\mathbf{r},\ldots, {}^s\mathbf{r}, {}^1\mathbf{p},\ldots, {}^s\mathbf{p}; t) = \int f_N F_1 \cdots F_s \, d\mathbf{r}_1 \cdots d\mathbf{p}_N \tag{2.3}$$

where we used in a slight modification of (1.1)

$$F_i({}^i\mathbf{r}, {}^i\mathbf{p}) = \delta({}^i\mathbf{r} - \mathbf{r}_i(t)) \, \delta({}^i\mathbf{p} - \mathbf{p}_i(t)) \tag{2.4}$$

To derive the BBGKY hierarchy which governs the specific distribution functions f_s we multiply the equation (1.9) with the phase space density (2.1) and integrate over the Γ space

$$\int f_N \Bigg\{ \left(\frac{\partial F_i}{\partial t}\right)_{{}^i\mathbf{r},{}^i\mathbf{p}} + \frac{{}^i\mathbf{p}}{m_i} \cdot \left(\frac{\partial F_i}{\partial\, {}^i\mathbf{r}}\right)_{{}^i\mathbf{p},t}$$

$$- e_i \frac{\partial}{\partial\, {}^i\mathbf{p}} \cdot \left\{ \int d^j\mathbf{r}' \int d^j\mathbf{p}' \left(\frac{\partial}{\partial\, {}^i\mathbf{r}} \frac{1}{|\,{}^i\mathbf{r} - {}^j\mathbf{r}'\,|}\right) F_i \sum_j{}' F_j \, e_j \right\}_{{}^i\mathbf{r},t} \Bigg\} \, d\mathbf{r}_1 \cdots d\mathbf{p}_N = 0 \tag{2.5}$$

The phase-space density $f_N(\mathbf{r}_1, \ldots, \mathbf{r}_N, \mathbf{p}_1, \ldots, \mathbf{p}_N; t)$ does not depend on the coordinates ${}^i\mathbf{r}$, ${}^i\mathbf{p}$, and consequently the integration in the second and third term on the left-hand side can readily be carried out.

The first term requires additional consideration. Let us first recall that

$$\begin{aligned} f_N \left(\frac{\partial F_i}{\partial t}\right)_{{}^i\mathbf{r},{}^i\mathbf{p}} &= \left(\frac{\partial f_N F_i}{\partial t}\right)_{{}^i\mathbf{r},{}^i\mathbf{p}} - F_i \left(\frac{\partial f_N}{\partial t}\right)_{{}^i\mathbf{r},{}^i\mathbf{p}} \\ &= \left(\frac{\partial f_N F_i}{\partial t}\right)_{{}^i\mathbf{r},{}^i\mathbf{p}} - F_i \frac{d f_N}{dt} = \left(\frac{\partial f_N F_i}{\partial t}\right)_{{}^i\mathbf{r},{}^i\mathbf{p}} \end{aligned} \tag{2.6}$$

holds.

Using the relation

$$\left(\frac{\partial f_N F_i}{\partial t}\right)_{{}^i\mathbf{r},{}^i\mathbf{p}} = \left(\frac{\partial f_N F_i}{\partial t}\right)_{{}^i\mathbf{r},{}^i\mathbf{p},\mathbf{r}_i,\mathbf{p}_i} + \sum_j \left(\frac{\mathbf{p}_j}{m_j} \cdot \frac{\partial}{\partial \mathbf{r}_j} + \dot{\mathbf{p}}_j \cdot \frac{\partial}{\partial \mathbf{p}_j}\right) (f_N F_i) \tag{2.7}$$

together with (2.6), we find

$$f_N \left(\frac{\partial F_i}{\partial t}\right)_{{}^i\mathbf{r},{}^i\mathbf{p}} = \left(\frac{\partial f_N F_i}{\partial t}\right)_{{}^i\mathbf{r},{}^i\mathbf{p},\mathbf{r}_i,\mathbf{p}_i} + \sum_j \left(\frac{\mathbf{p}_j}{m_j} \cdot \frac{\partial}{\partial \mathbf{r}_j} + \dot{\mathbf{p}}_j \cdot \frac{\partial}{\partial \mathbf{p}_j}\right) (f_N F_i) \tag{2.8}$$

which yields

$$\int f_N \left(\frac{\partial F_i}{\partial t}\right)_{{}^i\mathbf{r},{}^i\mathbf{p}} d\mathbf{r}_1 \cdots d\mathbf{p}_N = \left(\frac{\partial f_1({}^i\mathbf{r}, {}^i\mathbf{p}; t)}{\partial t}\right)_{{}^i\mathbf{r},{}^i\mathbf{p}} \tag{2.9}$$

since the second term in (2.8) disappears by partial integration and application of the equations of motion.

With the definition (2.3) of the specific distribution functions f_s we consequently find for the average distributions

$$\frac{\partial f_1}{\partial t} + \frac{{}^i\mathbf{p}}{m_i} \cdot \frac{\partial f_1}{\partial\, {}^i\mathbf{r}}$$
$$- e_i \int d^j\mathbf{r}' \int d^j\mathbf{p}' \left(\frac{\partial}{\partial\, {}^i\mathbf{r}} \frac{1}{|\,{}^i\mathbf{r} - {}^j\mathbf{r}'\,|}\right) \cdot \frac{\partial}{\partial\, {}^i\mathbf{p}} \sum_j{}' f_2({}^i\mathbf{r}, {}^j\mathbf{r}', {}^i\mathbf{p}, {}^j\mathbf{p}'; t)\, e_j = 0 \tag{2.10}$$

This is the first equation of the hierarchy for the distribution functions. The next equation expresses f_2 through f_3.

Before deriving it, we suggest simplifying our problem by the following specification of the model:

1. We assume that we have only one particle component. This does not restrict generality. The extension of our formalism to many particle components is straightforward.

2. We request symmetry for the phase space density f_N. Since the Liouville equation is symmetric in the coordinates this is always a correct assumption if the initial distribution for f_N is symmetric in the particle coordinates. Nevertheless, we should note that interesting cases which do not satisfy this assumption may occur. For instance, the test particle problem would violate the symmetry requirement. On the other hand, having the formalism given in the following it is not difficult to write down the relations for nonsymmetric cases.

Under these circumstances we may replace $e_i e_j$ by e^2. With that, we find from (2.10)

$$\frac{\partial f_1}{\partial t} + \frac{{}^i\mathbf{p}}{m} \cdot \frac{\partial f_1}{\partial {}^i\mathbf{r}}$$
$$- e^2(N-1) \int d^j\mathbf{r}' \int d^j\mathbf{p}' \left(\frac{\partial}{\partial {}^i\mathbf{r}} \frac{1}{|\,{}^i\mathbf{r} - {}^j\mathbf{r}'\,|}\right) \cdot \frac{\partial}{\partial\, {}^i\mathbf{p}} f_2({}^i\mathbf{r}, {}^j\mathbf{r}', {}^i\mathbf{p}, {}^j\mathbf{p}'; t) = 0 \tag{2.11}$$

This is the first equation of the BBGKY hierarchy. It relates the specific one-particle distribution function to the specific pair distribution function.

To solve (2.11) we must derive an equation for the pair distribution function f_2. To this end we consider the two Klimontovich equations for the particles i and k in the form

$$\frac{\partial F_i}{\partial t} + \frac{{}^i\mathbf{p}}{m} \cdot \frac{\partial F_i}{\partial\, {}^i\mathbf{r}}$$

$$- e^2 \frac{\partial}{\partial\, {}^i\mathbf{p}} \cdot \left\{ \int d^j\mathbf{r}' \int d^j\mathbf{p}' \left(\frac{\partial}{\partial\, {}^i\mathbf{r}} \frac{1}{|\, {}^i\mathbf{r} - {}^j\mathbf{r}' \,|} \right) F_i({}^i\mathbf{r}, {}^i\mathbf{p}) \sum_{j \neq i} F_j({}^j\mathbf{r}', {}^j\mathbf{p}') \right\} = 0 \tag{2.12}$$

and

$$\frac{\partial F_k}{\partial t} + \frac{{}^k\mathbf{p}}{m} \cdot \frac{\partial F_k}{\partial {}^k\mathbf{r}}$$

$$- e^2 \frac{\partial}{\partial\, {}^k\mathbf{p}} \cdot \left\{ \int d^j\mathbf{r}' \int d^j\mathbf{p}' \left(\frac{\partial}{\partial\, {}^k\mathbf{r}} \frac{1}{|\, {}^k\mathbf{r} - {}^j\mathbf{r}' \,|} \right) F_k({}^k\mathbf{r}, {}^k\mathbf{p}) \sum_{j \neq k} F_j({}^j\mathbf{r}', {}^j\mathbf{p}') \right\} = 0 \tag{2.13}$$

We multiply (2.12) by $F_k f_N$ and (2.13) by $F_i f_N$, sum both equations, and integrate over the Γ space using the relations (2.3) to obtain

$$\frac{\partial f_2}{\partial t} + \frac{{}^i\mathbf{p}}{m} \cdot \frac{\partial f_2}{\partial\, {}^i\mathbf{r}} + \frac{{}^k\mathbf{p}}{m} \cdot \frac{\partial f_2}{\partial\, {}^k\mathbf{r}} - e^2 \int d^j\mathbf{r}' \int d^j\mathbf{p}' \int d\mathbf{r}_1 \cdots d\mathbf{p}_N$$

$$\cdot f_N \left(\frac{\partial}{\partial\, {}^i\mathbf{r}} \frac{1}{|\, {}^i\mathbf{r} - {}^j\mathbf{r}' \,|} \right) \cdot \frac{\partial}{\partial\, {}^i\mathbf{p}} F_k F_i \sum_{j \neq i} F_j({}^j\mathbf{r}', {}^j\mathbf{p}')$$

$$- e^2 \int d^j\mathbf{r}' \int d^j\mathbf{p}' \int d\mathbf{r}_1 \cdots d\mathbf{p}_N f_N \left(\frac{\partial}{\partial\, {}^k\mathbf{r}} \frac{1}{|\, {}^k\mathbf{r} - {}^j\mathbf{r}' \,|} \right)$$

$$\cdot \frac{\partial}{\partial\, {}^k\mathbf{p}} F_i F_k \sum_{j \neq k} F_j({}^j\mathbf{r}', {}^j\mathbf{p}') = 0 \tag{2.14}$$

Let us consider the two last terms on the left-hand side. Since the phase-space density f_N depends only on variables with indices at the lower right, we can shift f_N over the differentiation symbols. The two terms contain two different types of contributions.

There are those whose indices of the F-functions are all different. Here the integration over the Γ space can easily be carried out and produces the distribution function $f_3({}^i\mathbf{r}, {}^k\mathbf{r}, {}^l\mathbf{r}, {}^i\mathbf{p}, {}^k\mathbf{p}, {}^l\mathbf{p}; t)$.

Then there are the contributions where two of the F-functions have the same index. They are given by

$$-e^2 \int d^k\mathbf{r}' \int d^k\mathbf{p}' \int d\mathbf{r}_1 \cdots d\mathbf{p}_N \left(\frac{\partial}{\partial\, {}^i\mathbf{r}} \frac{1}{|\,{}^i\mathbf{r} - {}^k\mathbf{r}'\,|}\right)$$

$$\cdot \frac{\partial}{\partial\, {}^i\mathbf{p}} F_k({}^k\mathbf{r}, {}^k\mathbf{p}) F_k({}^k\mathbf{r}', {}^k\mathbf{p}') F_i({}^i\mathbf{r}, {}^i\mathbf{p}) f_N(\mathbf{r}_1, \ldots, \mathbf{p}_N\,;\, t)$$

$$-e^2 \int d^i\mathbf{r}' \int d^i\mathbf{p}' \int d\mathbf{r}_1 \cdots d\mathbf{p}_N \left(\frac{\partial}{\partial\, {}^k\mathbf{r}} \frac{1}{|\,{}^k\mathbf{r} - {}^i\mathbf{r}'\,|}\right)$$

$$\cdot \frac{\partial}{\partial\, {}^k\mathbf{p}} F_i({}^i\mathbf{r}, {}^i\mathbf{p}) F_i({}^i\mathbf{r}', {}^i\mathbf{p}') F_k({}^k\mathbf{r}, {}^k\mathbf{p}) f_N(\mathbf{r}_1, \ldots, \mathbf{p}_N\,;\, t) \tag{2.15}$$

Here we first integrate with respect to the coordinates ${}^k\mathbf{r}'$, ${}^k\mathbf{p}'$ and ${}^i\mathbf{r}'$, ${}^i\mathbf{p}'$, respectively, then with respect to the coordinates of the Γ space. This yields

$$-e^2 \left\{\left(\frac{\partial}{\partial\, {}^i\mathbf{r}} \frac{1}{|\,{}^i\mathbf{r} - {}^k\mathbf{r}\,|}\right) \cdot \frac{\partial}{\partial\, {}^i\mathbf{p}} f_2 + \left(\frac{\partial}{\partial\, {}^k\mathbf{r}} \frac{1}{|\,{}^i\mathbf{r} - {}^k\mathbf{r}\,|}\right) \cdot \frac{\partial}{\partial\, {}^k\mathbf{p}} f_2\right\} \tag{2.16}$$

Introducing these results into (2.14), we arrive at

$$\begin{aligned} &\frac{\partial f_2}{\partial t} + \frac{{}^i\mathbf{p}}{m} \cdot \frac{\partial f_2}{\partial\, {}^i\mathbf{r}} + \frac{{}^k\mathbf{p}}{m} \cdot \frac{\partial f_2}{\partial\, {}^k\mathbf{r}} \\ &\quad - e^2 \left\{\left(\frac{\partial}{\partial\, {}^i\mathbf{r}} \frac{1}{|\,{}^i\mathbf{r} - {}^k\mathbf{r}\,|}\right) \cdot \frac{\partial f_2}{\partial\, {}^i\mathbf{p}} + \left(\frac{\partial}{\partial\, {}^k\mathbf{r}} \frac{1}{|\,{}^i\mathbf{r} - {}^k\mathbf{r}\,|}\right) \cdot \frac{\partial f_2}{\partial\, {}^k\mathbf{p}}\right\} \\ &\quad - e^2(N-2) \int d^j\mathbf{r}' \int d^j\mathbf{p}' \left\{\left(\frac{\partial}{\partial\, {}^i\mathbf{r}} \frac{1}{|\,{}^i\mathbf{r} - {}^j\mathbf{r}'\,|}\right) \right. \\ &\quad \cdot \frac{\partial}{\partial\, {}^i\mathbf{p}} f_3({}^i\mathbf{r}, {}^k\mathbf{r}, {}^j\mathbf{r}', {}^i\mathbf{p}, {}^k\mathbf{p}, {}^j\mathbf{p}'; t) \\ &\quad \left. + \left(\frac{\partial}{\partial\, {}^k\mathbf{r}} \frac{1}{|\,{}^k\mathbf{r} - {}^j\mathbf{r}'\,|}\right) \cdot \frac{\partial}{\partial\, {}^k\mathbf{p}} f_3({}^i\mathbf{r}, {}^k\mathbf{r}, {}^j\mathbf{r}', {}^i\mathbf{p}, {}^k\mathbf{p}, {}^j\mathbf{p}'; t)\right\} = 0 \end{aligned} \tag{2.17}$$

This is the second equation of the BBGKY hierarchy which relates the pair distribution function to the third-order specific distribution function.

It is readily seen that this procedure can be continued giving for the general member of our hierarchy

$$\frac{\partial f_s}{\partial t} + \sum_{i=1}^{s} \frac{{}^i\mathbf{p}}{m} \cdot \frac{\partial f_s}{\partial\, {}^i\mathbf{r}} - e^2 \sum_{i,k}^{s}{}' \left(\frac{\partial}{\partial\, {}^i\mathbf{r}} \frac{1}{|\,{}^i\mathbf{r} - {}^k\mathbf{r}\,|}\right) \cdot \frac{\partial f_s}{\partial\, {}^i\mathbf{p}}$$

$$- (N - s)\, e^2 \int d^j\mathbf{r}' \int d^j\mathbf{p}' \sum_{i=1}^{s} \left(\frac{\partial}{\partial\, {}^i\mathbf{r}} \frac{1}{|\,{}^i\mathbf{r} - {}^j\mathbf{r}'\,|}\right)$$

$$\cdot \frac{\partial}{\partial\, {}^i\mathbf{p}} f_{s+1}(\ldots, {}^j\mathbf{r}', \ldots, {}^j\mathbf{p}'; t) = 0 \tag{2.18}$$

For realistic physical problems, we are usually not interested in the specific molecular distribution functions f_s since it is, in principle, impossible to distinguish between particles of the same kind. Rather we are interested in studying the general distribution functions $f^{(s)}$ which give the simultaneous average density of any s particles in the prescribed positions.† The general distribution functions are defined through

$$f^{(s)} = [N!/(N - s)!]\, f_s \tag{2.19}$$

Introducing these general distribution functions $f^{(s)}$ into the hierarchy (2.18), we arrive at the final hierarchy

$$\frac{\partial f^{(s)}}{\partial t} + \sum_{i=1}^{s} \frac{{}^i\mathbf{p}}{m} \cdot \frac{\partial f^{(s)}}{\partial\, {}^i\mathbf{r}} - e^2 \sum_{i,k}^{s}{}' \left(\frac{\partial}{\partial\, {}^i\mathbf{r}} \frac{1}{|\,{}^i\mathbf{r} - {}^k\mathbf{r}\,|}\right) \cdot \frac{\partial f^{(s)}}{\partial\, {}^i\mathbf{p}}$$

$$- e^2 \int d^j\mathbf{r}' \int d^j\mathbf{p}' \sum_{i=1}^{s} \left(\frac{\partial}{\partial^i\mathbf{r}} \frac{1}{|\,{}^i\mathbf{r} - {}^j\mathbf{r}'\,|}\right) \cdot \frac{\partial}{\partial\, {}^i\mathbf{p}} f^{(s+1)}(\ldots, {}^j\mathbf{r}', \ldots, {}^j\mathbf{p}'; t) = 0 \tag{2.20}$$

In the literature it is quite common to use instead of the specific distribution functions f_s the normalized distribution functions $\bar{f}_s$ defined by

$$\bar{f}_s = V^s f_s \tag{2.21}$$

† For comparison, see the equivalent discussion at the beginning of Section 4 in Chapter I.

With these functions the hierarchy has the form

$$\frac{\partial \bar{f}_s}{\partial t} + \sum_{i=1}^{s} \frac{{}^i\mathbf{p}}{m} \cdot \frac{\partial \bar{f}_s}{\partial\, {}^i\mathbf{r}} - e^2 \sum_{i,k}^{s}{}' \left(\frac{\partial}{\partial\, {}^i\mathbf{r}} \frac{1}{|\, {}^i\mathbf{r} - {}^k\mathbf{r}\, |} \right) \cdot \frac{\partial \bar{f}_s}{\partial\, {}^i\mathbf{p}}$$

$$- e^2 \frac{N-s}{V} \int d^j\mathbf{r}' \int d^j\mathbf{p}' \sum_{i=1}^{s} \left(\frac{\partial}{\partial\, {}^i\mathbf{r}} \frac{1}{|\, {}^i\mathbf{r} - {}^j\mathbf{r}'\, |} \right)$$

$$\cdot \frac{\partial}{\partial\, {}^i\mathbf{p}} \bar{f}_{s+1}(\ldots, {}^j\mathbf{r}', \ldots, {}^j\mathbf{p}'; t) = 0 \tag{2.22}$$

It should further be noted that all integrations involved in the derivation of the hierarchy are extended to infinity. If we are considering a finite volume, then we should have additional terms in our hierarchy equations resulting from the effects of the boundaries. Therefore the set of equations (2.20) and (2.22), respectively, tacitly implies the limiting process $V \to \infty$, $N \to \infty$, with $N/V =$ constant.

Clearly it is necessary to truncate the hierarchy to arrive at a finite set of equations. The various kinetic approaches can be distinguished by where and how this truncation is achieved. The simplest approach neglects particle correlations altogether.

REFERENCES AND SUPPLEMENTARY READING

Chappell, W. R. (1967). *J. Math. Phys.* **8**, 298.

Chappell, W. R. (1967). Microscopic kinetic theory. *In* "Lectures in Theoretical Physics" Vol. IX, C, "Kinetic Theory" (W. E. Brittin, ed.). Univ. of Colorado Press, Denver, Colorado.

Klimontovich, Yu. L. (1967). "The Statistical Theory of Non-Equilibrium Processes in a Plasma." Pergamon, New York.

Wu, C. S. (1967). Plasma kinetic theory in the Klimontovich formalism. *In* "Lectures in Theoretical Physics," Vol. IX, C, "Kinetic Theory" (W. E. Brittin, ed.). Univ. of Colorado Press, Denver, Colorado.

Chapter III Nonequilibrium States of the Coulomb System, Description without Individual Particle Correlations

1. The Vlasov Approach

1.1. The Vlasov Equation

Let us consider the first equation of the set (II.2.20) and generalize it—for the sake of the following discussion—to include several components (μ, ν,...)†

$$\frac{\partial f^{(1)}({}^{\mu}\mathbf{r}, {}^{\mu}\mathbf{p}; t)}{\partial t} + \frac{{}^{\mu}\mathbf{p}}{m_{\mu}} \cdot \frac{\partial f^{(1)}({}^{\mu}\mathbf{r}, {}^{\mu}\mathbf{p}; t)}{\partial\, {}^{\mu}\mathbf{r}}$$

$$- e_{\mu} \sum_{\nu} \int d\,{}^{\nu}\mathbf{r}' \int d\,{}^{\nu}\mathbf{p}' \left(\frac{\partial}{\partial\, {}^{\mu}\mathbf{r}} \frac{e_{\nu}}{|\,{}^{\mu}\mathbf{r} - {}^{\nu}\mathbf{r}'\,|}\right) \cdot \frac{\partial}{\partial\, {}^{\mu}\mathbf{p}} f^{(2)}({}^{\mu}\mathbf{r}, {}^{\nu}\mathbf{r}', {}^{\mu}\mathbf{p}, {}^{\nu}\mathbf{p}'; t) = 0 \qquad (1.1)$$

† Latin indices denote particle variables; greek indices denote particle components.

Introducing the correlation function $g_{\mu\nu}$ through[†]

$$f^{(2)}({}^{\mu}\mathbf{r}, {}^{\nu}\mathbf{r}, {}^{\mu}\mathbf{p}, {}^{\nu}\mathbf{p}; t) = f^{(1)}({}^{\mu}\mathbf{r}, {}^{\mu}\mathbf{p}; t) f^{(1)}({}^{\nu}\mathbf{r}, {}^{\nu}\mathbf{p}; t) + g_{\mu\nu}({}^{\mu}\mathbf{r}, {}^{\nu}\mathbf{r}, {}^{\mu}\mathbf{p}, {}^{\nu}\mathbf{p}; t) \tag{1.2}$$

we find from (1.1) for the general distribution function $f^{(1)}$ of the μth component

$$\frac{\partial f^{(1)}}{\partial t} + \frac{{}^{\mu}\mathbf{p}}{m_{\mu}} \cdot \frac{\partial f^{(1)}}{\partial\, {}^{\mu}\mathbf{r}} + e_{\mu}\langle \mathbf{E} \rangle \cdot \frac{\partial f^{(1)}}{\partial\, {}^{\mu}\mathbf{p}} = e_{\mu} \sum_{\nu} \int d\, {}^{\nu}\mathbf{r}' \int d\, {}^{\nu}\mathbf{p}' \left(\frac{\partial}{\partial\, {}^{\mu}\mathbf{r}} \frac{e_{\nu}}{|{}^{\mu}\mathbf{r} - {}^{\nu}\mathbf{r}'|} \right) \cdot \frac{\partial}{\partial\, {}^{\mu}\mathbf{p}} g_{\mu\nu}({}^{\mu}\mathbf{r}, {}^{\nu}\mathbf{r}, {}^{\mu}\mathbf{p}, {}^{\nu}\mathbf{p}; t) \tag{1.3}$$

with

$$\langle \mathbf{E} \rangle = -\sum_{\nu} \int d\, {}^{\nu}\mathbf{r}' \int d\, {}^{\nu}\mathbf{p}' \left(\frac{\partial}{\partial\, {}^{\mu}\mathbf{r}} \frac{e_{\nu}}{|{}^{\mu}\mathbf{r} - {}^{\nu}\mathbf{r}'|} \right) f^{(1)}({}^{\nu}\mathbf{r}', {}^{\nu}\mathbf{p}'; t) + \mathbf{E}_{\mathrm{ex}} \tag{1.4}$$

where $\mathbf{E}_{\mathrm{ex}}$ designates the external electric field. The "correlation-free equation" or the **Vlasov equation** follows with

$$g_{\mu\nu} \equiv 0 \tag{1.5}$$

which yields (Vlasov, 1945)

$$\frac{\partial f^{(1)}}{\partial t} + \frac{{}^{\mu}\mathbf{p}}{m_{\mu}} \cdot \frac{\partial f^{(1)}}{\partial\, {}^{\mu}\mathbf{r}} + e_{\mu}\langle \mathbf{E} \rangle \cdot \frac{\partial f^{(1)}}{\partial\, {}^{\mu}\mathbf{p}} = 0 \tag{1.6}$$

with $\langle \mathbf{E} \rangle$ still given by (1.4)

The request (1.5) limits the applicability of Vlasov's equation. In a real system, this condition is only approximately fulfilled. To decide how much the pair correlation term contributes we have to anticipate some of the results which we will derive in later sections where we study higher-order truncations of the hierarchy. There we will show that within certain limits the contribution of the right-hand side in (1.3) is of the order of magnitude of $f^{(1)}/\tau_{\mathrm{c}}$ where τ_{c} is the electron collision time,[‡]

$$\tau_{\mathrm{c}} \approx \frac{\Lambda}{\omega_{\mathrm{p}-} \ln \Lambda} \tag{1.7}$$

[†] Note that the correlation functions $g_{\mu\nu}$ in this chapter are defined differently from the use in the equilibrium section, (see the footnote on p. 44).

[‡] For comparison, see the definitions in the Appendix.

with

$$\omega_{p-} = (4\pi ne^2/m)^{1/2}; \qquad \Lambda = 3\Theta^{3/2}/(4\pi n)^{1/2}\, e^3 = 9\delta \tag{1.8}$$

ω_{p-} is the well-known electron plasma frequency and Λ the plasma parameter which is proportional to the number δ of particles in the Debye region.

The correlation-free approach can be applied to phenomena with a characteristic time τ satisfying

$$\tau \ll \tau_c \tag{1.9}$$

Sufficiently below the critical density where we have $\Lambda \gg 1$ we may neglect the correlations up to characteristic times of the order of the electron plasma oscillation time τ_{p-}. This is important since many of the collective phenomena described by the Vlasov equation have characteristic times of order τ_{p-}.

To judge the implications of Vlasov's approach to real physical systems, Table I compiles the characteristic data for a number of typical examples.

TABLE I

Plasma (fully ionized)	Density n[cm^{-3}]	Temperature Θ/κ_B[°K]	Parameter Λ	Oscillation time τ_{p-}[sec]	Collision time τ_c[sec]
Interstellar gas	1	10^2	10^7	10^{-4}	10
Solar corona	10^6	10^6	10^{10}	10^{-7}	10
Low pressure discharge	10^{11}	$5 \cdot 10^4$	$4 \cdot 10^5$	$4 \cdot 10^{-10}$	$2 \cdot 10^{-6}$
High pressure discharge	10^{15}	$5 \cdot 10^3$	10	$4 \cdot 10^{-12}$	$3 \cdot 10^{-12}$
Thermonuclear plasma	10^{16}	10^8	10^8	10^{-12}	10^{-6}

It is noteworthy that Vlasov himself suggested a fictitious system for which the Vlasov equation provides the exact description—the **dispersion model** (Vlasov, 1950).

The model is based on an unlimited subdivision of each particle into equal parts so that we have $n \to \infty$. This subdivision process is subject to the conditions

$$\lim_{n\to\infty} (ne) = \text{constant}, \qquad \lim_{n\to\infty} (nm) = \text{constant} \tag{1.10}$$

Equations (1.10) imply e/m = constant and, since the thermal velocity is unchanged, $\Theta \to 0$.

It is quite obvious that under the conditions (1.10) the Vlasov equation and the plasma oscillation time τ_{p-} remain unchanged. The same is true for the Debye length which is the product of the plasma oscillation time and the thermal velocity.

Therefore we have

$$\lim_{n\to\infty} \delta = \lim_{n\to\infty} \Lambda = \infty \tag{1.11}$$

With this it follows from (1.7) that $\tau_c \to \infty$, which means that the influence of individual particle correlations disappears.

1.2. General Properties

For the sake of simplicity we intend to study in the following a one-component system and remove the indices μ and ν in (1.4) and (1.6). In doing so, we introduce a compensating smeared-out uniform background, the total charge of which has the same value and opposite sign as that of the component studied. This avoids field divergencies in the limit $N, V \to \infty$. For convenience, this background of the density n_c may be introduced into (1.4) through†

$$\nabla \cdot \mathbf{E}_{\text{ex}} = 4\pi n_c e_c = -(4\pi e/V) \int f^{(1)}(\mathbf{r}', \mathbf{v}'; t)\, d\mathbf{r}'\, d\mathbf{v}' \tag{1.12}$$

We therefore have for a single component the relations

$$\frac{\partial f^{(1)}}{\partial t} + \mathbf{v} \cdot \frac{\partial f^{(1)}}{\partial \mathbf{r}} + \frac{e}{m} \langle \mathbf{E} \rangle \cdot \frac{\partial f^{(1)}}{\partial \mathbf{v}} = 0 \tag{1.13}$$

$$\langle \mathbf{E} \rangle = -e \int d\mathbf{r}' \left(\frac{\partial}{\partial \mathbf{r}} \frac{1}{|\mathbf{r} - \mathbf{r}'|} \right) \left\{ \int d\mathbf{v}'\, f^{(1)}(\mathbf{r}', \mathbf{v}'; t) - n_c \right\} \tag{1.14}$$

1. The Vlasov equation is of the type of a Liouville equation in μ space

$$\frac{df^{(1)}}{dt} = \frac{\partial f^{(1)}}{\partial t} + \mathbf{v} \cdot \frac{\partial f^{(1)}}{\partial \mathbf{r}} + \dot{\mathbf{v}} \cdot \frac{\partial f^{(1)}}{\partial \mathbf{v}} = 0 \tag{1.15}$$

This is noteworthy since in the derivation of (1.13) and (1.14) individual particle interactions but not collective interactions of the particles have been neglected. Equation (1.15) shows that traveling along a trajectory with the particles the distribution function $f^{(1)}$ stays constant.

† We now change from the momentum to the velocity representation as this is common use in the literature on this subject.

2. Without lack of generality we may express the distribution function $f^{(1)}$ as a functional in the form $f^{(1)}(A_\nu)$. Introducing this into the Vlasov equation we find

$$\sum_\nu \frac{\partial f^{(1)}}{\partial A_\nu}\left\{\frac{\partial A_\nu}{\partial t} + \mathbf{v}\cdot\frac{\partial A_\nu}{\partial \mathbf{r}} + \dot{\mathbf{v}}\cdot\frac{\partial A_\nu}{\partial \mathbf{v}}\right\} = 0 \tag{1.16}$$

This shows that any function $f^{(1)}$ of the constants of motion A_ν is a solution of the Vlasov equation. For example, if the interaction between the particles is negligible and they move in a conservative external field, then any function of the Hamiltonian is a solution of Vlasov's equation.

3. Under the assumption that $f^{(1)}$ satisfies periodic boundary conditions in an arbitrary large but finite volume of μ space we calculate readily the time derivative of the H function

$$\begin{aligned}\frac{dH}{dt} = \frac{\partial H}{\partial t} &= \frac{\partial}{\partial t}\int f^{(1)} \ln f^{(1)}\, d\mathbf{r}\, d\mathbf{v}\\ &= \int \frac{\partial f^{(1)}}{\partial t} \ln f^{(1)}\, d\mathbf{r}\, d\mathbf{v} + \int \frac{\partial f^{(1)}}{\partial t}\, d\mathbf{r}\, d\mathbf{v} = 0\end{aligned} \tag{1.17}$$

The H function and with that the entropy $S = -\kappa_B H$ are constant. The equation and its solutions are time reversible, which is seen by replacing in (1.13) and (1.14) the time t by $-t$ and the velocity $\mathbf{v}$ by $-\mathbf{v}$.

4. If the initial distribution $f^{(1)}$ is positive definite, it stays positive definite for all times.

If $f^{(1)}$ attains negative values there should be a point $(\mathbf{r}_0, \mathbf{v}_0)$ and a time t_0 where and when $f^{(1)}$ has its first zero. Since the function is positive in the whole environment, we should have at this point the relations

$$\begin{aligned}&\frac{\partial^\nu f^{(1)}}{\partial q_i \cdots \partial q_j} = 0 \qquad \text{for all} \quad \nu < a\\ &\left(\Delta\mathbf{q}\cdot\frac{\partial}{\partial \mathbf{q}}\right)^a f^{(1)} = \Delta^a f^{(1)} > 0\end{aligned} \tag{1.18}$$

with

$$\mathbf{q} = \begin{pmatrix}\mathbf{r}\\ \mathbf{v}\end{pmatrix} \tag{1.19}$$

On the other hand, Vlasov's equation may be written in operator form as

$$\frac{\partial f^{(1)}}{\partial t} = \mathbf{c}(\mathbf{q}, t)\cdot\frac{\partial f^{(1)}}{\partial \mathbf{q}} \tag{1.20}$$

which yields for the time derivatives of higher order

$$\frac{\partial^\nu f^{(1)}}{\partial t^\nu} = (\mathbf{c}(\mathbf{q}, t) \cdot \partial/\partial \mathbf{q})^\nu f^{(1)}$$

$$+ \text{ terms of lower order than } \nu \text{ in the derivative } \partial/\partial \mathbf{q} \quad (1.21)$$

Due to (1.18), it follows that

$$\frac{\partial^\nu f^{(1)}}{\partial t^\nu} = \begin{cases} 0 & \text{for } \nu < a \\ (\mathbf{c}(\mathbf{q}, t) \cdot \partial/\partial \mathbf{q})^a f^{(1)} & \text{for } \nu = a \quad \text{positive definite} \end{cases} \quad (1.22)$$

where it should be observed that the operator $\partial/\partial \mathbf{q}$ acts also on $\mathbf{c}(\mathbf{q}, t)$. However, the corresponding terms do not contribute due to the first line in (1.18). Combination of (1.21) and (1.18) shows that $\partial^a f^{(1)}/\partial t^a$ is positive definite.

5. Any function

$$f^{(1)}(\mathbf{r}, \mathbf{v}, t) = h(\mathbf{v}) \quad (1.23)$$

is a solution of Vlasov's equation provided that the relation $\langle \mathbf{E} \rangle = 0$ holds. The latter is true, of course, in Landau's linear theory, where the relation

$$\int h(\mathbf{v})\, d\mathbf{v} = n_c \quad (1.24)$$

holds.

1.3. The Linear Approximation

The nonlinear Vlasov equation can only be solved in a few particular cases. Therefore the main effort in the past was concentrated on the solution of the linearized Vlasov equation. We introduce

$$f^{(1)} = {}^{(0)}f^{(1)} + {}^{(1)}f^{(1)} \quad (1.25)$$

with the assumption

$${}^{(1)}f^{(1)} \ll {}^{(0)}f^{(1)} \quad (1.26)$$

We also assume that the functions ${}^{(0)}f^{(1)}$ and ${}^{(1)}f^{(1)}$ are smooth, so that differential and integral operations do not change the order of magnitude. The zero-order distribution ${}^{(0)}f^{(1)}$ is homogeneous and stationary, and we have no external fields. To simplify the terminology, we use in the remainder of this chapter

$${}^{(0)}f^{(1)} = f_0, \qquad {}^{(1)}f^{(1)} = f \quad (1.27)$$

We then find the linear relations

$$\frac{\partial f}{\partial t} + \mathbf{v} \cdot \frac{\partial f}{\partial \mathbf{r}} + \frac{e}{m} \mathbf{E} \cdot \frac{\partial f_0}{\partial \mathbf{v}} = 0$$
$$\mathbf{E} = -\nabla\phi, \qquad \nabla^2\phi = -4\pi e \int f \, d\mathbf{v} \tag{1.28}$$

This is an integro-differential equation for f. Various procedures have been applied to solve this set of equations.

2. Solutions of the Linear Vlasov Equation

2.1. The Eigensolution Method

To find the eigensolutions of (1.28), we first separate the space dependence from the velocity and time dependence. We use

$$f = f(\mathbf{r}) f(\mathbf{v}, t) \tag{2.1}$$

Since we will need in the following a large number of different functions it seems impossible to distinguish them by different symbols. Rather it should be observed that different functions are characterized by their independent variables with the exception of the Fourier transform, which is distinguished by the symbol $\sim$.

Introducing (2.1) into (1.28) and applying well-known techniques, we conclude for the space dependence

$$f(\mathbf{r}) = A \exp(\boldsymbol{\lambda} \cdot \mathbf{r}) \tag{2.2}$$

To ensure that $f(\mathbf{r})$ stays finite for all values of $\mathbf{r}$, we have to choose the separation parameter $\boldsymbol{\lambda}$ as a purely imaginary quantity

$$f(\mathbf{r}) = A \exp(i\mathbf{k} \cdot \mathbf{r}), \qquad \mathbf{k} \text{ real} \tag{2.3}$$

We recognize that our normal mode solutions allow a Fourier analysis of any phenomenon in $\mathbf{r}$ space in accordance with

$$\tilde{\Omega}(\mathbf{k}, \mathbf{v}, t) = (2\pi)^{-3/2} \int \Omega(\mathbf{r}, \mathbf{v}, t) \, e^{-i\mathbf{k}\cdot\mathbf{r}} \, d\mathbf{r}$$
$$\Omega(\mathbf{r}, \mathbf{v}, t) = (2\pi)^{-3/2} \int \tilde{\Omega}(\mathbf{k}, \mathbf{v}, t) \, e^{i\mathbf{k}\cdot\mathbf{r}} \, d\mathbf{k} \tag{2.4}$$

Fourier transformation of (1.28) leads to the differential equations

$$(\partial/\partial t + i\mathbf{k}\cdot\mathbf{v})\tilde{f}(\mathbf{k}, \mathbf{v}, t) = i(e/m)\tilde{\Phi}(\mathbf{k}, t)\,\mathbf{k}\cdot\partial f_0(\mathbf{v})/\partial\mathbf{v} \tag{2.5}$$

and

$$\begin{aligned}\tilde{\mathbf{E}}(\mathbf{k}, t) &= -i\mathbf{k}\tilde{\Phi}(\mathbf{k}, t)\\ \tilde{\Phi}(\mathbf{k}, t) &= (4\pi e/k^2)\int \tilde{f}(\mathbf{k}, \mathbf{v}', t)\,d\mathbf{v}'\end{aligned} \tag{2.6}$$

Note that the relation for the field **E** can also be derived directly from (1.14) and (1.24) using the convolution theorem of Fourier analysis and recalling that

$$\left(\frac{1}{|\mathbf{r}-\mathbf{r}'|}\right)^{\sim} = \frac{4\pi}{k^2}\frac{1}{(2\pi)^{3/2}} \tag{2.7}$$

holds.

It is obvious that in (2.5) and (2.6), the velocity perpendicular to the wave vector **k**

$$\mathbf{v}_\perp = \mathbf{v} - \hat{\mathbf{k}}(\mathbf{v}\cdot\hat{\mathbf{k}}), \qquad \hat{\mathbf{k}} = \mathbf{k}/k \tag{2.8}$$

enters the equations only as a parameter. Since its specification is not of interest in general we transform all quantities in (2.5) and (2.6) according to the relation

$$\tilde{\Omega}(\mathbf{k}, u, t) = \int \tilde{\Omega}(\mathbf{k}, \mathbf{v}, t)\,\delta(u - \hat{\mathbf{k}}\cdot\mathbf{v})\,d\mathbf{v} \tag{2.9}$$

Here we have designated the component of the velocity parallel to **k** by u. We arrive at

$$\left(\frac{\partial}{\partial t} + iku\right)\tilde{f}(\mathbf{k}, u, t) = \frac{i}{k}\frac{4\pi e^2}{m}\left(\int \tilde{f}(\mathbf{k}, u', t)\,du'\right)\frac{\partial f_0(\hat{\mathbf{k}}, u)}{\partial u} \tag{2.10}$$

To find the eigenvalue solutions we apply

$$\tilde{f}(\mathbf{k}, u, t) = \tilde{f}(\mathbf{k}, u)\,e^{-i\omega t} \tag{2.11}$$

which yields

$$\tilde{f}(u) = \frac{1}{k^2}\frac{4\pi e^2}{m}\frac{\partial f_0(u)/\partial u}{u - u_{\mathrm{p}}}\int \tilde{f}(u')\,du' \tag{2.12}$$

with the complex phase velocity

$$u_{\mathrm{p}} = \omega/k \tag{2.13}$$

where here and in the following, for reasons of simplicity, we do not always mark the dependence on **k**.

The phase velocity is in general complex because ω is a complex quantity. The reason is that (2.12) is not self-adjoint. Equation (2.12) immediately gives the distribution function $\tilde{f}(u)$ since the integral on the right-hand side is only a constant determined by the normalization which can be chosen conveniently in the linear system.

Integration of (2.12) yields the **dispersion relation**

$$k^2 = \frac{4\pi e^2}{m} \int \frac{\partial f_0(u)/\partial u}{(u - u_{\mathrm{p}})}\, du \tag{2.14}$$

which determines the phase velocity $u_{\mathrm{p}}(k)$ or $\omega(k)$ as a function of the wavenumber vector. In general, the integration in (2.14) is problematic, since the value of the integral is not defined due to the pole at $u = u_{\mathrm{p}}$. We will see in the following how to deal with this problem.

Here we first study the two situations where the existence problem does not arise. This is the case if either $\partial f_0(u)/\partial u = 0$ at $u = u_{\mathrm{p}}$ holds, or if u_p is a complex quantity. These situations are met in many situations of practical interest.

Before we enter the discussion of these problems, we introduce the distribution $F(u)$ through

$$f_0(u) = F(u) \int f_0(u')\, du' = F(u) n \tag{2.15}$$

Allowing for the possibility that the total distribution function is composed of various components ν with

$$f_{0\nu}(u) = F_\nu(u) \int f_{0\nu}(u')\, du' = F_\nu(u) n_\nu \tag{2.16}$$

and the plasma frequencies

$$\omega_{\mathrm{p}\nu}^2 = 4\pi e_\nu{}^2 n_\nu / m_\nu \tag{2.17}$$

we find for the dispersion relation

$$1 - \sum_{\nu=1}^{N} \frac{\omega_{\mathrm{p}\nu}^2}{k^2} \int_{-\infty}^{+\infty} \frac{\partial F_\nu(u)/\partial u}{u - u_{\mathrm{p}}}\, du =: \varepsilon(k, \omega) = 0 \tag{2.18}$$

Of course, we have now as many equations as components, but each of them produces the same dispersion relation.

In the following sections we discuss the meaning of (2.18) for various important examples.

2.2. Particle Beams with a Sharp Velocity

We describe a set of charged particle beams penetrating each other with definite but different velocities in the $\hat{\mathbf{k}}$ direction by

$$F_\nu(u) = \delta(u - u_\nu) \tag{2.19}$$

None of these beams has a temperature spread. Introducing (2.19) into (2.18) and making use of the relations (II.1.3), we find

$$k^2 = \sum_\nu \omega_{\mathrm{p}\nu}^2/(u_\nu - u_{\mathrm{p}})^2 \tag{2.20}$$

It can be shown that our condition $f(u) \ll f_0(u)$ can still be satisfied with the Dirac functions of (2.19) except when the beam velocity comes close to the phase velocity. To prove this it is advisable to consider the Dirac function as a limiting case of proper functions.

a. $u_\nu = 0$. The somewhat academic case where the beam velocity of all charged particles is identically zero is particularly simple. Our dispersion relation reads

$$\omega^2 = \sum_\nu \omega_{\mathrm{p}\nu}^2 = \omega_{\mathrm{p}0}^2 \tag{2.21}$$

All particles oscillate with the same frequency $\omega_{\mathrm{p}0}$ which we call the plasma frequency of the system. ω is independent of k. Consequently the group velocity

$$u_{\mathrm{g}} = \partial\omega/\partial k \tag{2.22}$$

within the system is zero. No signal can be transmitted through this system. In fact, we do not deal with waves but rather with oscillations.

b. $u_\nu \neq 0$. In this case, k is a function of the phase velocity u_{p} in the form

$$k^2 = \sum_\nu \omega_{\mathrm{p}\nu}^2/(u_\nu - u_{\mathrm{p}})^2 = G(u_{\mathrm{p}}) \tag{2.23}$$

Since a complex phase velocity means a complex frequency, and a complex frequency implies an exponential damping or amplification of our mode, we are interested in understanding under what circumstances we have complex solutions of (2.23).

Equation (2.23) is a polynomial of $2N$th order in u_{p}. Therefore it always has $2N$ solutions for u_{p}. They may be real or complex. The complex ones are conjugated.

In the following, we discuss the number of real solutions which allow us to conclude how many complex solutions we have. Let us consider the function $G(u_p)$ for real values of u_p. We note that with $u_p \to \infty$ the function $G(u_p)$ tends to zero, whereas it tends to infinity in the neighborhood of the N values u_ν. Since the function is positive definite, it has $(N - 1)$ minima between these values u_ν, which correspond to wavenumbers $k_1^2, k_2^2, \ldots, k_{N-1}^2$, arranged in increasing order.

The above discussion of the curve shows that we have in the interval $k_{\nu-1}^2 < k^2 < k_\nu^2$ a total of 2ν real solutions and with that

$$z_c = 2(N - \nu) \tag{2.24}$$

complex solutions. Here we have

$$N \geqslant \nu \geqslant 1, \qquad k_0^2 = 0, \qquad k_N^2 = \infty \tag{2.25}$$

Since the complex solutions occur always in conjugate sets, our system is unstable if there is a complex solution anywhere in the range between k_0^2 and k_N^2.

A Single Beam

For the single beam we have $N = 1$, $\nu = 1$, and consequently $z_c = 0$ in the whole area k_0^2 to k_N^2. Our system is stable. The dispersion relation reads

$$\omega = ku_1 \pm \omega_p \tag{2.26}$$

This relation can be understood as a plasma oscillation with a Doppler shift superimposed. The group velocity

$$u_g = \partial\omega/\partial k = u_1 \tag{2.27}$$

is independent of k and identical with the beam velocity.

The above result may be interpreted in the form that the particles of the beam in a coordinate system moving with the beam show the normal plasma oscillations of case (a). The translational motion with the group velocity $u_g = u_1$ superimposes the Doppler effect.

Two or More Beams

For $N \geqslant 2$, we always have at least two complex solutions in the range $k_{N-2}^2 < k^2 < k_{N-1}^2$.

Consequently a system of two or more beams penetrating each other is always unstable. The number of complex modes decreases with increasing k.

For very large values of k, we have only real solutions close to

$$\omega_\nu = ku_\nu \pm \omega_{\mathrm{p}} \tag{2.28}$$

These solutions for very small wavelengths are decoupled. This means—as in the single-beam case—decoupled plasma oscillations of the beam superimposed with the corresponding Doppler shift.

The Electron–Ion Beam System

As a system of practical interest, we study an electron beam and an ion beam characterized by the relations

$$\left(\frac{\omega_{\mathrm{p}+}}{\omega_{\mathrm{p}-}}\right)^{2/3} \ll 1, \qquad \frac{u_+}{u_-}\left(\frac{\omega_{\mathrm{p}-}}{\omega_{\mathrm{p}+}}\right)^{2/3} \ll 1 \tag{2.29}$$

The first condition is secured through the mass ratio of electrons and ions. The second condition prescribes that in our model the ion velocity should be very small in comparison to the electron velocity, a situation which is given in many practical cases. The dispersion relation of our system reads

$$k^2 = \frac{\omega_{\mathrm{p}-}^2}{(u_- - u_{\mathrm{p}})^2} + \frac{\omega_{\mathrm{p}+}^2}{(u_+ - u_{\mathrm{p}})^2} \tag{2.30}$$

To find the critical value k_{cr} we derive

$$\frac{dk^2}{du_{\mathrm{p}}} = \frac{2\omega_{\mathrm{p}-}^2}{(u_- - u_{\mathrm{p}})^3} + \frac{2\omega_{\mathrm{p}+}^2}{(u_+ - u_{\mathrm{p}})^3} = 0 \tag{2.31}$$

or applying (2.29)

$$u_{\mathrm{p}} = u_-\left(\frac{\omega_{\mathrm{p}+}}{\omega_{\mathrm{p}-}}\right)^{2/3} \tag{2.32}$$

Introducing this into (2.30) and again making use of (2.29), we find for the critical wavevector

$$k_{\mathrm{cr}}^2 = (\omega_{\mathrm{p}-}/u_-)^2 \tag{2.33}$$

The electron–ion beam system is stable for $k > k_{\mathrm{cr}}$ and unstable for $k < k_{\mathrm{cr}}$.

2.3. Spherical Distribution of Monoenergetic Particles

Here we consider charged particles with a spherical monoenergetic velocity distribution moving in a charge compensating immobile background.

The dispersion relation is given according to (2.18) in the form

$$k^2 = \omega_p^2 \int \frac{\partial F(u)/\partial u}{u - u_p}\, du \tag{2.34}$$

with

$$F(u) = (1/n) \int f_0(\mathbf{v})\, \delta(u - \hat{\mathbf{k}} \cdot \mathbf{v})\, d\mathbf{v} \tag{2.35}$$

and in this case

$$f_0(\mathbf{v}) = An\delta(v^2 - v_0^2); \qquad A = 1/2\pi \mid v_0 \mid \tag{2.36}$$

It follows that

$$\begin{aligned} F(u) &= A \int \delta(v_\perp^2 + v_\parallel^2 - v_0^2)\, \delta(u - v_\parallel)\, v_\perp\, dv_\perp\, dv_\parallel\, d\phi \\ &= A\pi \int_0^\infty \delta[v_\perp^2 - (v_0^2 - u^2)]\, dv_\perp^2 \end{aligned} \tag{2.37}$$

or after simple substitution

$$F(u) = A\pi \int_{-\infty}^{v_0^2 - u^2} \delta(y)\, dy = A\{\Delta_0(v_0^2 - u^2) + \pi/2\} \tag{2.38}$$

where

$$\Delta(x, 0) = \Delta_0(x) = \pi \int_{-\infty}^{x} \delta(t)\, dt - \pi/2 \tag{2.39}$$

is a special value of Dirichlet's function

$$\Delta(x, y) = \int_0^\infty t^{-1} \sin xt \cos yt\, dt \tag{2.40}$$

As a reminder the values of Dirichlet's function are shown in Figure III.1. According to (2.38) and Figure III.1, the distribution $F(u)$ has the form shown in Figure III.2. This is a so called **packet distribution**. It may also be represented in the form

$$\begin{aligned} F(u) &= A[\Delta_0(u + v_0) + \pi/2], \qquad & u < 0 \\ &= A[-\Delta_0(u - v_0) + \pi/2], \qquad & u > 0 \end{aligned} \tag{2.41}$$

and remembering

$$d\Delta_0(x)/dx = \pi\delta(x) \tag{2.42}$$

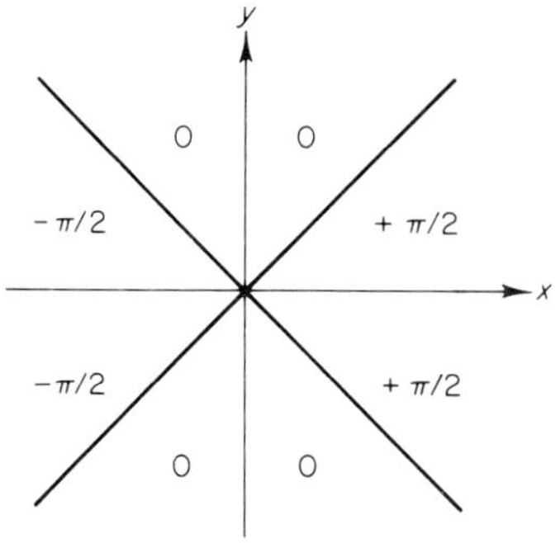

FIGURE III.1. Schematic presentation of the Dirichlet function.

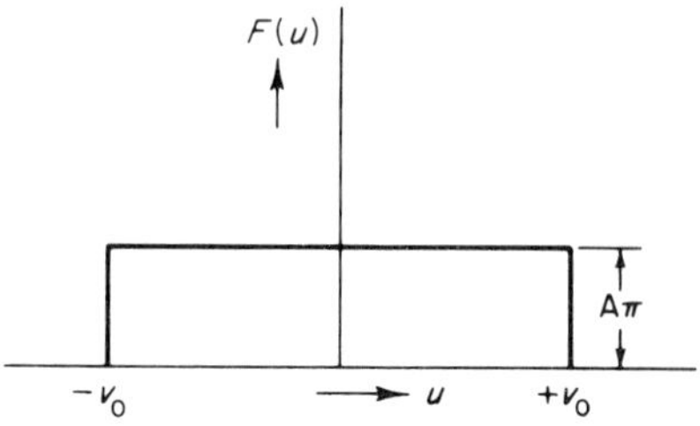

FIGURE III.2. Distribution function of a spherical distribution of monoenergetic particles of one velocity component u.

we find

$$dF(u)/du = A\pi[\delta(u + v_0) - \delta(u - v_0)] \tag{2.43}$$

This yields the dispersion relation

$$\begin{aligned} k^2 &= A\pi{\omega_\mathrm{p}}^2 \int \frac{\delta(u + v_0) - \delta(u - v_0)}{u - u_\mathrm{p}}\, du \\ &= -A\pi{\omega_\mathrm{p}}^2 \left(\frac{1}{v_0 + u_\mathrm{p}} + \frac{1}{v_0 - u_\mathrm{p}}\right) \end{aligned} \tag{2.44}$$

or since $F(u)$ is normalized to one

$$k^2 = -\frac{A\pi{\omega_\mathrm{p}}^2 2v_0}{{v_0}^2 - {u_\mathrm{p}}^2} = -\frac{{\omega_\mathrm{p}}^2}{{v_0}^2 - {u_\mathrm{p}}^2} \tag{2.45}$$

or

$$\omega^2 = k^2{v_0}^2 + {\omega_\mathrm{p}}^2 \tag{2.46}$$

We conclude that a system of spherically expanding monoenergetic particles is always stable. The product of the group velocity and the phase velocity is identical with the square of the particle velocity, $u_\mathrm{g}u_\mathrm{p} = {v_0}^2$.

2.4. THE PACKET DISTRIBUTIONS

In this section, we would like to consider a more general system of the packet distributions. Let $u_{1\nu}$ and $u_{2\nu}$ be the lowest and the highest velocity u of the νth packet distribution. Then it follows from (2.41) and (2.42) that we have

$$dF(u)/du = \sum_{\nu} A_\nu \pi \{\delta(u - u_1) - \delta(u - u_{2\nu})\} \tag{2.47}$$

and, with that, the dispersion relation

$$\begin{aligned} k^2 &= \sum_{\nu} \pi A_\nu \omega_{\mathrm{p}\nu}^2 \left(\frac{1}{u_{1\nu} - u_{\mathrm{p}}} - \frac{1}{u_{2\nu} - u_{\mathrm{p}}}\right) \\ &= \sum_{\nu} \pi A_\nu \omega_{\mathrm{p}\nu}^2 \Delta u_\nu \frac{1}{(\langle u \rangle_\nu - u_{\mathrm{p}})^2 - (\Delta u_\nu/2)^2} \end{aligned} \tag{2.48}$$

where we introduced the relations

$$\langle u \rangle_\nu = \tfrac{1}{2}(u_{1\nu} + u_{2\nu}), \qquad u_{2\nu} - u_{1\nu} = \Delta u_\nu \tag{2.49}$$

and applying the normalization condition, we find

$$k^2 = \sum_{\nu} \frac{\omega_{\mathrm{p}\nu}^2}{(\langle u \rangle_\nu - u_{\mathrm{p}})^2 - (\Delta u_\nu/2)^2} \tag{2.50}$$

As an example, we study the situation shown in Figure III.3. The characteristic function $G(u_{\mathrm{p}})$ is very similar to that of a system of beams,

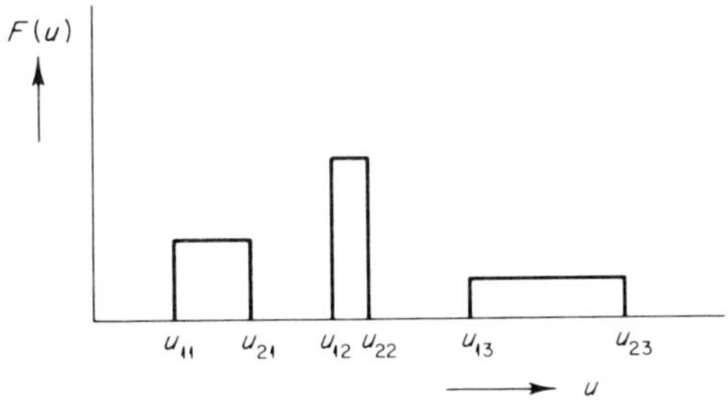

FIGURE III.3. Nonoverlapping packets.

except that in the neighborhood of each average velocity $\langle u \rangle_\nu$ within the range $u_{1\nu} \leqslant u \leqslant u_{2\nu}$, the characteristic function assumes negative values, whereas it is positive definite otherwise. Figure III.4 demonstrates schematically the characteristic function for the situation presented in Figure III.3.

A discussion analogous to that given for the case of multiple beams shows that such a combination of nonoverlapping packet distributions is always unstable, provided the number of packets is larger than one.

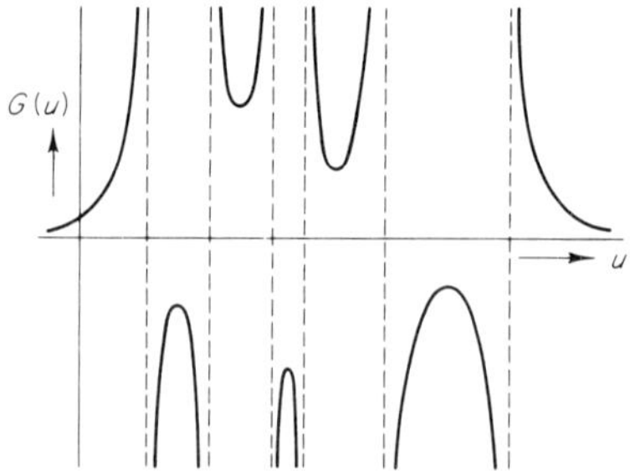

FIGURE III.4. Characteristic function for the distribution of Figure III.3, schematically.

2.5. THE GAP DISTRIBUTIONS

In this section we want to study distributions which are arbitrary except for the fact that they have a specified gap, where $F(u) \equiv 0$ holds. We consider phenomena for which u_p is within this gap.

Langmuir Waves of Long Wavelengths

We have a set of N arbitrary distributions $F_\nu(u)$, each of them having a gap extending from u_ν to ∞. The maximum value of all u_ν is u_m. The following derivation deals with the case $u_p \gg u_m$.

We rewrite the dispersion relation (2.18) after partial integration as

$$k^2 = \sum_{\nu=1}^{N} \omega_{p\nu}^2 \int_{-\infty}^{+\infty} \frac{\partial F_\nu(u)/\partial u}{u - u_p} du = \sum_{\nu=1}^{N} \omega_{p\nu}^2 \int_{-\infty}^{+\infty} \frac{F_\nu(u)}{(u - u_p)^2} du \tag{2.51}$$

The integrand may be developed into a series

$$\frac{1}{u_p{}^2[1 - (u/u_p)]^2} = \frac{1}{u_p{}^2} \sum_{\mu=0}^{\infty} (1 + \mu) \left(\frac{u}{u_p}\right)^\mu \tag{2.52}$$

With $u_p k = \omega$, we find from (2.51)

$$\begin{aligned} \omega^2 &= \sum_{\nu=1}^{N} \sum_{\mu=0}^{\infty} \omega_{p\nu}^2 \int_{-\infty}^{+\infty} (1 + \mu) \left(\frac{u}{u_p}\right)^\mu F_\nu(u)\, du \\ &= \sum_{\nu=1}^{N} \sum_{\mu=0}^{\infty} \omega_{p\nu}^2 (1 + \mu) \left\langle\left(\frac{u}{u_p}\right)^\mu\right\rangle_\nu \end{aligned} \tag{2.53}$$

respectively

$$\omega^2 = \sum_{\nu=1}^{N} \sum_{\mu=0}^{\infty} \omega_{p\nu}^2 (1 + \mu)(k/\omega)^\mu \langle u^\mu \rangle_\nu \tag{2.54}$$

If we truncate the series at $\mu = m$ we have an implicit equation of order $m + 2$ relating ω to k.

To arrive at the Langmuir result—which is meaningful in the above range $u_p \gg u_m$—we truncate the series at $\mu = 2$. Applying the relations

$$\sum_\nu \omega_{p\nu}^2 =: \omega_{p0}^2\,; \qquad \sum_\nu \omega_{p\nu}^2 \langle u^2 \rangle_\nu =: \omega_{p0}^2 \langle u^2 \rangle \tag{2.55}$$

and the condition

$$\sum_\nu \omega_{p\nu}^2 \langle u \rangle_\nu = 0 \tag{2.56}$$

we arrive at

$$\omega^2 = \omega_{p0}^2 \{1 + 3(k/\omega)^2 \langle u^2 \rangle\} \tag{2.57}$$

The relation (2.56) can always be established through an appropriate choice of the coordinate system.

Since we have neglected fourth-order terms in (2.56), it is consistent to use $\omega = \omega_{p0}$ on the right-hand side of (2.57)

$$\omega^2 = \omega_{p0}^2 + 3k^2 \langle u^2 \rangle \tag{2.58}$$

Obviously (2.58) expresses a dispersion depending on the average velocity square. The group velocity is given by

$$u_g = \partial\omega/\partial k = 3(k/\omega)\langle u^2 \rangle = 3\langle u^2 \rangle / u_p = 3(k/\omega_{p0})\langle u^2 \rangle \tag{2.59}$$

Our treatment is consistent for values of k which satisfy

$$|\, k \,| \ll \omega_{p0}/u_m \tag{2.60}$$

Ion Acoustic Waves

We study a system of cold ions and hot electrons. The ions are represented by a Dirac function in the form

$$F_+(u) = \delta(u) \tag{2.61}$$

Their "gap" extends from zero to infinity. The electrons are represented by the distribution function

$$\begin{aligned} F_-(u) &= c \exp(-m_- u^2/2\Theta_-) \qquad &\text{for } |\, u \,| > \bar{u} \\ F_-(u) &= 0 &\text{for } |\, u \,| < \bar{u} \end{aligned} \tag{2.62}$$

where c is a normalization constant. The electron gap extends over the range $|u| < \bar{u}$ and therewith $u_p < \bar{u}$. A simple transformation yields, for the dispersion relation,

$$k^2 = -\frac{\omega_{p-}^2 m_-}{\Theta_-} \left\{1 + u_p \int_{|u|>\bar{u}} c \exp\left(-\frac{m_- u^2}{2\Theta_-}\right) \frac{1}{u - u_p} \, du\right\} + \frac{\omega_{p+}^2}{u_p^2} \tag{2.63}$$

We now develop $(1 - u_p/u)^{-1}$ in terms of powers of (u_p/u).

At this stage, we choose $\bar{u}$ in accordance with

$$(m_-/m_+)^{1/2}\langle u\rangle_- \ll \bar{u} \ll \langle u\rangle_- \tag{2.64}$$

This guarantees on the one hand that most of the electrons are included in our considerations, and on the other hand that $\bar{u} \gg u_p$ holds. Observing the fact that the distribution function F_- is even, we find from (2.63)

$$k^2 = -\frac{\omega_{p-}^2 m_-}{\Theta_-} + \frac{\omega_{p+}^2}{u_p^2} \tag{2.65}$$

It follows that

$$\omega^2 = \omega_{p+}^2 \frac{(k\lambda_{D-})^2}{1 + (k\lambda_{D-})^2} \tag{2.66}$$

where λ_{D-} designates the Debye length of the electrons.

For very large values of $(k\lambda_{D-})$, we find decoupled oscillations with the ion plasma frequency ω_{p+}. The interesting case is $(k\lambda_{D-}) \ll 1$. Our dispersion relation then produces

$$u_p^2 = \Theta_-/m_+ \tag{2.67}$$

This phase velocity is identical with that of an acoustic wave in a system of particles of mass m_+ and thermal energy Θ_-, assuming an isothermal process. For this reason, the waves with $(k\lambda_{D-}) \ll 1$ are often called **ion acoustic waves**.

At first sight, it may seem surprising that we find a structure similar to an acoustic wave, since we have no collisions or correlations in our Vlasov description. On the other hand, the occurrence of the ion acoustic velocity is plausible from the following physical reasoning.

The ion waves are—so far as the electrons are concerned—a slow process. The electrons may therefore be considered as in equilibrium with the potential

$$n_- = n_+ = c \exp(e\Phi/\Theta_-) \tag{2.68}$$

Consequently, we have

$$\nabla n_+/n_+ = \nabla n_-/n_- = -(e\mathbf{E}/\Theta_-) \tag{2.69}$$

and introducing this into the equation of collision-free motion of the ions, we find

$$m_+ n_+ \, d\mathbf{v}_+/dt = en_+\mathbf{E} = -n_+\Theta_- \, \nabla n_+/n_+ = -\nabla(n_+\Theta_-) \tag{2.70}$$

Comparing this with the basic equation of sound propagation we understand why we have the "acoustic behavior."

2.6. Arbitrary Distributions—Penrose Criterion

So far we have been able to avoid complications stemming from the zero of the denominator of the dispersion relation by restricting our discussion to special distributions. We now consider arbitrary distributions. Since the value of the integral in the dispersion relation is not defined for real values of u_p, we first concentrate on complex phase velocities. This allows us to draw conclusions with respect to the stability of systems with arbitrary distributions. We investigate whether the dispersion relation (2.18) rewritten for a single component ($\nu = 0$) with smeared-out neutralizing background

$$k^2 = G(u_p) = \omega_p{}^2 \int_{-\infty}^{+\infty} \frac{F_0'(u)}{u - u_p} \, du = \omega_p{}^2 \int_{-\infty}^{+\infty} \frac{F_0(u)}{(u - u_p)^2} \, du \tag{2.71}$$

has a solution with $\mathrm{Im}(u_p) > 0$.

Exponentially growing modes exist if the function $G(u_p)$ takes a real positive value somewhere in the upper half of the complex u_p plane. They are the most important cause of instabilities, so that we shall derive a necessary and sufficient criterion for exponentially growing instability and a necessary one for stability. The analysis to be given will hold only for fairly smooth functions $F_0(u)$ with suitable behavior at infinity. Mathematically speaking, we require for real u

$$\begin{gathered}\int_{-\infty}^{+\infty} F_0'(u)^2 \, du < \infty, \qquad \int_{-\infty}^{+\infty} |F_0'(u)| \, du < \infty, \qquad |F_0'(u)| \leqslant M_1 \\ \int_{-\infty}^{+\infty} F_0''(u)^2 \, du < \infty, \qquad |F_0''(u)| \leqslant M_2\end{gathered} \tag{2.72}$$

From (2.71) and (2.72) we conclude that $G(u_p)$ is a holomorphic function in the u_p plane cut along the real axis.

Its behavior on the upper boundary of the cut is given by

$$G(u_{\mathrm{p}} + i0) = \omega_{\mathrm{p}}^2 \left\{ \mathcal{P} \int_{-\infty}^{+\infty} \frac{F_0'(u)}{u - u_{\mathrm{p}}}\, du + i\pi F_0'(u_{\mathrm{p}}) \right\} \tag{2.73}$$

due to Plemelj's formula

$$\frac{1}{x \pm i0} = \mathcal{P} \frac{1}{x} \mp i\pi\delta(x) \tag{2.74}$$

($\mathcal{P}$ denotes the Cauchy principal value.)

The validity of (2.74) is evident from

$$\int_{-\infty}^{+\infty} \frac{f(x)}{x \pm i0}\, dx = \int_{|x|>\epsilon} \frac{f(x)}{x}\, dx + \int_{C_\pm} \frac{f(z)}{z}\, dz \tag{2.75}$$

where $C_\pm$ is the semicircle with radius ε around the origin in the upper respectively lower half plane in the limit $\varepsilon \to 0$. We find

$$\begin{aligned} \int_{-\infty}^{+\infty} \frac{f(x)}{x \pm i0}\, dx &= \mathcal{P} \int_{-\infty}^{+\infty} \frac{f(x)}{x}\, dx + \int_{\pm\pi}^{0} \frac{f(0)}{\varepsilon e^{i\phi}}\, i\varepsilon e^{i\phi}\, d\phi \\ &= \mathcal{P} \int_{-\infty}^{+\infty} \frac{f(x)}{x}\, dx \mp i\pi f(0) \end{aligned} \tag{2.76}$$

We consider the conditions (2.72). The boundedness of $F_0'(u)$ implies a Lipschitz condition on $F_0(u)$ and by that the boundedness and continuity of $G(u_{\mathrm{p}} + i0)$.

(The boundedness can be easily derived from (2.72):

$$\left| \mathcal{P} \int_{-\infty}^{+\infty} \frac{F_0'(u)}{u - u_{\mathrm{p}}}\, du \right| \leqslant \frac{1}{\Delta} \int_{-\infty}^{+\infty} |F_0'(u)|\, du + \left| \mathcal{P} \int_{u_{\mathrm{p}}-\Delta}^{u_{\mathrm{p}}+\Delta} \frac{F_0'(u)}{u - u_{\mathrm{p}}}\, du \right|$$

$$\begin{aligned} \left| \mathcal{P} \int_{u_{\mathrm{p}}-\Delta}^{u_{\mathrm{p}}+\Delta} \frac{F_0'(u)}{u - u_{\mathrm{p}}}\, du \right| &\leqslant \left| \mathcal{P} \int_{u_{\mathrm{p}}-\Delta}^{u_{\mathrm{p}}+\Delta} \frac{F_0'(u_{\mathrm{p}})}{u - u_{\mathrm{p}}}\, du \right| + \int_{u_{\mathrm{p}}-\Delta}^{u_{\mathrm{p}}+\Delta} \frac{M_2(u - u_{\mathrm{p}})}{u - u_{\mathrm{p}}}\, du \\ &= 0 + 2\Delta M_2 \end{aligned}$$

So $G(u_{\mathrm{p}} + i0)$ is bounded, since we chose Δ arbitrary but fixed.)

The image $G(u_{\mathrm{p}} + i0)$ of the upper boundary of the real u_{p} axis reveals those properties of $G(u_{\mathrm{p}})$ which are of relevance here.

Since $G(\mathbb{R} + i0)$ is bounded and continuous, the image will be a directed curve. This curve starts and finishes at the origin $G = 0$, since $G(\pm\infty) = 0$.

Now, if G_0 is any point not on $G(\mathbb{R} + i0)$, then the curve $G(\mathbb{R} + i0)$ winds anticlockwise around G_0 as many times as $G(u_p)$ takes the value G_0 in the upper half-plane. This can be seen by application of Cauchy's theorem

$$(1/2\pi i) \oint [F'(u_p)/F(u_p)]\, du_p = N_0 \tag{2.77}$$

where N_0 is the number of zeros in the domain enclosed by the integration contour to $F(u_p) := G(u_p) - G_0$. We get

$$\frac{1}{2\pi i} \oint \frac{G'(u_p)}{G(u_p) - G_0}\, du_p = \frac{1}{2\pi i} \oint \frac{dG}{G - G_0} = \frac{2\pi i M}{2\pi i} = M \tag{2.78}$$

where we take $\oint (G - G_0)^{-1}\, dG$ as a Stieltjes integral, and where M is the number of times the image contour winds around G_0. So we get $M = N_0$.

Hence the image of any point in the upper half-plane (of the u_p-plane) must be either inside $G(\mathbb{R} + i0)$ or upon it; therefore the interior of $G(\mathbb{R} + i0)$ is the image of the upper half-plane. Consequently $G(u_p)$ takes positive values somewhere in this half-plane if and only if $G(\mathbb{R} + i0)$ encloses some parts of the positive real G axis. This happens if and only if $G(\mathbb{R} + i0)$ crosses the positive real G axis. Since the point $G(u_p + i0)$ moving along $G(\mathbb{R} + i0)$ encloses the points inside in an anticlockwise sense, it must be moving upward in its right-hand-most crossing of the positive real G axis.

This is equivalent with a change in sign to $\operatorname{Im} G(u_p + i0) = \omega_p{}^2 \pi F_0'(u_p)$ corresponding to a minimum of $F_0(u)$. Also, $\operatorname{Re} G(u_p + i0)$ can, by (2.73) be expressed in the form

$$\operatorname{Re} G(u_p + i0) = \omega_p{}^2\, \mathcal{P} \int \frac{F_0'(u)}{u - u_p}\, du \tag{2.79}$$

We may therefore formulate the **Penrose criterion** (Penrose, 1960):

A necessary and sufficient condition for the existence of exponentially growing modes is the requirement that:

a. the distribution function has at least one minimum; and
b. that in one of these minimum values the condition

$$\mathcal{P} \int \frac{F_0'(u)}{u - u_p}\, du > 0 \tag{2.80}$$

is satisfied.

It is interesting to remark that the Penrose criterion holds identically for the existence of exponentially damped modes. This can easily be seen by taking the complex conjugate of the dispersion relation

$$\frac{k^2}{\omega_p{}^2} = \int_{-\infty}^{+\infty} \frac{F_0'(u)}{u - u_p{}^*}\, du \tag{2.81}$$

which yields the result that u_p is a solution if and only if $u_p{}^*$ is a solution.

(Using Penrose's treatment, we look for the image $G(\mathbb{R} - i0)$ of the lower boundary of the real u_p axis. We find, due to Plemelj's formula, that $G(\mathbb{R} - i0)$ coincides with $G(\mathbb{R} + i0)$. It is directed, however, in the opposite sense. $G(u_p)$ is a conformal mapping, and therefore the image of any domain stays at the same side of the boundary. Consequently the lower half-plane again is mapped into the interior of $G(\mathbb{R} \pm i0)$. The inspection of the right-hand-most crossing of $G(\mathbb{R} - i0)$ with the real k^2 axis again yields (2.80), since we have changed the sense of direction of $G(\mathbb{R} + i0)$ and the sign of the imaginary part too.)

The application of these results will be illustrated by two examples.

The Single-Humped Distributions

The simplest case is a distribution with a single maximum. According to the first criterion, we find Nyquist's result that a single humped distribution cannot lead to exponentially growing modes. We show in Figure III.5 the curve $G(\mathbb{R} + i0)$ for the Maxwellian distribution function $F_0(u) = C \exp(-mu^2/2\Theta)$. The image of the upper half-plane is shaded and includes no positive real values. Possible $G(\mathbb{R} + i0)$ curves which do enclose positive real values are shown in Figure III.6.

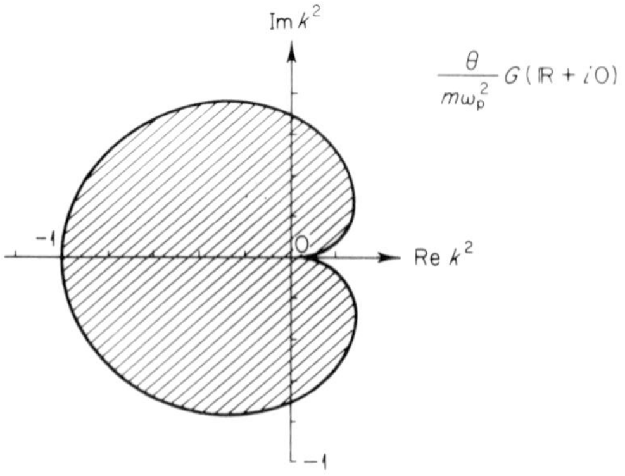

FIGURE III.5. Image of the upper half u_p plane in the k^2 plane for a Maxwellian distribution.

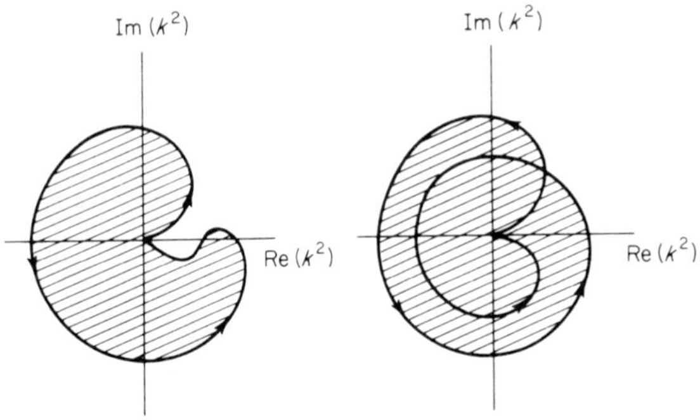

FIGURE III.6. Two examples of $u_p - k^2$ mappings for unstable distributions.

Stability of an Isotropic Distribution

We consider an isotropic distribution $f_0(v^2)$ which is arbitrary

$$F(u) = \int f_0(v^2)\,\delta(u - \hat{\mathbf{k}} \cdot \mathbf{v})\,d\mathbf{v} = \pi \int_0^\infty f_0(v_\perp{}^2 + u^2)\,dv_\perp{}^2$$
$$= \pi \int_{u^2}^\infty f_0(x)\,dx \qquad (2.82)$$

producing the derivative

$$F'(u) = -2\pi u f_0(u^2) \qquad (2.83)$$

Applying the Penrose criterion (a.) we see that such an isotropic distribution is stable.

3. Fourier Transformation in Time

In the preceding section, we studied the eigensolution method for the solution of the linearized Vlasov equation. This procedure—sometimes called normal mode analysis—is appropriate for solving the initial value problem if the set of the eigensolutions is complete. For certain special distributions we have found such complete sets. For arbitrary distributions we discussed those eigensolutions which have a complex frequency or phase velocity. In Section 5 we shall study the "Van Kampen modes," which supplement the complex solutions to a complete set.

3.1. The External Perturbation Problem

We study the case where we apply a perturbation given by an external potential disturbance $\Phi_{ex}(\mathbf{r}, t)$ at the time $t = 0$. For $t < 0$, the perturbation $\Phi_{ex} \equiv 0$, and we assume that f disappears in this region. This means that f is a so called "causal function." This assumption destroys the reversible character of our description, whereas the Vlasov equation—as we proved above—is reversible.

According to (2.5), (2.6) and (2.15), the linearized Vlasov equation reads after Fourier analysis in configuration space

$$(\partial/\partial t + iku)\tilde{f}(u, k, t) = i(e/m)\, kn[\partial F_0(u, \hat{\mathbf{k}})/\partial u]\, \tilde{\Phi}(k, t)$$
$$\tilde{\Phi}(k, t) = (4\pi e/k^2) \int \tilde{f}(u, k, t)\, du + \tilde{\Phi}_{ex}\,; \tag{3.1}$$

where $\tilde{\Phi}_{ex}$ represents the external perturbation.

Before applying Fourier transformation with respect to time, we have to make sure that the integrals of the Fourier transformation exist. The functions to be transformed must be quadratic integrable. If the plasma is unstable—and this is frequently the case—the functions increase exponentially in time and cannot fulfill the above condition. However, it is not difficult to overcome this problem. Let η_m be the maximum growth rate of a certain function Ω of the unstable plasma. We then consider the Fourier transform of the function

$$\Omega'(\mathbf{r}, t) = \Omega(\mathbf{r}, t)\, e^{-\omega_i t} \tag{3.2}$$

with $\omega_i > \eta_m$. This function Ω' is quadratic integrable, since we require $\Omega \equiv 0$, for $t < 0$. We may apply real Fourier transformation to Ω'. This yields

$$\tilde{\Omega}'(\mathbf{r}, \omega_r) = (2\pi)^{-1/2} \int \Omega(\mathbf{r}, t)\, e^{i\omega t}\, dt = \tilde{\Omega}(\mathbf{r}, \omega) \tag{3.3}$$

with

$$\omega = \omega_r + i\omega_i\,, \qquad \omega_i = \text{constant} > \eta_m \tag{3.4}$$

and

$$\Omega(\mathbf{r}, t) = (2\pi)^{-1/2} \int_{-\infty}^{+\infty} \tilde{\Omega}(\mathbf{r}, \omega)\, e^{-i\omega t}\, d\omega_r = (2\pi)^{-1/2} \int_{i\omega_i-\infty}^{i\omega_i+\infty} \tilde{\Omega}(\mathbf{r}, \omega)\, e^{-i\omega t}\, d\omega \tag{3.5}$$

Note that now in contrast to the real Fourier transformation the contour of integration in (3.5) is shifted by ω_i into the upper half of the ω plane.

Applying the transformation (3.3) to (3.1), we find†

$$
\begin{aligned}
(-i\omega + iku)\tilde{\tilde{f}}(u, k, \omega) &= i(e/m)\, kn[\partial F_0(u, \hat{\mathbf{k}})/\partial u]\, \tilde{\tilde{\Phi}}(k, \omega) \\
\tilde{\tilde{\Phi}}(k, \omega) &= (4\pi e/k^2) \int \tilde{\tilde{f}}(u, k, \omega)\, du + \tilde{\tilde{\Phi}}_{\mathrm{ex}}(k, \omega)
\end{aligned}
\tag{3.6}
$$

It follows that

$$
\tilde{\tilde{f}}(u, k, \omega) = \frac{ne}{m} \frac{\partial F_0(u, \hat{\mathbf{k}})/\partial u}{u - (\omega/k)} \tilde{\tilde{\Phi}}(k, \omega) \tag{3.7}
$$

and

$$
\tilde{\tilde{\Phi}}(k, \omega) = \frac{\tilde{\tilde{\Phi}}_{\mathrm{ex}}(k, \omega)}{1 - (\omega_{\mathrm{p}}{}^2/k^2) \int_{-\infty}^{+\infty} du\, F_0{}'(u, \hat{\mathbf{k}})/[u - (\omega/k)]} \tag{3.8}
$$

Observe that the denominators in (3.7) and (3.8) cause no difficulties since ω is a complex quantity.

For a stable plasma we may now consider the limit $\omega_{\mathrm{i}} = +0$ yielding together with (2.74)

$$
\tilde{\tilde{\Phi}}(k, \omega) = \frac{\tilde{\tilde{\Phi}}_{\mathrm{ex}}(k, \omega)}{1 - (\omega_{\mathrm{p}}{}^2/k^2) \int F_0{}'(u, \hat{\mathbf{k}}) \left\{ \mathcal{P} \left(\dfrac{1}{u - \omega/k} \right) + i\pi\delta \left(u - \dfrac{\omega}{k} \right) \right\} du} \tag{3.9}
$$

In principle, the problem is solved. Since $\tilde{\tilde{\Phi}}_{\mathrm{ex}}(k, \omega)$ is given we find $\tilde{\tilde{\Phi}}(k, \omega)$ from (3.8) and introducing this into (3.7) we may calculate the perturbed distribution function.

The linear response of the system to the external perturbation is reflected in the denominator of (3.9) which reads in the generalization to many components

$$
\varepsilon(k, \omega) = 1 - \sum_{\nu} \frac{\omega_{\mathrm{p}_\nu}^2}{k^2} \int_{-\infty}^{+\infty} F'_{\nu}(u, \hat{\mathbf{k}}) \left\{ \mathcal{P} \left(\frac{1}{u - \omega/k} \right) + i\pi\delta \left(u - \frac{\omega}{k} \right) \right\} du \tag{3.10}
$$

The quantity $\varepsilon(k, \omega)$ is closely related to the dielectric constant of the plasma. This is readily seen if we assume an isotropic and homogeneous plasma where Poisson's equation reads

$$
\varepsilon(\omega)\, \Delta \tilde{\Phi} = -4\pi e \tilde{n}_{\mathrm{ex}} \tag{3.11}
$$

$$
\tilde{\tilde{\Phi}}(k, \omega) = 4\pi e \tilde{\tilde{n}}_{\mathrm{ex}}(k, \omega)/k^2 \varepsilon(\omega) = \tilde{\tilde{\Phi}}_{\mathrm{ex}}(k, \omega)/\varepsilon(\omega) \tag{3.12}
$$

† The symbol ≈ denotes the double Fourier transform with respect to the time and the space variable.

Comparing (3.12) with (3.9), we find that $\varepsilon(k, \omega)$ is the dielectric constant measured in a plasma when observing a single wave with wavevector $\mathbf{k}$ and frequency ω. It should be noted, however, that in general $\varepsilon(k, \omega)$ is not the Fourier transform of the space dependent dielectric constant $\varepsilon(\mathbf{r}, \omega)$.

In the limits of low and high frequencies we evaluate the results for Maxwellian distribution functions F_ν.

For **low frequencies** ($\omega_r \to 0$), our result (3.10) simplifies to

$$\varepsilon(k, \omega = 0) = 1 - \sum_\nu (\omega_{p\nu}^2/k^2) \int F'_\nu(u, \hat{\mathbf{k}}) \left[\mathcal{P}(1/u) + i\pi\delta(u)\right] du \tag{3.13}$$

We specify the distribution functions

$$F_\nu(u, \hat{\mathbf{k}}) = c_\nu \exp(-m_\nu u^2/2\Theta_\nu) \tag{3.14}$$

representing a system of different particle components each in thermal equilibrium. Introducing this into (3.13) we arrive at

$$\begin{aligned} \varepsilon &= 1 + \sum_\nu \omega_{p\nu}^2 m_\nu/k^2\Theta_\nu = 1 + (1/k^2\lambda_D^2) \\ 1/\lambda_D^2 &= \sum_\nu 4\pi n_\nu e^2/\Theta_\nu \end{aligned} \tag{3.15}$$

For the Fourier spectral function of the potential we have

$$\tilde{\tilde{\Phi}}(k, \omega = 0) = \frac{4\pi e}{k^2} \frac{\tilde{\tilde{n}}_{ex}(k, \omega = 0)}{1 + (k\lambda_D)^{-2}} = \frac{4\pi e \tilde{\tilde{n}}_{ex}(k, \omega = 0)}{k^2 + \lambda_D^{-2}} \tag{3.16}$$

Remembering the convolution theorem of the Fourier transformation, the inverse transformation of (3.16) is given as

$$\tilde{\Phi}(\mathbf{r}, \omega = 0) = e \int \tilde{n}_{ex}(\mathbf{r}', \omega = 0) \frac{\exp(-|\mathbf{r} - \mathbf{r}'|/\lambda_D)}{|\mathbf{r} - \mathbf{r}'|} d\mathbf{r}' \tag{3.17}$$

where we used

$$\frac{1}{r} \exp(-r/\lambda_D) = \int d\mathbf{k} \frac{e^{i\mathbf{k}\cdot\mathbf{r}}}{2\pi^2(k^2 + \lambda_D^{-2})} \tag{3.18}$$

We find that for a quasistatic perturbation the plasma affects the external influence through the well known Debye shielding.

In the limit of **high frequencies** $\omega_r \to \infty$, we may develop the denominator $(u - \omega/k)$ of (3.10) into a series

$$\frac{1}{u - \omega/k} = -\frac{1}{\omega/k} \left\{ \sum_{\nu=0}^{\infty} \left(\frac{ku}{\omega}\right)^\nu \right\} \tag{3.19}$$

and after introducing the distribution functions (3.14) into (3.10), we arrive at

$$\varepsilon(k, \omega \to \infty) = 1 - \omega_p{}^2/\omega^2 \tag{3.20}$$

As should be expected for reasons of inertia we find that the dielectric susceptibility which is proportional to $\varepsilon - 1$ disappears for very high frequencies as $\omega_p{}^2/\omega^2$.

3.2. The Initial Value Problem

To study the initial value problem, we again consider equation (3.1) without the external contribution Φ_{ex}. We then have

$$\begin{aligned}(\partial/\partial t + iku)\tilde{f}(u, k, t) &= i(e/m)\, kn[\partial F_0(u, \hat{\mathbf{k}})/\partial u]\, \tilde{\Phi}(k, t)\\ \tilde{\Phi}(k, t) &= (4\pi e/k^2) \int \tilde{f}(u, k, t)\, du\end{aligned} \tag{3.21}$$

We multiply these equations by $\Delta_1 \exp(i\omega t)$ and integrate from $t = -\infty$ to $t = +\infty$. The function Δ_1 is related to the Dirichlet function through

$$\Delta_1 = \Delta(x, 0)/\pi + \tfrac{1}{2} \tag{3.22}$$

and according to (2.39) has the derivative

$$d\Delta_1/dx = \delta(x) \tag{3.23}$$

It follows that

$$\int_{-\infty}^{+\infty} \Delta_1 e^{i\omega t}(\partial\tilde{f}/\partial t)\, dt = -\int_{-\infty}^{+\infty} \tilde{f}(u, k, t)[\delta(t) + i\omega\Delta_1]\, e^{i\omega t}\, dt \tag{3.24}$$

respectively

$$\int_{-\infty}^{+\infty} \Delta_1 e^{i\omega t}(\partial\tilde{f}/\partial t)\, dt = -\tilde{f}(u, k, 0) - i\omega \int_{-\infty}^{+\infty} \Delta_1 e^{i\omega t}\tilde{f}(u, k, t)\, dt \tag{3.25}$$

Using instead of (3.3), the relation†

$$\widehat{\Omega}(u, \mathbf{r}, \omega) := (2\pi)^{-1/2} \int_{-\infty}^{+\infty} \Omega(u, \mathbf{r}, t)\, \Delta_1 e^{i\omega t}\, dt \tag{3.26}$$

† The symbol $\frown$ denotes the spectral function of the Laplace transformation in the following.

we find for (3.21), accounting for (3.25),

$$i(ku-\omega)\tilde{\hat{f}}(u,k,\omega) - i(e/m)\,kn[\partial F_0(u,\hat{\mathbf{k}})/\partial u]\,\tilde{\hat{\Phi}}(k,\omega) = (2\pi)^{-1/2}\tilde{f}(u,k,t=0)$$

$$\tilde{\hat{\Phi}}(k,\omega) = (4\pi e/k^2)\int_{-\infty}^{+\infty}\tilde{\hat{f}}(u,k,\omega)\,du \tag{3.27}$$

Dividing the first of these relations by $i(ku-\omega)$, integrating from $u=-\infty$ to $u=+\infty$, and substituting the second equation of (3.27), we have the potential $\tilde{\hat{\Phi}}(k,\omega)$ expressed in terms of the given initial value. With this we can also find $\tilde{\hat{f}}(u,k,\omega)$ from the first equation of (3.27). The inverse transformation in space and time yields the desired functions. So, in principle, the problem is solved.

Before we get involved with the evaluation one more remark: If we use the relations

$$\omega = ip, \qquad p_r = \omega_i\,, \quad p_i = -\omega_r \tag{3.28}$$

then we have

$$\hat{\Omega}(u,\mathbf{r},p) = (2\pi)^{-1/2}\int_0^{\infty} e^{-pt}\Omega(u,\mathbf{r},t)\,dt \tag{3.29}$$

where $\hat{\Omega}(u,\mathbf{r},p)$ is now the Laplace transform of $\Omega(u,\mathbf{r},t)$. Of course, we have to observe that now in the p plane the integration contour of the inverse transformation is different from that in the ω plane. It is given by $p = p_r + ip_i$, where p_i varies from $-\infty$ to $+\infty$, and p_r is a constant larger than the maximum growth rate η_m.

4. Landau's Treatment of the Vlasov Equation

4.1. Landau's Solution

Introducing (3.28) and (3.29) into (3.27) and eliminating $\tilde{\hat{f}}(u,k,p)$, we find for the potential

$$\tilde{\hat{\Phi}}(k,p) = \frac{4\pi e}{k^2}\,\frac{(2\pi)^{-1/2}\int_{-\infty}^{+\infty}[\tilde{f}(u,k,t=0)/(iku+p)]\,du}{1 - i\omega_p^2/k\int_{-\infty}^{+\infty}[F_0'(u)/(iku+p)]\,du} \tag{4.1}$$

and therewith

$$\tilde{\Phi}(k,t) = \frac{1}{2\pi i}\int_{\sigma-i\infty}^{\sigma+i\infty} dp\; e^{+pt}\,\frac{4\pi e}{k^2}\left\{\frac{\displaystyle\int_{-\infty}^{+\infty}\frac{\tilde{f}(u,k,t=0)}{iku+p}\,du}{\displaystyle 1-\frac{i\omega_p^2}{k}\int_{-\infty}^{+\infty}\frac{F_0'(u)}{iku+p}\,du}\right\} \tag{4.2}$$

As indicated, the contour of integration is a vertical line in the complex p plane with $\sigma > \eta_m$ to the right of all singularities of the integrand.

We now restrict our considerations to distribution functions $F_0(u)$ which according to the Penrose criterion are stable and therefore do not yield complex roots u_p of the dispersion relation. Consequently the integrand of (4.2) is regular in both half-planes $p_r > 0$ and $p_r < 0$, so that we can choose $\sigma > 0$ arbitrarily.

Straightforward evaluation of (4.2) is precluded because the integrand of (4.2) is discontinous on the imaginary p axis, as is readily seen from the application of Plemelj's formula (2.74). In this situation, we follow a method suggested by Landau based on analytic continuation. In addition to the requirement already imposed on $F_0(u)$, we demand that $F_0(u)$ and $\tilde{f}(u, k, t = 0)$ allow analytic continuation and $\tilde{\tilde{\Phi}}(k, p)$ vanishes sufficiently for $|p| \to \infty$. The intention is to apply residue calculus to the evaluation of the integral (4.2) closing the integration path by a semicircle in the negative half-plane. As it is, the discontinuity prohibits this. However, the integral value with the prescribed path ($\sigma - i\infty$ to $\sigma + i\infty$) is independent of the values of the integrand outside the path. Therefore if we can find an analytic continuation of the integrand into the region at the left of the integration path, the calculus of residues is applicable.

Numerator and denominator of the integrand of (4.2) are both defined by complex integrals. Cauchy's theorem shows that the integrals behave analytically if we deform the path of integration as represented in Figure III.7 when we turn from positive to negative p_r values. Consequently we have for the denominator

$$D(k, p) = 1 - \frac{\omega_p{}^2}{k^2} \int_{-\infty}^{+\infty} \frac{F_0'(u)}{u - ip/k}\, du \qquad p_r > 0 \tag{4.3}$$

$$D(k, p) = 1 - \frac{\omega_p{}^2}{k^2} \left\{ \mathcal{P} \int_{-\infty}^{+\infty} \frac{F_0'(u)}{u - ip/k}\, du + \pi i F_0'(ip/k) \right\} \qquad p_r = 0 \tag{4.4}$$

$$D(k, p) = 1 - \frac{\omega_p{}^2}{k^2} \left\{ \int_{-\infty}^{+\infty} \frac{F_0'(u)}{u - ip/k}\, du + 2\pi i F_0'(ip/k) \right\} \qquad p_r < 0 \tag{4.5}$$

The analytic continuation of the numerator is performed in the same way.

By construction the zeros of $D(k, p)$ can lie in the open left p half-plane

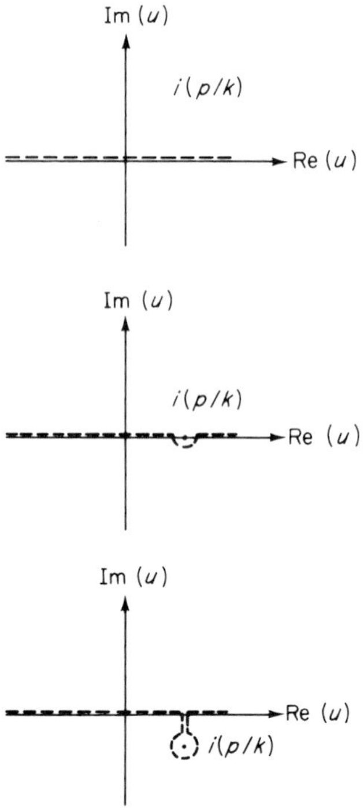

FIGURE III.7. Path deformation used in the process of analytic continuation.

only.† They are isolated since the existence of a finite cluster point would imply $D(k, p) \equiv 0$.‡

† Note that this analytic continuation $D(k, p)$ can have zeros, although the original denominator was supposed to have no zeros.

‡ For a Maxwellian $F_0(u)$, the assumption $\tilde{\Phi}(k, p) \to 0$ for $|p| \to \infty$ does not hold. In this case $D(k, p)$ has an infinite number of zeros in the left p half-plane (Denavit, 1965) which cluster at infinity and are located close to the lines $\arg(p) = \pm 3\pi/4$. Any vertical strip contains only a finite number of zeros of $D(k, p)$ (Saenz, 1965). This has the consequence that the inverse transformation is determined through

$$\tilde{\Phi}(k, t) = \sum_{p_{\nu r} > \rho} c_\nu e^{p_\nu t} + O(e^{\rho t})$$

where the p_ν are the poles in the strip $\rho < p_{\nu r} < 0$. In addition it can be shown that the contribution of an integration path shifted to infinity vanishes for arbitrary times $t > 0$ if the initial disturbance is a Maxwellian of a temperature not lower than that of the main plasma. For arbitrary initial distributions, the problem of this contribution is not yet solved in full.

In application of the theorem of residues, we shift the integration path of the inverse Laplace transform to $\sigma = -\alpha$ and account for the poles by the sum of the residues

$$\tilde{\Phi}(k, t) = \sum_{\nu} c_\nu e^{p_\nu t} + (1/2\pi i) \int_{-\alpha-i\infty}^{-\alpha+i\infty} \bar{\tilde{\Phi}}'(k, p)\, e^{pt}\, dp \tag{4.6}$$

where $\bar{\tilde{\Phi}}'(k, p)$ is the analytic continuation of $\bar{\tilde{\Phi}}(k, p)$ and c_ν is the residue of $\bar{\tilde{\Phi}}'(k, p)$ at the pole p_ν. The sum is extended over all zeros of $D(k, p)$. We assume that the integral in (4.6) vanishes for $\alpha \to \infty$, which yields

$$\tilde{\Phi}(k, t) = \sum_{\nu} c_\nu e^{p_\nu t} \underset{t\to\infty}{\simeq} c_0 e^{p_0 t} \tag{4.7}$$

Since $p_{\nu\mathrm{r}} < 0$ the initial perturbation is damped. This damping is called **Landau damping** (Landau, 1946).

As (4.7) demonstrates only the contribution with the smallest modulus of the real part prevails in the limit $t \to \infty$. We therefore look for the root p_0 of $D(k, p) = 0$ with the smallest real part $|p_\mathrm{r}|$. For this aim we develop $D(k, p)$ on the p_i axis into a power series.

This can be done by a generalization of Plemelj's formula (Jackson, 1960). Let a function $g(u)$ have all the properties we required for $F_0(u)$ then

$$I(z) := \int_{-\infty}^{+\infty} \frac{g(u)}{u - z}\, du \tag{4.8}$$

defines a function regular in the upper z half-plane (and, of course, in the lower, too). This function can be continued analytically into the other half-plane by the development into a power series

$$I(z) = \sum_{n=0}^{\infty} (1/n!)\, I^{(n)}(z_0)(z - z_0)^n \tag{4.9}$$

For z_0 in the upper half-plane we find from (4.8)

$$I'(z_0) = \int_{-\infty}^{+\infty} \frac{g(u)}{(u - z_0)^2}\, du = \int_{-\infty}^{+\infty} \frac{g'(u)}{u - z_0}\, du \tag{4.10}$$

By induction, we conclude

$$I^{(n)}(z_0) = \int_{-\infty}^{+\infty} \frac{g^{(n)}(u)}{u - z_0}\, du \tag{4.11}$$

Due to Plemelj's formula (2.74) we get, with $z_0 = x_0 + iy_0$,

$$\lim_{y_0 \to +0} I^{(n)}(z_0) = \mathcal{P} \int_{-\infty}^{+\infty} \frac{g^{(n)}(u)}{u - x_0}\, du + i\pi g^{(n)}(x_0) \tag{4.12}$$

and, therefore, with $x_0 = x$, $y_0 = 0$,

$$I(x + iy) = \sum_{n=0}^{\infty} \frac{(iy)^n}{n!} \left\{ \mathcal{P} \int_{-\infty}^{+\infty} \frac{g^{(n)}(u)}{u - x} \, du + i\pi g^{(n)}(x) \right\} \tag{4.13}$$

With $g(u) = F_0'(u)$, $x = -p_i/k$, $y = p_r/k$, this yields for the development of $D(k, p)$ on the imaginary p axis

$$D(k, p) = 1 - \frac{\omega_p^2}{k^2} \sum_{n=0}^{\infty} \frac{1}{n!} \left(\frac{ip_r}{k} \right)^n \left\{ \mathcal{P} \int_{-\infty}^{+\infty} \frac{F_0^{(n+1)}(u)}{u + p_i/k} \, du + i\pi F_0^{(n+1)} \left(-\frac{p_i}{k} \right) \right\} \tag{4.14}$$

Since we look for the zero with the smallest $|p_r|$, we neglect higher powers of p_r/k, and find from $D(k, p) = 0$ after separation of the real and imaginary parts

$$\frac{k^2}{\omega_p^2} = \mathcal{P} \int_{-\infty}^{+\infty} \frac{F_0'(u)}{u + p_i/k} \, du + O(p_r^2/k^2) \tag{4.15}$$

$$0 = \pi F_0'(-p_i/k) + \frac{p_r}{k} \mathcal{P} \int_{-\infty}^{+\infty} \frac{F_0''(u)}{u + p_i/k} \, du + O(p_r^3/k^3) \tag{4.16}$$

Differentiating (4.15) with respect to k, we have

$$\frac{2k}{\omega_p^2} = \frac{p_i - k(dp_i/dk)}{k^2} \mathcal{P} \int_{-\infty}^{+\infty} \frac{F_0''(u)}{u + p_i/k} \, du \tag{4.17}$$

Elimination of the integrals from (4.16) and (4.17) yields

$$p_r = (\pi/2)\, p_i (\omega_p/k)^2 F_0'(-p_i/k) \left(\frac{k}{p_i} \frac{dp_i}{dk} - 1 \right) \tag{4.18}$$

We now have to insert p_i from (4.15). To simplify this, we restrict ourselves to small values of k. Then only the range $|u| \ll p_i/k$ (low temperature case) contributes essentially to the integral, so that we may develop the denominator into a power series with respect to uk/p_i. The singularity of the integrand is situated outside the circle of convergence of this series where we do not expect essential contributions to the integral. This allows us to use for the principal part

$$\frac{k^2}{\omega_p^2} \approx \frac{k}{p_i} \int_{-\infty}^{+\infty} F_0'(u) \left(1 - \frac{ku}{p_i} + \frac{k^2u^2}{p_i^2} - \frac{k^3u^3}{p_i^3} + \cdots \right) du \tag{4.19}$$

which yields after partial integration

$$\frac{k^2}{\omega_p^2} \approx -\frac{k}{p_i}\int_{-\infty}^{+\infty} F_0(u)\left(-\frac{k}{p_i} + 2u\frac{k^2}{p_i^2} - 3u^2\frac{k^3}{p_i^3} + \cdots\right) du \qquad (4.20)$$

With $\langle u\rangle = 0$, we arrive at

$$\frac{1}{\omega_p^2} \approx \frac{1}{p_i^2}\left(1 + 3\langle u^2\rangle \frac{k^2}{p_i^2}\right) \qquad (4.21)$$

The zero-order solution is $p_i^2 = \omega_p^2$, and therefore we get in the first order

$$p_i \approx \pm\omega_p[1 + 3\langle u^2\rangle(k^2/\omega_p^2)]^{1/2} \approx \pm\omega_p[1 + \tfrac{3}{2}\langle u^2\rangle(k^2/\omega_p^2)] \approx \pm\omega_p \qquad (4.22)$$

This result is identical with (2.58), and indeed the approximation (4.15) brings us back to the imaginary p axis (i.e., the real u_p axis). Further, the neglect of the singularity reduces (4.15) to the dispersion equation of a distribution with a gap. Here and there we have assumed small values of k. To find the corresponding expression for p_r, we differentiate with respect to k. Simple calculations yield

$$1 - \frac{k}{p_i}\frac{dp_i}{dk} = \frac{1 - \frac{3}{2}\langle u^2\rangle(k^2/\omega_p^2)}{1 + \frac{3}{2}\langle u^2\rangle(k^2/\omega_p^2)} \approx 1 - 3\langle u^2\rangle(k^2/\omega_p^2) \approx 1 \qquad (4.23)$$

which we insert into (4.18)

$$-p_r = \pm(\pi/2)(\omega_p/k)^3\, kF_0'(\mp(\omega_p/k)[1 + \tfrac{3}{2}\langle u^2\rangle(k^2/\omega_p^2)]) \qquad (4.24)$$

We now take the Maxwellian velocity distribution

$$F_0(u) = (m/2\pi\Theta)^{1/2}\exp[-mu^2/2\Theta] \qquad (4.25)$$

With $m\langle u^2\rangle = \Theta$, we obtain for the derivative

$$F_0'\left(\pm\frac{\omega_p}{k}[\cdots]\right) \approx \pm(2\pi)^{-1/2}(m/\Theta)^{3/2}(\omega_p/k)\exp[-(m\omega_p^2/2k^2\Theta) - \tfrac{3}{2}] \qquad (4.26)$$

Thus we get for the damping decrement

$$-p_r = \omega_p(\pi/8)^{1/2}(k\lambda_D)^{-3}\exp(-\tfrac{1}{2}(k\lambda_D)^{-2} - \tfrac{3}{2}) \qquad (4.27)$$

using again $\lambda_D^{-2} = m\omega_p^2/\Theta$.

The relative damping rate $-p_r/\omega_p$ computed from (4.27) is plotted in Figure III.8. We see that the damping is very weak for $k\lambda_D \ll 1$. An

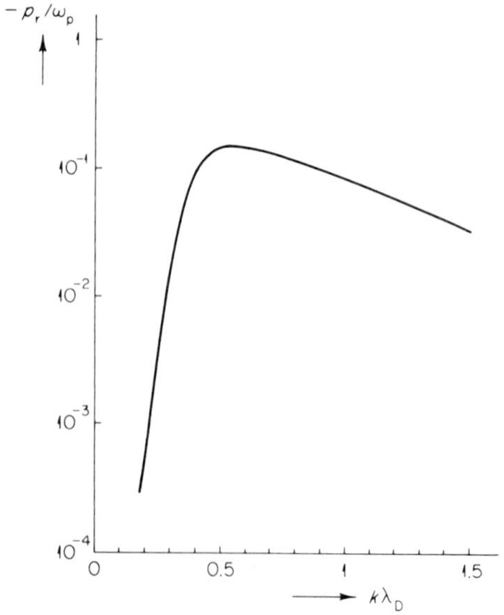

FIGURE III.8. Damping constant as calculated from Landau's approximation.

essential damping can be expected only when $k\lambda_D \sim 1$. Here, however, our result becomes inconsistent since we assumed $k\lambda_D \ll 1$ for the evaluation.

It is interesting to consider the possibility of observing Landau damping. On the one hand there is the difficulty in experimental setups to distinguish between Landau damping and collision damping. Landau damping prevails when the damping decrement $|p_r|$ is larger than the collision frequency ν_c. That means

$$|p_r| \gg \nu_c \approx \frac{\ln \Lambda}{\Lambda} \omega_p \tag{4.28}$$

or

$$\frac{\Lambda}{\ln \Lambda} \frac{|p_r|}{\omega_p} \gg 1 \tag{4.29}$$

From the nomogram on p. 330, we see that in typical plasmas $\Lambda/\ln \Lambda$ hardly will exceed the value 10^4. This implies $|p_r|/\omega_p \gg 10^{-4}$ and therefore $k\lambda_D \gtrsim 0.2$.

On the other hand, to allow observation, the damping rate should not be too large, so that the oscillations will survive at least for some, say ten, periods. Figure III.8 therewith imposes the condition $k\lambda_D \leqslant 0.4$.

Consequently, one can hope to find Landau damping only in the small range $0.2 \lesssim k\lambda_D \lesssim 0.4$. Even if we admit a much larger $\Lambda/\ln \Lambda$, the range does not grow essentially because $|p_r|/\omega_p$ falls very rapidly with decreasing $k\lambda_D$.

The preceding evaluation is only approximate, and one would like to know the rigorous results.

Introducing (4.25) into the relation (4.5), the equation $D(k, p) = 0$ reads

$$\frac{k^2}{\omega_p{}^2} = -\frac{(m/\Theta)^{3/2}}{(2\pi)^{1/2}}\left\{\int_{-\infty}^{+\infty}\frac{u \exp[-mu^2/2\Theta]}{u - ip/k}\,du + 2\pi i\,\frac{ip}{k}\exp\left[-\frac{m}{2\Theta}\left(\frac{ip}{k}\right)^2\right]\right\} \tag{4.30}$$

Using the substitution $t = u[m/(2\Theta)]^{1/2}$ it follows that

$$-k^2\lambda_D{}^2 = \pi^{-1/2}\int_{-\infty}^{+\infty}\frac{te^{-t^2}}{t-\zeta}\,dt + 2i\pi^{1/2}\,\zeta e^{-\zeta^2} =: J(\zeta) \tag{4.31}$$

with the abbreviation

$$\zeta := (ip/k)(m/2\Theta)^{1/2} \tag{4.32}$$

We look for solutions with $p_r < 0$, i.e., $\zeta_i < 0$.

Multiplying numerator and denominator in (4.31) by $(t + \zeta)$, observing when the function is even or odd, adding one, and subtracting the quantity

$$\pi^{-1/2}\int_{-\infty}^{+\infty} e^{-t^2}\,dt = 1 \tag{4.33}$$

we find

$$\begin{aligned} J(\zeta) &= 1 + \pi^{-1/2}\int_{-\infty}^{+\infty}\frac{\zeta^2 e^{-t^2}}{t^2-\zeta^2}\,dt + 2i\pi^{1/2}\,\zeta e^{-\zeta^2} \\ &= 1 + \pi^{-1/2}\zeta^2 e^{-\zeta^2}\left\{\int_{-\infty}^{+\infty}\frac{e^{-(t^2-\zeta^2)}}{t^2-\zeta^2}\,dt + \frac{2\pi i}{\zeta}\right\} \end{aligned} \tag{4.34}$$

With the identity

$$e^{-A}/A = -\int_0^1 e^{-As}\,ds + 1/A, \qquad A = t^2 - \zeta^2 \tag{4.35}$$

this can be written as

$$J(\zeta) = 1 + \pi^{-1/2}\zeta^2 e^{-\zeta^2}\left\{\int_{-\infty}^{+\infty} dt\left[-\int_0^1 e^{-(t^2-\zeta^2)s}\,ds + (t^2-\zeta^2)^{-1}\right] + 2\pi i/\zeta\right\} \tag{4.36}$$

Now we integrate with respect to t. Due to Cauchy's formula we find

$$\int_{-\infty}^{+\infty} \frac{dt}{t^2 - \zeta^2} = \pm \frac{i\pi}{\zeta} \quad \text{for} \quad \zeta_i \gtrless 0 \tag{4.37}$$

and therefore, since we consider $\zeta_i < 0$,

$$J(\zeta) = 1 + \zeta^2 e^{-\zeta^2} \left\{ -\int_0^1 s^{-1/2} e^{\zeta^2 s} \, ds + \frac{i\pi^{1/2}}{\zeta} \right\} \tag{4.38}$$

Substituting $\nu = i\zeta s^{1/2}$, it follows that

$$\begin{aligned} J(\zeta) &= 1 + \zeta^2 e^{-\zeta^2} \left\{ -\frac{2}{i\zeta} \int_0^{i\zeta} e^{-\nu^2} \, d\nu + \frac{i\pi^{1/2}}{\zeta} \right\} \\ &= 1 + 2i\zeta e^{-\zeta^2} \left\{ \tfrac{1}{2}\pi^{1/2} + \int_0^{i\zeta} e^{-\nu^2} \, d\nu \right\} \end{aligned} \tag{4.39}$$

so that we finally have

$$J(\zeta) = 1 + 2i\zeta e^{-\zeta^2} \int_{-\infty}^{i\zeta} e^{-\nu^2} \, d\nu = -k^2 \lambda_D^2 \tag{4.40}$$

Using the dispersion function

$$Z(\zeta) := 2i e^{-\zeta^2} \int_{-\infty}^{i\zeta} e^{-\nu^2} \, d\nu \tag{4.41}$$

the equation for the zeros of $D(k, p)$ simply reads

$$k^2 \lambda_D^2 = -[1 + \zeta Z(\zeta)] = \tfrac{1}{2} Z'(\zeta) \tag{4.42}$$

The dispersion function $Z(\zeta)$ and its derivative $Z'(\zeta)$ have been evaluated and tabulated by Fried and Conte (1961).

Their results allow us to present the solutions of (4.42) in the ζ plane. We use the notation $\zeta = v + iw$. Choosing a fixed value of w, we may plot the imaginary and real part of the function Z versus v. In Figure III.9 this is demonstrated for the value $w = -2$. The imaginary part of Z' shows a number of zero values. At these zeros the real part may be either positive or negative. If it is negative we have to discard the corresponding solutions; if it is positive it provides us with a possible value $2k^2\lambda_D^2$. Applying this procedure to arbitrary values of w we can find the loci of all solutions in the ζ plane. To plot these curves, asymptotic solutions may be helpful. For large values of k the roots of the dispersion

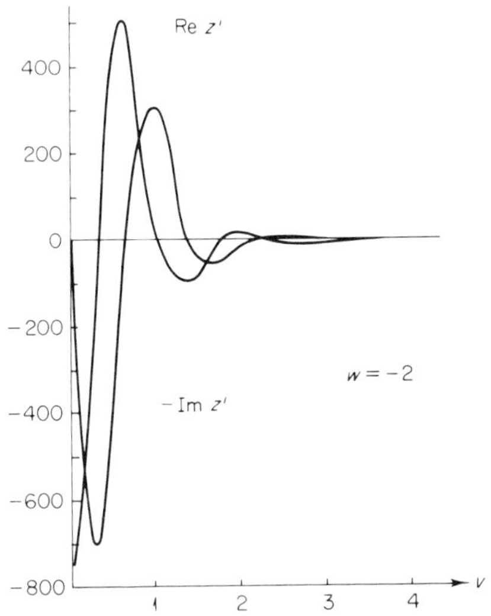

FIGURE III.9. Real and imaginary part of the dispersion function for a constant imaginary part $w = -2$.

relation have a large modulus so that an asymptotic expansion of $Z(\zeta)$ may be used yielding

$$v_n w_n = \frac{\arctan(w_n/v_n) - (2n + \frac{1}{2})\pi}{2}, \quad (v_n^2 + w_n^2)^{1/2} \exp(w_n^2 - v_n^2) = \frac{(k/k_D)^2}{2\pi^{1/2}}$$

$$k/k_D \gg 1, \qquad n = 0, 1, \ldots \tag{4.43}$$

Here we introduced $k_D = 1/\lambda_D$. For $n = 0$ the result may be approximated by

$$v_0 = -\pi/2w_0, \qquad -w_0 e^{w_0^2} = \tfrac{1}{2}\pi^{-1/2}(k/k_D)^2, \qquad k/k_D \gg 1 \tag{4.43a}$$

For large numbers n, all zeros are situated close to the line $w = -v$ and can be represented by

$$\rho_n^2 = (2n + \tfrac{3}{4})\pi, \qquad \delta_n = \frac{1}{2\rho_n^2} \ln \frac{k^2\lambda_D^2}{2\pi^{1/2}\rho_n}$$

$$k/k_D \gg 1, \quad n \gg 1, \quad \zeta_n = \rho_n \exp(-i(\pi/4 + \delta_n)) \tag{4.43b}$$

For small values of k Landau's result ($n = 0$)

$$v_0^2 = \frac{k^2 + 3k_D^2}{2k_D^2}, \qquad w_0 = -\pi^{1/2}\, v_0^4 e^{-v_0^2} \qquad \text{for} \quad k/k_D \ll 1 \quad (4.44)$$

is reproduced.

In Figure III.10 the heavy lines give the physically reasonable solutions with positive values for k_D^2. The light lines correspond to the negative values of k_D^2. The dashed curves represent the asymptotic calculations.

Since there is a coordination of k values to the points of each curve, we may replot the values of w and v versus k/k_D. This is shown in Figure III.11. In this figure the dashed line again gives the result of our

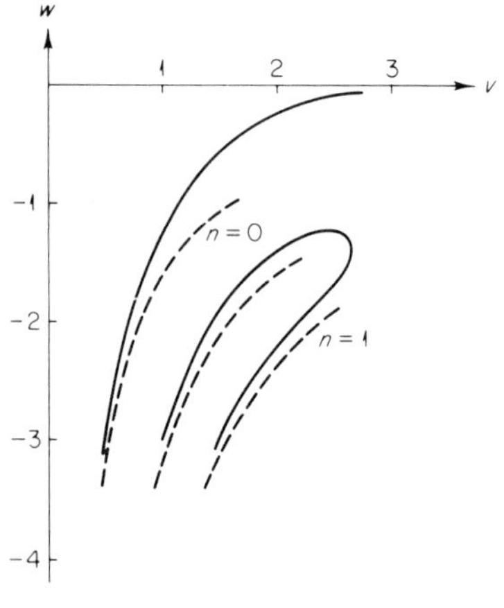

FIGURE III.10. Solution of the dispersion relation; n designates the different modes.

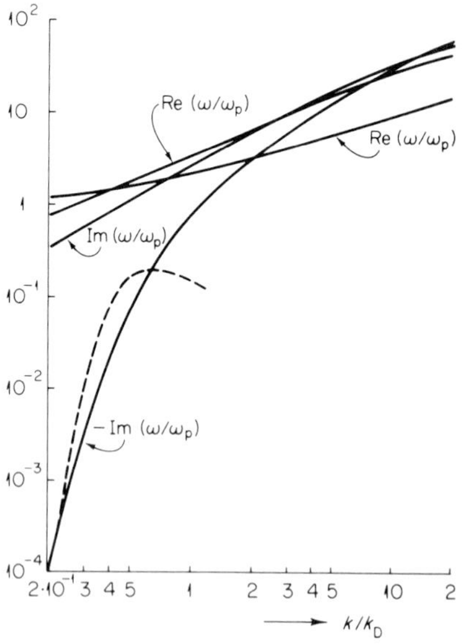

FIGURE III.11. Frequency and damping of longitudinal plasma waves as a function of k/k_D. Heavy lines distinguish the first mode from the second.

approximate Landau solution (see Figure III.8). It demonstrates the breakdown of this solution for large values of k. Further, we see from this figure that, as we expect from Figure III.10, there are multiple solutions for each value of k.† The Landau mode is the one with the smallest damping decrement.

It is interesting to note that the function $Z(\zeta)$ obeys the relation

$$Z(\zeta^*) = -Z^*(-\zeta) \tag{4.45}$$

Applying this result to (4.42) we immediately see that $-\zeta^*$ is a solution if ζ is one. This means that waves propagating in the direction $\mathbf{k}$ and $-\mathbf{k}$ exhibit the same damping.

4.2. General Aspects of Landau Damping

Damping and amplification of longitudinal collective modes is already familiar to us from our treatment of the normal modes with complex frequencies. There we learned that to each complex solution of the dispersion relation another one is given as the conjugate complex quantity. Therefore we have amplified and damped modes in pairs.

It is interesting to note that Landau damping does not fall into this group, since there is no amplification process as a counterpart to the Landau damping.

Analytically the fact that the Landau poles do not occur in pairs ζ, ζ^*, follows from the fact that the Landau solutions are derived from the analytic continuation of the dispersion function into the negative p_r plane (see (4.3) and (4.5))

$$\begin{aligned} D(k, p) &= 1 - \frac{\omega_p^2}{k^2} \int_{-\infty}^{+\infty} \frac{F_0'(u)}{u - ip/k}\, du \qquad \text{for } p_r > 0 \\ D(k, p) &= 1 - \frac{\omega_p^2}{k^2} \left\{ \int_{-\infty}^{+\infty} \frac{F_0'(u)}{u - ip/k}\, du + 2\pi i F_0'(ip/k) \right\} \qquad \text{for } p_r < 0 \end{aligned} \tag{4.46}$$

whereas the complex pairs are deduced from the dispersion relation given by the discontinuous function

$$\varepsilon(k, p) = 1 - \frac{\omega_p^2}{k^2} \int_{-\infty}^{+\infty} \frac{F_0'(u)}{u - ip/k}\, du \tag{4.47}$$

The Landau procedure therefore yields all the poles of the complex solutions of the dispersion relation and some additional ones which are the "Landau poles."

† This may easily be proved with the help of a theorem by Picard as has been shown by Hayes (1961, 1963) and by Saenz (1964, 1965).

To come closer to a physical interpretation of the Landau damping phenomenon, let us observe that our system of complex eigensolutions—as we discussed at the beginning of Section 3—is not complete. We have to add the modes with real values of ω—the Van Kampen modes. This we will discuss in detail in the following section. Anticipating that the set consisting of the complex solutions of the dispersion relation and the Van Kampen modes is a complete one, we can present any solution of the Vlasov equation in the form

$$\tilde{\Phi}(k, t) = (4\pi e/k^2) \int_{-\infty}^{+\infty} \tilde{\bar{h}}(\omega, k)\, e^{-i\omega t}\, d\omega + \sum_{\nu} A_\nu e^{-i\omega_\nu t} \tag{4.48}$$

On the other hand, applying the Landau procedure to an unstable plasma, we will find a solution of the form

$$\tilde{\Phi}(k, t) = \sum \text{Landau} + \sum_{\nu} A_\nu e^{-i\omega_\nu t} \tag{4.49}$$

where the first term on the right-hand side is the sum from (4.7) and the second term is identical with the second term of the right-hand side of (4.48). It follows that the first term of (4.49), the Landau contribution, must be identical with the contribution of superimposed Van Kampen modes. Since the Van Kampen modes are the normal modes for real values of ω, each of them is undamped. But, of course, superposition of undamped oscillations can be attenuated by interference. We conclude that Landau damping is a result of **phase mixing of Van Kampen modes**.

Another question of physical interest with respect to the Landau damping phenomenon is the question of energy dissipation. If we consider a stable plasma which confronts us only with the damped Landau solutions, the final state of the system is free of all collective electric phenomena, and we ask in what energy form the electrical energy has been converted. Since our system is a closed one, the only possible transition is into the kinetic energy. The total energy of our system neglecting correlations is given by

$$W = \int \tfrac{1}{2} m v^2 f^{(1)}(\mathbf{r}, \mathbf{v}, t)\, d\mathbf{r}\, d\mathbf{v} + \int (8\pi)^{-1} \,|\, E \,|^2\, d\mathbf{r} \tag{4.50}$$

and therefore the change of the kinetic energy is related to the change of the electrical energy by

$$\int \tfrac{1}{2} m v^2 [\partial f(\mathbf{r}, \mathbf{v}, t)/\partial t]\, d\mathbf{r}\, d\mathbf{v} = -(8\pi)^{-1} \int [\partial \,|\, E \,|^2/\partial t]\, d\mathbf{r} \tag{4.51}$$

If we calculate the electric field to the first order, the corresponding energy term is of the second order, and consequently we require knowledge of $f^{(1)}$ to the second order if we want to present the energy correctly. This shows that the linearized theory presented so far is inadequate to the solution of the energy problem.

Nevertheless, it is revealing to have another look at our linear approximation. The inverse Laplace transformation of $\tilde{\bar{f}}(u, k, p)$ produces (see 3.27)

$$\tilde{f}(u, k, t) = (2\pi)^{-1/2} \int_{\sigma-i\infty}^{\sigma+i\infty} \frac{i(e/m)\, knF_0'(u)\, \tilde{\bar{\Phi}}(k, p) + (2\pi)^{-1/2} \tilde{f}(u, k, t=0)}{iku + p}\, e^{pt}\, dp \tag{4.52}$$

In contrast to the electric field and the corresponding energy this contribution does not approach zero for $t \to \infty$. The Fourier Laplace transform $\tilde{\bar{f}}(u, k, p)$ has in addition to the poles of the Fourier Laplace transform of $\tilde{\bar{\Phi}}$ one pole on the imaginary axis which results in an oscillating term $A(u, k)\, e^{-ikut}$ which shows no damping.

It is particularly noteworthy that the frequency of this phenomenon depends on the velocity u. If we integrate f with respect to u over an arbitrary interval again damping occurs as a phase mixing of contributions with different velocities u.

Any attempt to arrive at more detailed statements about the energy requires nonlinear theory (see Einaudi and Sudan, 1969, and the references cited there).

5. Van Kampen's Treatment of the Vlasov Equation

We shall describe in this section an alternative method for solving Vlasov's equation. This method is due to Van Kampen (1955, 1957) who tackled the difficulty arising from the pole at $u = u_p$ in (2.14) by methods of the theory of generalized functions (Mushkelishvili, 1953).

Let us start from the Fourier-transformed linearized Vlasov equation without external perturbation [see (3.6)]

$$[u - (\omega/k)]\, \tilde{\tilde{f}}(u, k, \omega) - (\omega_p^2/k^2)\, F_0'(u) \int_{-\infty}^{+\infty} \tilde{\tilde{f}}(u, k, \omega)\, du = 0 \tag{5.1}$$

We now want to study its normal mode solutions with real $\omega/k =: \nu$, a problem which Landau evaded by analytic continuation. We have to solve the eigenvalue problem for real values of ν. To stress this aspect we write ν as an index and suppress the k dependence. Equation

(5.1) is homogeneous so that the function is determined only up to an arbitrary factor. We are therefore free to require an additional normalization condition

$$\int_{-\infty}^{+\infty} \tilde{\tilde{f}}_\nu(u)\, du = 1 \tag{5.2}$$

thus transforming (5.1) into

$$(u - \nu)\, \tilde{\tilde{f}}_\nu(u) = (\omega_p{}^2/k^2)\, F_0{}'(u) \tag{5.3}$$

We note that the solution of this equation cannot be written simply in the form

$$\tilde{\tilde{f}}_\nu(u) = (\omega_p{}^2/k^2)\, F_0{}'(u)[1/(u - \nu)] \tag{5.4}$$

We need an unambiguous rule stating the contribution of the singularity.

Further, it should be observed that it is sufficient to define $f_\nu(u)$ as a generalized function in the sense of Schwartz, since measurable quantities are all averages calculated by integration over $f_\nu(u)$.

Using the results of functional analysis we arrive at

$$\tilde{\tilde{f}}_\nu(u) = \frac{\omega_p{}^2}{k^2} F_0{}'(u)\, \mathcal{P} \frac{1}{u - \nu} + A(\nu)\, \delta(u - \nu) \tag{5.5}$$

where the second term arises from the fact that we have to add an arbitrary solution of the associated homogeneous equation

$$(u - \nu)\, \tilde{\tilde{f}}_\nu(u) = 0 \tag{5.6}$$

The normalization condition leads to

$$A(\nu) = 1 - \frac{\omega_p{}^2}{k^2}\, \mathcal{P} \int_{-\infty}^{+\infty} \frac{F_0{}'(u)}{u - \nu}\, du \tag{5.7}$$

so that we get

$$\tilde{\tilde{f}}_\nu(u) = \frac{\varepsilon_2(u)}{\pi}\, \mathcal{P} \frac{1}{u - \nu} + \varepsilon_1(u)\, \delta(u - \nu) \tag{5.8}$$

Here we have introduced the abbreviations ε_1, $-\varepsilon_2$, for the real and imaginary part of the dielectric constant (3.10) which now can be written in the limits $\nu_i = \pm 0$ as

$$\begin{aligned} \varepsilon_\pm(k, \nu) &= \varepsilon_1(k, \nu) \mp i\varepsilon_2(k, \nu) \\ \varepsilon_1(k, \nu) &:= 1 - \frac{\omega_p{}^2}{k^2}\, \mathcal{P} \int_{-\infty}^{+\infty} \frac{F_0{}'(u)}{u - \nu}\, du \\ \varepsilon_2(k, \nu) &:= \pi \frac{\omega_p{}^2}{k^2} F_0{}'(\nu) \end{aligned} \tag{5.9}$$

In (5.8) we have the surprising result that one can find an eigenfunction $\tilde{\tilde{f}}_\nu(u)$ for any value of ν and for any value of the wavenumber k. The spectrum of eigenvalues of the Vlasov equation is the whole real axis. This means that there exists no dispersion relation $\nu = \nu(k)$.

5.1. The Adjoint Vlasov Equation

In order to solve the initial value problem, we have to expand a given initial disturbance into a series of eigenfunctions

$$\tilde{f}(u, t = 0) = \int_{-\infty}^{+\infty} K(\nu)\, \tilde{\tilde{f}}_\nu(u)\, d\nu \tag{5.10}$$

Since Vlasov's equation is not Hermitian we need a system of functions $\tilde{\tilde{f}}{}^{\dagger}_{\nu'}(u)$ which are orthogonal to Van Kampen's modes $\tilde{\tilde{f}}_\nu(u)$

$$\int_{-\infty}^{+\infty} \tilde{\tilde{f}}_\nu(u)\, \tilde{\tilde{f}}{}^{\dagger}_{\nu'}(u)\, du = C(\nu)\, \delta(\nu - \nu') \tag{5.11}$$

in order to compute the coefficients of the expansions in accord with

$$K(\nu) = [1/C(\nu)] \int_{-\infty}^{+\infty} \tilde{f}(u, t = 0)\, \tilde{\tilde{f}}{}^{\dagger}_{\nu'}(u)\, du \tag{5.12}$$

This means that we have to consider the adjoint Vlasov equation. More generally the operator $L^\dagger$ which is adjoint to a linear operator L is defined by the equation

$$\int [f_1 L f_2 - f_2 L^\dagger f_1]\, dx = 0 \tag{5.13}$$

This definition ensures that the respective eigenfunctions of L and $L^\dagger$ are orthogonal:

$$\int [f_1{}^\dagger \lambda_2 f_2 - f_2 \lambda_1 f_1{}^\dagger]\, dx = (\lambda_2 - \lambda_1) \int f_1{}^\dagger f_2\, dx = 0 \tag{5.14}$$

The Vlasov operator contains an integral operator

$$I = b(x) \int_{-\infty}^{+\infty} d\xi \tag{5.15}$$

the adjoint of which can be found easily

$$\int_{-\infty}^{+\infty} dx\, f_1(x)\, b(x) \int_{-\infty}^{+\infty} d\xi\, f_2(\xi) = \int_{-\infty}^{+\infty} dx\, f_2(x) \int_{-\infty}^{+\infty} d\xi\, b(\xi)\, f_1(\xi) \tag{5.16}$$

which means that

$$I^{\dagger} = \int_{-\infty}^{+\infty} d\xi\, b(\xi) \tag{5.17}$$

The adjoint Vlasov equation therefore reads

$$(u - \nu)\tilde{f}_{\nu}^{\dagger}(u) - (\omega_{\mathrm{p}}^2/k^2) \int_{-\infty}^{+\infty} F_0'(u)\tilde{f}_{\nu}^{\dagger}(u)\, du = 0 \tag{5.18}$$

This is a homogeneous equation of the same type as (5.1) and can be solved by the same method. We again have the freedom to impose a normalization condition; it is convenient to take it in the form

$$(\omega_{\mathrm{p}}^2/k^2) \int_{-\infty}^{+\infty} F_0'(u)\tilde{f}_{\nu}^{\dagger}(u)\, du = 1 \tag{5.19}$$

By the same arguments as before, the solution of (5.18) is

$$\tilde{f}_{\nu}^{\dagger}(u) = \mathcal{P}\,[1/(u - \nu)] + B(\nu)\,\delta(u - \nu) \tag{5.20}$$

where $B(\nu)$ is determined by the normalization condition (5.19) as

$$B(\nu) = [\varepsilon_1(\nu)/\varepsilon_2(\nu)] \tag{5.21}$$

The orthogonality relation (5.11) has been proved in all generality by (5.14). The determination of $C(\nu)$ shall be postponed until we have shown the completeness of Van Kampen's modes. The straightforward evaluation of $C(\nu)$ is difficult because care must be taken not to exchange the order of multiple integrations in terms involving products of principal parts.

5.2. Completeness of Van Kampen's Modes

The expansion (5.10) is possible only if the set $\tilde{f}_{\nu}(u)$ is complete. This problem may be stated in the following form: What are the conditions under which (5.10) has a solution for any given initial disturbance $\tilde{f}(u, t = 0)$?

For the initial disturbance which is defined only on the real u axis, we shall require that it can be split into two parts, a "plus function" and a "minus function." The plus function has an analytic continuation into the upper complex half-plane and the minus function one into the lower one. The decomposition is unique if we further require $\tilde{f}_+(\infty) = 0$. The function

$$\frac{1}{2\pi i} \int_{-\infty}^{+\infty} \frac{\tilde{f}(u)}{u - z}\, du =: \begin{cases} \tilde{f}_+(z) & \text{for} \quad z_i > 0 \\ \tilde{f}_-(z) & \text{for} \quad z_i < 0 \end{cases} \tag{5.22}$$

defines analytic functions in these half planes under the condition that $f(u)$ is integrable. Due to Plemelj's formula, we find in the limit $z_i = \pm 0$

$$\tilde{f}_\pm(u) = \frac{1}{2\pi i} \mathcal{P} \int_{-\infty}^{+\infty} \frac{\tilde{f}(x)}{x - u}\, dx \pm \frac{1}{2} \tilde{f}(u) \tag{5.23}$$

We therefore get

$$\tilde{f} = \tilde{f}_+ - \tilde{f}_-$$

$$\frac{1}{\pi i} \mathcal{P} \int_{-\infty}^{+\infty} \frac{\tilde{f}(x)}{x - u}\, dx = \tilde{f}_+ + \tilde{f}_- \tag{5.24}$$

We now impose on $F_0(u)$ the conditions which we needed when deriving Penrose's criterion.

From (5.10) we get, substituting (5.8) for $\tilde{f}_\nu(u)$, the equivalent relation

$$\tilde{f}(u, t = 0) = \varepsilon_1(u)\, K(u) - \frac{\varepsilon_2(u)}{\pi} \mathcal{P} \int_{-\infty}^{+\infty} \frac{K(\nu)}{\nu - u}\, d\nu \tag{5.25}$$

or using the decomposition (5.24) for $K(u)$ and $\tilde{f}(u, t = 0)$

$$\tilde{f}_+ - \tilde{f}_- = \varepsilon_1(K_+ - K_-) - i\varepsilon_2(K_+ + K_-) \tag{5.26}$$

Together with (5.9) this yields

$$\tilde{f}_+ - \tilde{f}_- = \varepsilon_+ K_+ - \varepsilon_- K_- \tag{5.27}$$

Equation (5.27) is a statement for the function $\varepsilon K - \tilde{f}$, which is composed of functions which are analytic outside the real axis, and has the step zero across the real axis. The only such function is a constant, in this case zero, for

$$\varepsilon(u)\, K(u) - \tilde{f}(u) \to 0 \qquad \text{as} \quad |u| \to \infty \tag{5.28}$$

We therefore have

$$\varepsilon_\pm K_\pm = \tilde{f}_\pm \tag{5.29}$$

This procedure was allowed if and only if $K_\pm = \tilde{f}_\pm / \varepsilon_\pm$ is a plus (minus) function. Therefore it is necessary and sufficient that $\varepsilon_+(z) \neq 0$ for $z_i \geqslant 0$. Then by $\varepsilon_-(z) = \varepsilon_+{}^*(z) = \varepsilon_+(z^*)$, we have $\varepsilon_-(z) \neq 0$ for

$z_i \leqslant 0$, too. From our discussion of the curves $G(\mathbb{R} \pm i0)$ when deriving the Penrose criterion, it is clear that this situation is equivalent with the stability of the zero-order distribution $F_0(u)$. We thus get the theorem that the stability of $F_0(u)$ implies the completeness of the system of Van Kampen's eigensolutions $\tilde{f}_\nu(u)$ for real ν.

Case has shown how to generalize the previous procedure in order to include unstable situations too. One has to add to the continuous real spectrum a set of discrete eigenvalues which are the zeros of $\varepsilon_\pm(z)$ for $z_i \gtrless 0$. The corresponding normalized eigenfunctions are

$$\tilde{f}_j(u) = [\varepsilon_2(u)/\pi][1/(u - z_j)] \tag{5.30}$$

The adjoint eigenfunctions are given by

$$\tilde{f}_j^\dagger(u) = 1/(u - z_j) \tag{5.31}$$

and are normalized according to (5.19). The closure relation then is modified by an additional sum over the discrete spectrum

$$\int_{-\infty}^{+\infty} [1/C(\nu)]\, \tilde{f}_\nu(u_1)\, \tilde{f}_\nu^\dagger(u_2)\, d\nu + \sum_j (1/C_j)\, \tilde{f}_j(u_1)\, \tilde{f}_j'(u_2) = \delta(u_1 - u_2) \tag{5.32}$$

We again consider only stable situations and determine the normalization constant $C(\nu)$ by the evaluation of $K(\nu)$ according to (5.12) and comparison with our result

$$K(\nu) = \frac{\tilde{f}_+(\nu)}{\varepsilon_+(\nu)} - \frac{\tilde{f}_-(\nu)}{\varepsilon_-(\nu)} \tag{5.33}$$

Introducing $\tilde{f}_\nu^\dagger(u)$ from (5.20) and (5.21), we get by simple transformations

$$\begin{aligned} K(\nu) &= \frac{\pi}{C(\nu)\,\varepsilon_2(\nu)} \int_{-\infty}^{+\infty} \tilde{f}(u, t = 0) \left\{ \frac{\varepsilon_2(\nu)}{\pi}\, \mathcal{P} \frac{1}{u - \nu} + \varepsilon_1(\nu)\, \delta(u - \nu) \right\} du \\ &= \frac{\pi}{C(\nu)\,\varepsilon_2(\nu)} \{\tilde{f}_+(\nu)\, \varepsilon_-(\nu) - \tilde{f}_-(\nu)\, \varepsilon_+(\nu)\} \end{aligned} \tag{5.34}$$

This can be written

$$K(\nu) = \frac{\pi\varepsilon_+(\nu)\, \varepsilon_-(\nu)}{C(\nu)\, \varepsilon_2(\nu)} \left\{ \frac{\tilde{f}_+(\nu)}{\varepsilon_+(\nu)} - \frac{\tilde{f}_-(\nu)}{\varepsilon_-(\nu)} \right\} \tag{5.35}$$

Comparison with (5.33) finally gives

$$C(\nu) = \pi \mid \varepsilon_\pm(\nu)|^2/\varepsilon_2(\nu) \tag{5.36}$$

5.3. Solution of the Initial Value Problem by Eigensolution Expansion–Comparison with Landau's Result

Any initial disturbance of the form (5.10) develops in time according to

$$\tilde{f}(k, u, t) = \int_{-\infty}^{+\infty} K(\nu) \tilde{\tilde{f}}_\nu(u)\, e^{-ik\nu t}\, d\nu \tag{5.37}$$

since we have the normal mode type time dependence for Van Kampen's modes.

We are mainly interested in the time dependence of the potential which by (3.27) only involves the first-order distribution integrated with respect to the velocity u. Because of the normalization of Van Kampen's modes, we get

$$\tilde{h}(k, t) := \int_{-\infty}^{+\infty} \tilde{f}(k, u, t)\, du = \int_{-\infty}^{+\infty} K(\nu)\, e^{-ik\nu t}\, d\nu \tag{5.38}$$

and introducing (5.33)

$$\tilde{h}(k, t) = \int_{-\infty}^{+\infty} e^{-ik\nu t} \left\{ \frac{\tilde{\tilde{f}}_+(\nu)}{\varepsilon_+(\nu)} - \frac{\tilde{\tilde{f}}_-(\nu)}{\varepsilon_-(\nu)} \right\} d\nu \tag{5.39}$$

For $t > 0$, the contour of integration can be closed from $+\infty$ to $-\infty$ in the lower ν half-plane. There $\tilde{\tilde{f}}_-/\varepsilon_-$ is regular so that we have for $t > 0$

$$\tilde{h}(k, t) = \int_{-\infty}^{+\infty} e^{-ik\nu t} \frac{\tilde{\tilde{f}}_+(\nu)}{\varepsilon_+(\nu)}\, d\nu \tag{5.40}$$

Landau obtained for $\tilde{h}(k, t)$ [see (4.2)]

$$\tilde{h}(k, t) = \frac{1}{2\pi i} \int_{\sigma - i\infty}^{\sigma + i\infty} e^{pt} \frac{\int_{-\infty}^{+\infty} \tilde{f}(u, k, t = 0)/(p + iku)\, du}{\varepsilon_+(k, p)}\, dp \tag{5.41}$$

By the substitution $p = -i\omega$ we get

$$\tilde{h}(k, t) = \frac{1}{2\pi} \int_{-\infty + i\sigma}^{+\infty + i\sigma} e^{-i\omega t} \frac{\int_{-\infty}^{+\infty} \tilde{f}(u, k, t = 0)/[i(ku - \omega)]\, du}{\varepsilon_+(k, \omega)}\, d\omega \tag{5.42}$$

Recalling that we can take for a stable plasma $\sigma = +0$ and therefore substituting $\omega = k(\nu + i0)$, it follows that

$$\tilde{h}(k, t) = \frac{1}{2\pi} \int_{-\infty}^{+\infty} \frac{e^{-ik\nu t}}{\varepsilon_+(k, \nu)} \left[\int_{-\infty}^{+\infty} \frac{\tilde{f}(u, k, t = 0)}{ik(u - \nu - i0)}\, du \right] k\, d\nu \tag{5.43}$$

Using (5.22), we see that this expression is identical with the result (5.40). This demonstrates the complete equivalence of Landau's and Van Kampen's treatment of the initial value problem.

There is no contradiction between the fact that Van Kampen's modes are undamped and do not obey a dispersion relation and the fact that Landau's oscillations are damped: In a stable plasma any physical distribution function is a superposition of Van Kampen's singular eigensolutions and thus damped by "phase mixing."

REFERENCES AND SUPPLEMENTARY READING

General References

Case, K. M., and Zweifel, P. F. (1967). "Linear Transport Theory." Addison-Wesley, Reading, Massachusetts.

Grad, H., ed, (1967). Magneto-Fluid and Plasma Dynamics. *Proc. Symp. Appl. Math.* **18**, Amer. Math. Soc., Providence, Rhode Island.

Jackson, J. D. (1962). "Classical Electrodynamics." Wiley, New York.

Montgomery, D., and Tidman, D. (1964). "Plasma Kinetic Theory." McGraw-Hill, New York.

Roos, B. W. (1969). "Analytic Functions and Distributions in Physics and Engineering." Wiley, New York.

Stix, T. H. (1962). "The Theory of Plasma Waves." McGraw-Hill, New York.

van Kampen, N. G., and Felderhoff, B. U. (1967). "Theoretical Methods in Plasma Physics." Wiley, New York.

Wu, T. Y. (1966). "Kinetic Equations of Gases and Plasmas." Addison-Wesley, Reading, Massachusetts.

Section 1

Vlasov, A. A. (1938). *Zh. Eksp. Teor. Fiz.* **8**, 291.

Vlasov, A. A. (1945). *J. Phys. (U.S.S.R.)* **9**, 25.

Vlasov, A. A. (1950; 1961). "Many-Particle Theory and its Application to Plasma," Engl. Transl. Gordon and Breach. (Originally published in Russian by State Publishing House for Technical-Theoretical Literature, Moscow and Leningrad, 1950.)

Section 2

Bohm, D., and Gross, E. P. (1949). *Phys. Rev.* **75**, 1851.

Bohm, D., and Gross, E. P. (1949). *Phys. Rev.* **75**, 1864.

Fowler, T. K. (1961). *Phys. Fluids*, **4**, 1393.

Fowler, T. K. (1962). *J. Nucl. Energy Part C* **4**, 391.

Harrison, E. R. (1962). *Proc. Phys. Soc. London* **79**, 317.

Harrison, E. R. (1962). *Proc. Phys. Soc. London* **80**, 432.

Penrose, O. (1960). *Phys. Fluids* **3**, 258.

Tonks, L., and Langmuir, I. (1929). *Phys. Rev.* **33**, 196.

SECTION 3 and 4

Altshul, L. M., and Karpman, V.I. (1966). *Sov. Phys. JETP.* **22**, 361.
Backus, G. (1960) *J. Math. Phys.* **1**, 178.
Bernstein, I. B., Greene, J. M., and Kruskal, M. D. (1957). *Phys. Rev.* **108**, 546.
Denavit, J. (1965). *Phys. Fluids* **8**, 471.
Denavit, J. (1966). *Phys. Fluids* **9**, 134.
Derfler, H., and Simonen, T. C. (1969). *Phys. Fluids* **12**, 269.
Drummond, W. E., and Pines, D. (1962). *Nucl. Fusion Suppl.* Pt. 2, 1049.
Einaudi, F., and Sudan, R. N. (1969). *Plasma Phys.* **11**, 359.
Fried, B. D., and Conte, S. D. (1961). "The Plasma Dispersion Function." Academic Press, New York.
Fried, B. D., and Gould, R. W. (1961). *Phys. Fluids* **4**, 139.
Gary, S. P. (1967). *Phys. Fluids* **10**, 570.
Hayes, J. (1961). *Phys. Fluids* **4**, 1387.
Hayes, J. (1963). *Nuovo Cimento* **30**, 1048.
Jackson, J. D. (1960). *J. Nucl. Energy Part C* **1**, 171.
Källén, G. (1966). Intuitive Analyticity. *In* "Preludes in Theoretical Physics in Honor of V. F. Weisskopf" (A. Shalit, ed.). North-Holland Publ., Amsterdam.
Landau, L. (1946). *J. Phys. (U.S.S.R.)* **10**, 25.
McGune, J. E. (1966). *Phys. Fluids* **9**, 2082.
Saenz, A. W. (1964). Rep. NRL-6125. U.S. Naval Res. Lab., Baltimore, Maryland.
Saenz, A. W. (1965). *J. Math. Phys.* **6**, 859.
Taylor, E. C. (1965). *Phys. Fluids* **8**, 2250.
Turski, A. J. (1965). *Ann. Phys.* **35**, 240.
Vedenov, A. A., Velikov, E. P., and Sagdeev, R. Z. (1961). *Sov. Phys. Usp.* **4**, 332.
Weitzner, H. (1963). *Phys. Fluids* **6**, 1123.
Weitzner, H. (1964). *Phys. Fluids* **7**, 476.
Weitzner, H. (1965). *Comm. Pure Appl. Math.* **18**, 307.

SECTION 5

Case, K. M. (1959). *Ann. Phys. (New York)* **7**, 349.
McGune, J. E. (1966). *Phys. Fluids* **9**, 1788.
Mushkelishvili, N. T. (1953). "Singular Integral Equations," Engl. Transl. P. Noordhoff N.V., Groningen, Holland. Originally published in Russian by Institut of Mathematics, Tiflis, U.S.S.R., Moscow, 1946.
van Kampen, N. G. (1955). *Physica (Utrecht)* **21**, 949.
van Kampen, N. G. (1957). *Physica (Utrecht)* **23**, 647.
Zelazny, R. S. (1962). *Ann. Phys.* **19**, 177.

Chapter IV Nonequilibrium States of the Coulomb System with Individual Particle Correlations

1. Derivation of Kinetic Equations from the BBGKY Hierarchy

1.1. General Basis

We again start from the BBGKY hierarchy for the specific distribution functions as given in (II.2.18). Since we are interested in the internal structure of the Coulomb system, we omit external forces. In the subcritical region we have $\Lambda \gg 1$, and therefore we can restrict our interest to the cases $s \ll N$. The hierarchy is then given by the set of equations

$$\frac{\partial f_s}{\partial t} + \sum_{i=1}^{s} \mathbf{v}_i \cdot \frac{\partial f_s}{\partial \mathbf{r}_i} - \frac{1}{m} \sum_{i,k}^{s}{}' \frac{\partial}{\partial \mathbf{r}_i} \phi(\mathbf{r}_i, \mathbf{r}_k) \cdot \frac{\partial}{\partial \mathbf{v}_i} f_s$$

$$= \frac{N}{m} \int d\mathbf{r}' \int d\mathbf{v}' \sum_{i=1}^{s} \frac{\partial}{\partial \mathbf{r}_i} \phi(\mathbf{r}_i, \mathbf{r}') \cdot \frac{\partial}{\partial \mathbf{v}_i} f_{s+1}(\cdots \mathbf{r}', \mathbf{v}'; t) \tag{1.1}$$

where $\phi(\mathbf{r}_i, \mathbf{r}_k)$ is the potential energy of interaction depending only on $|\mathbf{r}_i - \mathbf{r}_k|$.

Since the distinction of the μ space variables (${}^i\mathbf{r}$, ${}^i\mathbf{v}$) and the Γ space variables ($\mathbf{r}_i$, $\mathbf{v}_i$) was of importance only with respect to the derivation of the Klimontovich formalism, we return here to the Γ space variables ($\mathbf{r}_i$, $\mathbf{v}_i$) generally used in the literature.

The hierarchy (1.1) presents the well-known problem that there is always one more unknown function than equations. To overcome this difficulty, we introduce an assumption expressing the three-particle correlations in terms of binary correlations. In the following sections, we will discuss the solution of the remaining two equations of the hierarchy (1.1) by appropriate expansions in terms of smallness parameters.

As a byproduct, we will find the inherent limitations of the correlationless approach (see Chapter III)

$$f_2(\mathbf{r}_i , \mathbf{v}_i , \mathbf{r}_j , \mathbf{v}_j ; t) = f_1(\mathbf{r}_i , \mathbf{v}_i ; t) f_1(\mathbf{r}_j , \mathbf{v}_j ; t) \tag{1.2}$$

Classification in Terms of the Coupling and Density Parameters

In this discussion, it is advisable to consider first the case of a system with a particle interaction law which permits a sensible definition of a characteristic value of the potential energy ϕ_c and of a characteristic interaction range r_c . We also consider distribution functions which allow the introduction of a representative value of the velocity v_c .

Of course, we will discuss later how the conclusions derived on this basis can be applied to the plasma. Normalizing (1.1) with r_c , v_c , ϕ_c , we arrive at

$$\frac{\partial \mathring{f}_s}{\partial \mathring{t}} + \sum_{i=1}^{s} \mathring{\mathbf{v}}_i \cdot \frac{\partial \mathring{f}_s}{\partial \mathring{\mathbf{r}}_i} - \Pi_{co} \sum_{i,k}^{s}{}' \frac{\partial}{\partial \mathring{\mathbf{r}}_i} \mathring{\phi} \cdot \frac{\partial \mathring{f}_s}{\partial \mathring{\mathbf{v}}_i}$$

$$= \Pi_{co}\Pi_d \int d\mathring{\mathbf{r}}' \int d\mathring{\mathbf{v}}' \sum_{i=1}^{s} \frac{\partial}{\partial \mathring{\mathbf{r}}_i} \mathring{\phi}(\mathring{\mathbf{r}}_i , \mathring{\mathbf{r}}') \cdot \frac{\partial}{\partial \mathring{\mathbf{v}}_i} \mathring{f}_{s+1} \tag{1.3}$$

with the reduced variables defined by

$$\mathring{\mathbf{r}} = \frac{\mathbf{r}}{r_c}, \qquad \mathring{\mathbf{v}} = \frac{\mathbf{v}}{v_c}, \qquad \mathring{t} = t\frac{v_c}{r_c}, \qquad \mathring{\phi} = \frac{\phi}{\phi_c}, \qquad \mathring{f}_s = v_c^{3s} V^s f_s \tag{1.4}$$

The dimensionless equation (1.3) contains only two typical parameters

$$\Pi_{co} = \frac{\phi_c}{m v_c^{2}}, \qquad \Pi_d = n r_c^{3} \tag{1.5}$$

We call Π_{co} the coupling parameter and Π_d the density parameter These parameters are suitable to characterize our expansion procedures.

If $\varepsilon \ll 1$ is a smallness parameter, then we cover all situations if we study all possible combinations of the following values of Π_{co} and Π_d

$$\begin{array}{llll} (1) & \Pi_{co} = O(\varepsilon) & (1') & \Pi_d = O(\varepsilon) \\ (2) & \Pi_{co} = O(1) & (2') & \Pi_d = O(1) \\ (3) & \Pi_{co} = O(\varepsilon^{-1}) & (3') & \Pi_d = O(\varepsilon^{-1}) \end{array} \tag{1.6}$$

The four combinations of Π_{co} and Π_d resulting from the groups (2), (3) and (2′), (3′), respectively, provide no basis for an expansion in terms of smallness parameters. These four combinations correspond to critical or above-critical situations where we have strong coupling with one or more partners. Such situations are of interest for fluid and solid states but not for our considerations, since the basic supposition of our model was subcritical behavior. We therefore have to omit these combinations.

One of the remaining five combinations is precluded because in the case

$$(3) \quad \Pi_{co} = O(\varepsilon^{-1}) \quad (1') \quad \Pi_d = O(\varepsilon) \tag{1.7}$$

we have a strictly binary character of the interactions because of $\Pi_d = 0(\varepsilon)$, and consequently—for reasons of energy conservation—Π_{co} cannot exceed the value one.

So we are left with the following four cases:

$$\begin{array}{llll} (a) & \Pi_{co} = O(1), & \Pi_d = O(\varepsilon), & \text{dilute gas} \\ (b) & \Pi_{co} = O(\varepsilon), & \Pi_d = O(\varepsilon), & \text{weak coupling} \\ (c) & \Pi_{co} = O(\varepsilon), & \Pi_d \geqslant O(1), & \text{long–range forces} \end{array} \tag{1.8}$$

This terminology is self explanatory and conforms with that introduced by Frieman (1962).

There are three standard kinetic equations corresponding to the cases given in (1.8):

Dilute gas	—	**Boltzmann equation**
Weak coupling	—	**Landau–Fokker–Planck equation**
Long range forces	—	**Bogolubov–Lenard–Balescu equation**

Applicability of the Standard Kinetic Approaches to the Description of the Plasma

The discussion in the preceding section was carried out under the provision that it is sensible to introduce a characteristic interaction length r_c and a characteristic value ϕ_c for the potential energy.

In a fully ionized plasma, this is not sensible due to the Coulomb interactions. No single value r_c and ϕ_c, respectively, can be given which could be applied in the whole range of interaction distances.

We may, however, overcome this difficulty by subdividing the whole range of interaction distances into three parts each of them being open to the application of one of the standard equations:

$0 < r \leqslant O(r_w)$

r_w designates the classical interaction radius† and is given by

$$r_w = e^2/\Theta \tag{1.9}$$

In this range, we introduce the characteristic values

$$r_c = r_w, \qquad \phi_c = \Theta, \qquad v_c = (\Theta/m)^{1/2} \tag{1.10}$$

since binary interactions $r \ll r_w$ do not occur because of the argument of energy conservation given above. It follows that

$$\Pi_{co} = 1, \qquad \Pi_d = 9/4\pi\Lambda^2 \tag{1.11}$$

Now due to our general assumption of subcritical behavior, we have

$$\Lambda = 12\pi n\lambda_D{}^3 = \frac{3\Theta^{3/2}}{(4\pi n)^{1/2}\, e^3} \gg 1 \tag{1.12}$$

This means that, in this case, ε is given by

$$\varepsilon = 9/4\pi\Lambda^2 \ll 1 \tag{1.13}$$

and that the Boltzmann equation (B) is the appropriate kinetic equation in this range.

$O(r_w) < r < O(r_0)$

To state the limitations of the parameters Π_{co} and Π_d in this range,

† In the following discussion, we refer to the terminology and the relations used in the appendix: Characteristic Quantities of the Plasma.

we first calculate the corresponding values for $r = O(r_0)$, where r_0 designates the average particle distance and is given by

$$r_0 = (3/4\pi n)^{1/3} \tag{1.14}$$

The characteristic values in this case are

$$r_c = r_0, \qquad \phi_c = e^2/r_0, \qquad v_c = (\Theta/m)^{1/2} \tag{1.15}$$

which yield the coupling and density parameters

$$\Pi_{co} = (3/\Lambda^2)^{1/3}, \qquad \Pi_d = 3/4\pi = O(1) \tag{1.16}$$

With the preceding results for r_0 and the corresponding results for r_w of the previous section, we have

$$\begin{aligned} (3/\Lambda^2)^{1/3} < \Pi_{co} < 1 \\ 1 > \Pi_d > 9/4\pi\Lambda^2 \end{aligned} \tag{1.17}$$

These relations show—with $\Lambda \gg 1$—that the Landau–Fokker–Planck equation (LFP) is the adequate description in the central part of the region $r_w < r < r_0$ breaking down in both limits.

$O(r_0) < r < \infty$

To give the limitation for the parameters Π_{co} and Π_d, we first state that, in the limit $r \to \infty$, we have

$$\Pi_{co} \to 0, \qquad \Pi_d \to \infty \tag{1.18}$$

With the results of the previous section, we find

$$(3/\Lambda^2)^{1/3} > \Pi_{co} > 0, \qquad 3/4\pi < \Pi_d < \infty \tag{1.19}$$

which shows—again for $\Lambda \gg 1$—that the Bogolubov–Lenard–Balescu equation (BLB) should be applied to this region.

The applicability of the standard kinetic procedures as concluded from the preceding discussion is demonstrated by the solid lines in Figure IV.1. Actually, however the range of validity of these approaches extends into the dotted ranges of Figure IV.1. The reasons are:

(B) The basic precondition of the Boltzmann approach is not

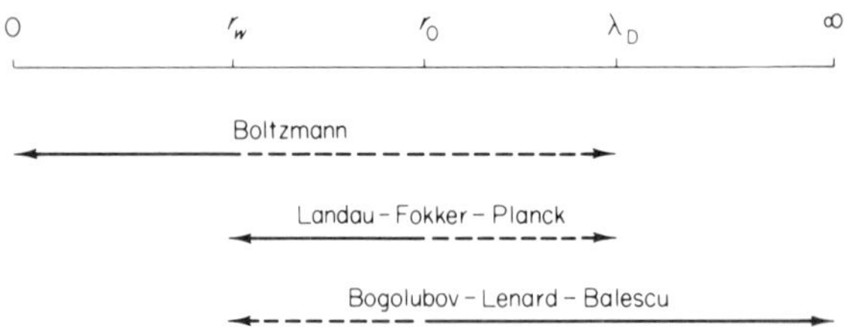

FIGURE IV.1. Range of applicability of the three standard kinetic equations (see text).

destroyed by multiple collisions in the range $r_w \lesssim r \lesssim \lambda_D$ since there occur only weak deflections which—like binary collisions—superimpose themselves independently within a linear approximation. Therefore Boltzmann's approach is valid up to the limit $r = O(\lambda_D)$.

(BLB) The Bogolubov–Lenard–Balescu procedure—with the cutoff $r_{\min} = r_0$—treats correctly the collective shielding effects. Its essential limitation is the restriction to weak deflections, which breaks down only for $r < r_w$. The Bogolubov–Lenard–Balescu approach is therefore a correct description if we apply it with a cutoff $r_{\min} = r_w$.

(LFP) Landau claimed in his original paper that his treatment without screening could and should be extended up to the limit $r = O(\lambda_D)$. We will show in the following (see Section 1.4) on the basis of the Bogolubov–Lenard–Balescu equation that Landau's claim was correct.

Judging by Figure IV.1, one should conclude that none of the three standard procedures is apt to describe the plasma throughout the whole range of interaction distances. In principle this conclusion is correct. However, we will show now that some intervals of the interaction distances contribute only negligibly, a fact which strongly influences our conclusion.

We claim: The contributions of the ranges $0 \leqslant r \leqslant r_w$ and $\lambda_D \leqslant r < \infty$ are negligible in comparison to the contributions of $r_w \leqslant r \leqslant \lambda_D$.

To prove this claim we rewrite the first two equations of the hierarchy (1.1)

$$\frac{\partial f_1}{\partial t} + \mathbf{v}_1 \cdot \frac{\partial f_1}{\partial \mathbf{r}_1} + \frac{e}{m} \langle \mathbf{E} \rangle \cdot \frac{\partial f_1}{\partial \mathbf{v}_1}$$

$$= \left(\frac{\delta f_1}{\delta t}\right)_{\text{coll}} = \frac{N}{m} \frac{\partial}{\partial \mathbf{v}_1} \cdot \int d\mathbf{r}_j \int d\mathbf{v}_j \frac{\partial \phi(\mathbf{r}_1, \mathbf{r}_j)}{\partial \mathbf{r}_1} g_{1j}(\mathbf{r}_1, \mathbf{r}_j, \mathbf{v}_1, \mathbf{v}_j; t) \tag{1.20}$$

and

$$\left(\frac{\partial}{\partial t} + \mathbf{v}_1 \cdot \frac{\partial}{\partial \mathbf{r}_1} + \mathbf{v}_2 \cdot \frac{\partial}{\partial \mathbf{r}_2}\right) g_{12}$$

$$- \frac{1}{m} \frac{\partial \phi_{12}}{\partial \mathbf{r}_1} \cdot \left(\frac{\partial}{\partial \mathbf{v}_1} - \frac{\partial}{\partial \mathbf{v}_2}\right) [f_1(\mathbf{r}_1, \mathbf{v}_1; t) f_1(\mathbf{r}_2, \mathbf{v}_2; t) + g_{12}]$$

$$- \frac{N}{m} \frac{\partial f_1(\mathbf{r}_1, \mathbf{v}_1; t)}{\partial \mathbf{v}_1} \cdot \int d\mathbf{r}_3 \int d\mathbf{v}_3 \frac{\partial \phi_{13}}{\partial \mathbf{r}_1} g_{23}$$

$$- \frac{N}{m} \frac{\partial g_{12}}{\partial \mathbf{v}_1} \cdot \int d\mathbf{r}_3 \int d\mathbf{v}_3 \frac{\partial \phi_{13}}{\partial \mathbf{r}_1} f_1(\mathbf{r}_3, \mathbf{v}_3; t)$$

$$- \frac{N}{m} \frac{\partial f_1(\mathbf{r}_2, \mathbf{v}_2; t)}{\partial \mathbf{v}_2} \cdot \int d\mathbf{r}_3 \int d\mathbf{v}_3 \frac{\partial \phi_{23}}{\partial \mathbf{r}_2} g_{13}$$

$$- \frac{N}{m} \frac{\partial g_{12}}{\partial \mathbf{v}_2} \cdot \int d\mathbf{r}_3 \int d\mathbf{v}_3 \frac{\partial \phi_{23}}{\partial \mathbf{r}_2} f_1(\mathbf{r}_3, \mathbf{v}_3; t)$$

$$- \frac{N}{m} \int d\mathbf{r}_3 \int d\mathbf{v}_3 \left(\frac{\partial \phi_{13}}{\partial \mathbf{r}_1} \cdot \frac{\partial g_{123}}{\partial \mathbf{v}_1} + \frac{\partial \phi_{23}}{\partial \mathbf{r}_2} \cdot \frac{\partial g_{123}}{\partial \mathbf{v}_2}\right) = 0 \tag{1.21}$$

where we applied the expansions of f_2 and f_3 in terms of correlation functions g

$$\begin{aligned} f_2(\mathbf{r}_1, \mathbf{r}_2, \mathbf{v}_1, \mathbf{v}_2; t) &= f_1(\mathbf{r}_1, \mathbf{v}_1; t) f_1(\mathbf{r}_2, \mathbf{v}_2; t) + g_{12} \\ f_3(\mathbf{r}_1, \ldots, \mathbf{v}_3; t) &= f_1 f_1 f_1 + f_1 g_{23} + f_1 g_{13} + f_1 g_{12} + g_{123} \end{aligned} \tag{1.22}$$

$\langle \mathbf{E} \rangle$ is the selfconsistent field from the Vlasov approach. We now make use of the fact that the three standard kinetic procedures discussed so far neglect the triple correlation function g_{123}, and consequently we cancel the last term in (1.21). Further, we introduce the dimensionless variables (1.4) into (1.21) and recognize that for all combinations of the expansion parameters under consideration

$$g_{12} = O(\Pi_{\mathrm{co}} f_1 f_1) \tag{1.23}$$

holds. This relation is physically reasonable, since the correlations are due to the interactions. Inserting (1.23) into (1.20), we find as the basic equation for our estimate

$$\left(\frac{\delta f_1}{\delta t}\right)_{\mathrm{coll}} = O\left(\frac{N}{m v_{\mathrm{c}}} f_1(\mathbf{r}_1, \mathbf{v}_1; t) \int d\mathbf{r}_2 \int d\mathbf{v}_2 \frac{\partial \phi_{12}}{\partial r_1} \Pi_{\mathrm{co}} f_1(\mathbf{r}_2, \mathbf{v}_2; t)\right) \tag{1.24}$$

Using the normalization condition

$$\int f_1(\mathbf{r}_2, \mathbf{v}_2; t)\, d\mathbf{v}_2 = 1/V \tag{1.25}$$

and transforming to the relative coordinate $r = |\mathbf{r}_1 - \mathbf{r}_2|$, we find

$$\left(\frac{\delta f_1}{\delta t}\right)_{\text{coll}} = O\left(\frac{n}{v_{\text{c}}\, m} f_1 \int_0^\infty \frac{\partial \phi_{12}}{\partial r}\, \Pi_{\text{co}}\, r^2\, dr\right) \tag{1.26}$$

We now split our integration in (1.26) into three parts according to the three regions (B), (LFP), and (BLB) defined above and write

$$\left(\frac{\delta f_1}{\delta t}\right)_{\text{coll}} = O[(C_{\mathbf{B}} + C_{\mathbf{LFP}} + C_{\mathbf{BLB}}) f_1] \tag{1.27}$$

where we have to estimate the magnitude of the coefficients.

(B) $r \leqslant O(r_w)$

The main contribution to the integral in (1.26) comes from the environment of r_w, so that we are justified in introducing $\Pi_{\text{co}} = 1$ from (1.11), and we have

$$C_{\mathbf{B}} = O\left(\frac{n}{m v_{\text{c}}} \int_0^{r_w} \frac{e^2}{r^2} r^2\, dr\right) = O(n v_{\text{c}}\, r_w^2) \tag{1.28}$$

or

$$C_{\mathbf{B}} = O(\omega_{\text{p}}/\Lambda) \tag{1.29}$$

where ω_{p} is the plasma frequency of the system. $C_{\mathbf{B}}$ characterizes the order of magnitude of the Boltzmann correlation term in region (B).

(BLB) $r \geqslant O(\lambda_{\text{D}})$

Due to the exponential decrease of the Debye screening law, the main contribution stems from the neighborhood of λ_{D} which allows us to use $\Pi_{\text{co}} = \Lambda^{-1}$. Here the estimate (1.26) yields

$$C_{\mathbf{BLB}} = O\left(\frac{n}{m v_{\text{c}}} \int_{\lambda_{\text{D}}}^\infty \frac{e^2 \exp(-r/\lambda_{\text{D}})}{r^2} \frac{1}{\Lambda} r^2\, dr\right) = O\left(\frac{n e^2}{m} \frac{\lambda_{\text{D}}}{v_{\text{c}}} \frac{1}{\Lambda}\right) \tag{1.30}$$

or

$$C_{\mathbf{BLB}} = O(\omega_{\text{p}}/\Lambda) \tag{1.31}$$

which means that the order of magnitude of the Bogolubov–Lenard–Balescu correlation term in region (BLB) is equal to the order of magnitude of the Boltzmann correlation term in region (B).

(LFP) $O(r_w) < r < O(\lambda_D)$

In this range the coupling parameter is given by

$$\Pi_{co} = O\left(\frac{e^2}{r\Theta}\right) \tag{1.32}$$

Then we find for the Landau–Fokker–Planck correlation term in region (LFP)

$$C_{LFP} = O\left(\frac{n}{mv_c}\int_{r_w}^{\lambda_D}\frac{e^2}{r^2}\frac{e^2}{r\Theta}r^2\,dr\right) = O\left(\frac{ne^2}{m}\frac{e^2}{v_c\Theta}\ln\frac{\lambda_D}{r_w}\right) \tag{1.33}$$

or

$$C_{LFP} = O\left(\frac{\omega_p \ln \Lambda}{\Lambda}\right) \tag{1.34}$$

This differs from the contributions of the regions (B) and (BLB) by the factor $\ln \Lambda$.

It is obvious from (1.29), (1.31), and (1.34) that the main effect on the distribution function originates from collisions with weak deflections in the range $r_w \lesssim r \lesssim \lambda_D$ if the condition

$$\ln \Lambda \gg 1 \tag{1.35}$$

is fulfilled. We require this postulate—more incisive than $\Lambda \gg 1$—for all following studies.

Therewith we have proved our claim.

Conclusions

The information of Figure IV.1 together with the estimate of the contributions which we have just given shows that each of the three standard kinetic approaches gives the full description of the plasma within the accuracy

$$\ln \Lambda \gg 1$$

provided that the inaccessible regions are excluded by appropriate cutoffs. These are:

(B) $r_{max} = \lambda_D$

(FLP) $r_{min} = r_w, \qquad r_{max} = \lambda_D$

(BLB) $r_{min} = r_w$

It is important to note that there exist combinations of the above procedures which remove the divergencies without introducing cutoffs. Such a combination may symbolically be written in the form

$$(\delta f_1/\delta t)_{\text{coll}} = I_{\text{B}} + I_{\text{BLB}} - I_{\text{LFP}} \tag{1.36}$$

The interpretation of this procedure is simple:

In the range $r \leqslant r_w$, the Landau–Fokker–Planck term (I_{LFP}) and the Bogolubov–Lenard–Balescu term (I_{BLB}) cancel, whereas the Boltzmann term (I_{B}) describes the situation correctly.

In the region $r_w \leqslant r \leqslant \lambda_{\text{D}}$, the three terms are all identical, and therefore only the contribution of one of them remains.

Finally, in the region $r > \lambda_{\text{D}}$, the Boltzmann and Landau–Fokker–Planck terms cancel leaving the Bogolubov–Lenard–Balescu contribution which is correct and nondivergent in this region.

Equations of the type symbolized in (1.36) have been derived by various authors (Frieman and Book, 1963; Hubbard, 1961; Kihara, Aono, 1963; Weinstock, 1963, 1964).

1.2. The Boltzmann Equation

Derivation of the Boltzmann Equation from the BBGKY Hierarchy

Boltzmann's equation for a dilute system of individuals with short range interaction can be derived in numerous ways. For the sake of continuity, we here prefer the derivation from the BBGKY hierarchy. With the conditions

$$\Pi_{\text{co}} = O(1), \qquad \Pi_{\text{d}} = O(\varepsilon) \ll 1 \tag{1.37}$$

omitting again external forces, the first two equations of the hierarchy read

$$\frac{\partial f_1}{\partial t} + \mathbf{v}_1 \cdot \frac{\partial f_1}{\partial \mathbf{r}_1} = -\frac{N}{m} \frac{\partial}{\partial \mathbf{v}_1} \cdot \int \frac{\partial \phi_{12}}{\partial \mathbf{r}} f_2(\mathbf{r}, \mathbf{r}_1, \mathbf{v}_1, \mathbf{v}_2; t)\, d\mathbf{r}\, d\mathbf{v}_2 \tag{1.38}$$

$$\left[\frac{\partial}{\partial t} + \mathbf{v}_1 \cdot \frac{\partial}{\partial \mathbf{r}_1} + (\mathbf{v}_2 - \mathbf{v}_1) \cdot \frac{\partial}{\partial \mathbf{r}} - \frac{1}{m} \frac{\partial \phi_{12}}{\partial \mathbf{r}} \cdot \left(\frac{\partial}{\partial \mathbf{v}_1} - \frac{\partial}{\partial \mathbf{v}_2}\right)\right] f_2(\mathbf{r}, \mathbf{r}_1, \mathbf{v}_1, \mathbf{v}_2; t)$$

$$= \frac{N}{m} \int \sum_{\nu=1}^{2} \left(\frac{\partial}{\partial \mathbf{r}_\nu} \phi_{\nu 3}\right) \cdot \frac{\partial}{\partial \mathbf{v}_\nu} f_3\, d\mathbf{r}_3\, d\mathbf{v}_3 \tag{1.39}$$

where we have introduced the relative coordinate $\mathbf{r} = \mathbf{r}_2 - \mathbf{r}_1$.

Note that (1.39) contains on the left-hand side terms identical with the integrand on the right-hand side of (1.38).

To eliminate these terms, we integrate (1.39) with respect to $\mathbf{v}_2$ and $\mathbf{r}$. If we restrict ourselves to linear contributions in $\Pi_{\rm d}$, we can neglect the right-hand side. Using Gauss's theorem for the velocity integration, we find

$$\left(\frac{\partial}{\partial t} + \mathbf{v}_1 \cdot \frac{\partial}{\partial \mathbf{r}_1}\right) \int f_2(\mathbf{r}, \mathbf{r}_1, \ldots)\, d\mathbf{v}_2\, d\mathbf{r} + \int (\mathbf{v}_2 - \mathbf{v}_1) \cdot \frac{\partial f_2(\mathbf{r}, \mathbf{r}_1, \ldots)}{\partial \mathbf{r}}\, d\mathbf{r}\, d\mathbf{v}_2$$

$$= \frac{1}{m} \frac{\partial}{\partial \mathbf{v}_1} \cdot \int \frac{\partial \phi_{12}}{\partial \mathbf{r}} f_2(\mathbf{r}, \mathbf{r}_1, \ldots)\, d\mathbf{v}_2\, d\mathbf{r} \tag{1.40}$$

We now introduce the domain of collisions between the first and the sth particle

$$D_s^\sigma := \{\mathbf{r}_s \mid |\mathbf{r}_s - \mathbf{r}_1| < \sigma\} \tag{1.41}$$

and the complement-region with respect to the configuration space γ_s of the sth particle

$$\bar{D}_s^\sigma := \{\mathbf{r}_s \mid |\mathbf{r}_s - \mathbf{r}_1| > \sigma\} = \gamma_s - D_s^\sigma \tag{1.42}$$

where σ denotes the range of interactions $\sigma \approx r_{\rm c}$.

Further, we define the **truncated distribution function** of order s:

$$f_s^\sigma := (1/V^{N-s}) \int_{\bar{D}_{s+1}^\sigma} d\mathbf{r}_{s+1}\, d\mathbf{v}_{s+1} \cdots \int_{\bar{D}_N^\sigma} d\mathbf{r}_N\, d\mathbf{v}_N\, f_N \tag{1.43}$$

From (1.42) and (1.43) we find the development

$$f_s^\sigma = f_s - N \int_{D_{s+1}^\sigma} d\mathbf{r}_{s+1}\, d\mathbf{v}_{s+1}\, f_{s+1}$$

$$+ (N^2/2) \int_{D_{s+1}^\sigma} d\mathbf{r}_{s+1}\, d\mathbf{v}_{s+1} \int_{D_{s+2}^\sigma} d\mathbf{r}_{s+2}\, d\mathbf{v}_{s+2}\, f_{s+2} - + \cdots \tag{1.44}$$

Since in the Boltzmann approach we neglect f_3, we get for the first two truncated distribution functions

$$f_1^\sigma = f_1 - N \int_{D_2^\sigma} d\mathbf{r}_2\, d\mathbf{v}_2\, f_2\,, \qquad f_2^\sigma \equiv f_2 \tag{1.45}$$

The truncated distribution function f_s^σ gives the probability density to find the particles $(1, 2, \ldots, s)$ at prescribed positions $(\mathbf{r}_1, \ldots, \mathbf{v}_s)$ and none of the particles $(s + 1, \ldots, N)$ within the collision range of the first particle. Especially $f_1^\sigma(\mathbf{r}_1, \mathbf{v}_1; t)$ is the probability density to find the particle 1 at $\mathbf{r}_1$, $\mathbf{v}_1$ not interacting with any other particle. Note that, due to the distinction of particle 1, the symmetry in the coordinates of f_s^σ is destroyed.

Introducing the truncated distribution functions into (1.38) and (1.40) we get

$$\left(\frac{\partial}{\partial t} + \mathbf{v}_1 \cdot \frac{\partial}{\partial \mathbf{r}_1}\right) f_1^\sigma = N \int_{D_2^\sigma} d\mathbf{r} \int d\mathbf{v}_2 (\mathbf{v}_2 - \mathbf{v}_1) \cdot \frac{\partial f_2^\sigma(\mathbf{r}, \mathbf{r}_1, \ldots)}{\partial \mathbf{r}} \tag{1.46}$$

or using Gauss's theorem

$$\left(\frac{\partial}{\partial t} + \mathbf{v}_1 \cdot \frac{\partial}{\partial \mathbf{r}_1}\right) f_1^\sigma = N \int d\mathbf{v}_2 \oint_S f_2^\sigma(\mathbf{r}, \mathbf{r}_1, \ldots)(\mathbf{v}_2 - \mathbf{v}_1) \cdot d\mathbf{S} \tag{1.47}$$

where S is the surface of the sphere D_2^σ.
This is the **second Grad equation** recovered from the BBGKY hierarchy. As shown by Grad (1958), this equation is exact and can be derived without neglecting f_3.

Due to the limitation $\Pi_\mathrm{d} \ll 1$, it is justified in the Boltzmann approach to identify f_2^σ and f_2. The resulting equation is

$$\frac{\partial f_1}{\partial t} + \mathbf{v}_1 \cdot \frac{\partial f_1}{\partial \mathbf{r}_1} = N \int d\mathbf{v}_2 \oint_S d\mathbf{S} \cdot (\mathbf{v}_2 - \mathbf{v}_1) f_2(\mathbf{r}, \mathbf{r}_1, \ldots) \tag{1.48}$$

This equation is not yet the Boltzmann equation, but we observe an important change from the general formulation toward Boltzmann's approach.

In the general description, the correlation term is provided by the smooth change of the number of particles of a given state under the continuous influence of the interaction forces. In (1.48), however, the change in the number of free particles (not interacting with others) represents itself as the difference of the loss of pairs of particles entering the interaction sphere and the corresponding gain of pairs leaving this interaction sphere. This is due to our truncation procedure and results from the development in Π_d.

To come closer to the formal similarity with the Boltzmann equation, let us introduce polar coordinates (b, φ) in a diametral plane C (see Figure IV.2) of the sphere S perpendicular to the relative velocity $\mathbf{g} = \mathbf{v}_2 - \mathbf{v}_1$. This description means a projection of the sphere S into the disc C. Since C is covered twice, we have to distinguish the two hemispheres

$$\begin{aligned} S^+ &:= \{\mathbf{r} \in S \mid \mathbf{g} \cdot d\mathbf{S} \geqslant 0\} =: \{\mathbf{r}^+\} \\ S^- &:= \{\mathbf{r} \in S \mid \mathbf{g} \cdot d\mathbf{S} < 0\} =: \{\mathbf{r}^-\} \end{aligned} \tag{1.49}$$

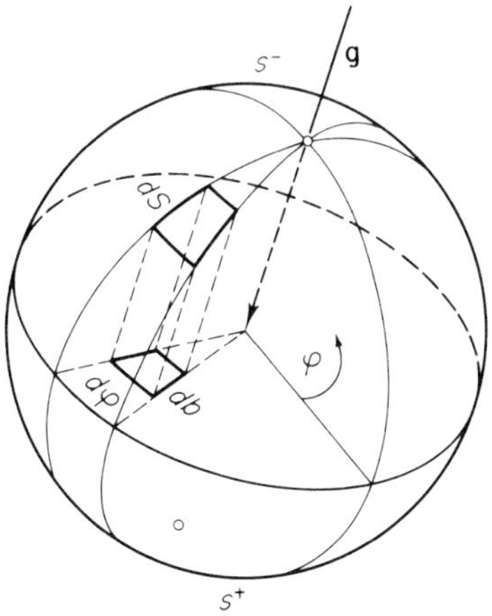

FIGURE IV.2. Collision sphere with impact parameters.

$S^{\pm}$ refer to receding and approaching particles, respectively (see Figure IV.2).

We may then rewrite (1.48) in the form

$$\frac{\partial f_1}{\partial t} + \mathbf{v}_1 \cdot \frac{\partial f_1}{\partial \mathbf{r}_1} = N \int d\mathbf{v}_2 \int_0^{\sigma} b\, db \int_0^{2\pi} d\varphi\, g\, [f_2(\mathbf{r}^+,...) - f_2(\mathbf{r}^-,...)] \quad (1.50)$$

The function $f_2(\mathbf{r}^{\pm}, \mathbf{r}_1, \mathbf{v}_1, \mathbf{v}_2 ; t)$ is still unknown. So far essentially the condition

$$\Pi_{\mathrm{d}} \ll 1 \quad (1.51)$$

has been used. One further conclusion can be drawn from the condition (1.51). The smallness of this parameter is the background of the restriction to binary collisions. Therefore each of these pair systems can be considered as closed, and Liouville's law may be applied in the form

$$f_2(\mathbf{r}^{-\prime}, \mathbf{r}_1, \mathbf{v}_1', \mathbf{v}_2'; t') = f_2(\mathbf{r}^+, \mathbf{r}_1, \mathbf{v}_1, \mathbf{v}_2 ; t) \quad (1.52)$$

where the primes designate precollision values corresponding to post-collision values $\mathbf{r}^+$, $\mathbf{v}_1$, $\mathbf{v}_2$, t.

We now need two further conditions: First,

$$\sigma \ll L, \qquad \sigma \ll v_{\mathrm{c}}\, T \quad (1.53)$$

which means that the characteristic length L and the characteristic time T of the variation of f_1 are large compared to the interaction distance and interaction time. If this condition is fulfilled, we may use

$$\begin{aligned} f_2(\mathbf{r}^{-\prime}, \mathbf{r}_1, \mathbf{v}_1', \mathbf{v}_2'; t') &= f_2(\mathbf{r}_1, \mathbf{r}_1, \mathbf{v}_1', \mathbf{v}_2'; t) \\ f_2(\mathbf{r}^{+}, \mathbf{r}_1, \mathbf{v}_1, \mathbf{v}_2; t) &= f_2(\mathbf{r}_1, \mathbf{r}_1, \mathbf{v}_1, \mathbf{v}_2; t) \end{aligned} \tag{1.54}$$

Second, we request molecular chaos at the surface S^- before the collision starts; that means we use

$$\begin{aligned} f_2(\mathbf{r}_1, \mathbf{r}_1, \mathbf{v}_1, \mathbf{v}_2; t) &= f_1(\mathbf{r}_1, \mathbf{v}_1; t) f_1(\mathbf{r}_1, \mathbf{v}_2; t) \\ f_2(\mathbf{r}_1, \mathbf{r}_1, \mathbf{v}_1', \mathbf{v}_2'; t) &= f_1(\mathbf{r}_1, \mathbf{v}_1'; t) f_1(\mathbf{r}_1, \mathbf{v}_2'; t) \end{aligned} \tag{1.55}$$

With this basic requirement which destroys the reversibility of the kinetic equation we find the **Boltzmann equation**

$$\begin{aligned} \frac{\partial f_1}{\partial t} + \mathbf{v}_1 \cdot \frac{\partial f_1}{\partial \mathbf{r}_1} = N \int [&f_1(\mathbf{r}_1, \mathbf{v}_1'; t) f_1(\mathbf{r}_1, \mathbf{v}_2'; t) \\ &- f_1(\mathbf{r}_1, \mathbf{v}_1; t) f_1(\mathbf{r}_1, \mathbf{v}_2; t)]\, g\, b\, db\, d\varphi\, d\mathbf{v}_2 \end{aligned} \tag{1.56}$$

Other Derivations of the Boltzmann Equation from the Liouville Equation

The derivation presented above corresponds to that given by Grad (1958). There are other derivations of the Boltzmann equation from the Liouville equation which use basically different approaches.

Probably the first serious mathematical attempt is due to Kirkwood (1947). The basic concept of his approach is a time-smoothing process, which in essence has the same effect as the introduction of the truncated distribution function by Grad. It, too, provides an equation in which only completed collisions occur. Of course, the assumption of molecular chaos is also basic in Kirkwood's approach.

The derivation of Bogolubov (1946) yields the Boltzmann equation as the first term of a general expansion. The basic idea of Bogolubov's procedure is the claim that the higher-order distribution functions f_s depend on time only via the distribution function f_1. The crucial assumption for the justification of this claim is the postulate that any s particles which are sufficiently dispersed satisfy the relation $f_s(\mathbf{r}_1, \mathbf{v}_1, ..., \mathbf{r}_s, \mathbf{v}_s) = f_1(\mathbf{r}_1, \mathbf{v}_1) \cdots f_1(\mathbf{r}_s, \mathbf{v}_s)$, and that any higher-order distribution function when followed back in time automatically reaches such a state of dispersion.

Similar to the procedure of Bogolubov is the approach used by Green (1956). There are also derivations which start from the master equation instead of the Liouville equation, which in the classical domain are, for instance, given by Prout (1956) and by Prout and Prigogine (1956). Since a critical survey of the features of the Boltzmann equation is not the aim of this investigation, we refer to the thorough discussion in Grad's paper.

Boltzmann's Original Approach (1872, 1875)

In contrast to all the derivations discussed above, the original study of Boltzmann considers a single system. Subsequent application of the ideas of Chapter II requires that the density has to be described in the Klimontovich form [see (II.1.1)]

$$F(\mathbf{r}, \mathbf{v}; t) = \sum_i \delta(\mathbf{r} - \mathbf{r}_i(t))\, \delta(\mathbf{v} - \mathbf{v}_i(t)) \tag{1.57}$$

and, omitting external forces for the sake of simplicity, we have

$$\frac{\partial F}{\partial t} + \mathbf{v} \cdot \frac{\partial F}{\partial \mathbf{r}} = - \frac{\partial}{\partial \mathbf{v}} \cdot (\mathbf{K}F) \tag{1.58}$$

where $\mathbf{K}$ represents all internal forces.

This true point density F is, however, not the object of Boltzmann's study. Rather, he uses the average particle number density resulting from a smoothing process in time and in the μ space

$$f_{\mathrm{B}} = \int \frac{d\mathbf{r}\, d\mathbf{v}}{\Delta\mu} \int \frac{dt}{\Delta T} F \tag{1.59}$$

After a suitable subdivision of the whole μ space into elements $\Delta\mu$ and of the time into intervals ΔT, the averaging process (1.59) provides a step function for f_{B}. Clearly, if this definition of the mean density is to be meaningful, we have the necessary condition

$$\Delta f_{\mathrm{B}} \ll f_{\mathrm{B}} \tag{1.60}$$

for the steps Δf_{B}. If this condition is fulfilled, we may replace the step function by a continuous function. To avoid strong deviations from the average values, we further must require

$$f_{\mathrm{B}}\, \Delta\mu \gg 1 \tag{1.61}$$

With these requirements we can write the left-hand side of (1.58) in the form

$$\frac{\partial f_{\mathrm{B}}}{\partial t} + \mathbf{v} \cdot \frac{\partial f_{\mathrm{B}}}{\partial \mathbf{r}} = \left(\frac{\delta f_{\mathrm{B}}}{\delta t}\right)_{\mathrm{coll}} \tag{1.62}$$

where the right-hand side represents the effect of the correlations.

The main contribution of Boltzmann is actually the approximation of the functional dependence of the correlation term on f_B using the basis of the completed collision concept in a dilute gas.

Boltzmann's collision concept incorporates several assumptions:

First, interactions are short, intensive, and rare ($\Pi_{co} = O(1)$, $\Pi_d \ll 1$). This means: (a) we can neglect higher-order correlations; (b) the binary interactions occur during a very short time interval in comparison to the free-flight time, so that all particles are practically noninteracting at all times; and (c) the effect of the correlations may be calculated on the basis of completed collisions.

Second, if T and L are the characteristic time and length of the convective μ-space variation of f_B, then Boltzmann postulates

$$T \gg \tau_c, \qquad L \gg \lambda_c \tag{1.63}$$

where τ_c is the mean collision time and λ_c the mean free path.

Third, if one introduces on the basis of the first postulate pre- and postcollision pair coordinates, then Boltzmann expects that the precollision pair variables are stochastically independent (molecular chaos).

On the basis of these postulates and with the help of Figure IV.2, it is easy to see that the collision term $(\delta f_1/\delta t)_{coll}$ presents itself as a difference of a production and destruction term for the state $\mathbf{r}_1$, $\mathbf{v}_1$ in the form of the right-hand side of (1.56).

Particular attention is drawn to the fact that, in principle, Boltzmann's result is quite different from the result (1.56) derived from the BBGKY hierarchy. Already the meaning of the f_1-functions is different. This is also reflected in the conditions (1.60) and (1.61) which—if violated—can render the Boltzmann equation—or more precisely, its application to a single system—invalid. These interesting problems and their consequences for experimental verification have been studied by several authors (Ludwig, Müller, Schröter 1964; Ecker, Hölling 1963).

1.3. The Landau–Fokker–Planck Equation

Landau Form of the Weak Coupling Equation

Within our specification (1.8), the Landau equation is characterized by

$$\Pi_{co} = O(\varepsilon), \qquad \Pi_d = O(\varepsilon) \tag{1.64}$$

Comparing these relations with the basic conditions (1.37) for the

Boltzmann equation, we recognize that it should be possible to derive the Landau equation from an expansion of the Boltzmann equation in the limiting case of weak interactions.

Such expansions are well known (see, e.g., van Kampen and Felderhof, 1967). However, in view of the general policy in this investigation, we prefer to deduce the Landau equation directly from the BBGKY hierarchy. This is also advised by the fact that the following derivation of the Lenard–Balescu equation will profit from the present approach.

Utilizing the expansions (1.22) and neglecting as always the third-order correlation function, we get from the two first equations of the hierarchy (1.20) and (1.21)[†]

$$\frac{\partial f_1}{\partial t} + \mathbf{v}_1 \cdot \frac{\partial f_1}{\partial \mathbf{r}_1} = \left(\frac{\delta f_1}{\delta t}\right)_{\text{coll}} = \frac{N}{m} \frac{\partial}{\partial \mathbf{v}_1} \cdot \int \frac{\partial \phi_{12}}{\partial \mathbf{r}_1} g_{12}(\mathbf{r}_1, \ldots, \mathbf{v}_2 ; t) \, d\mathbf{r}_2 \, d\mathbf{v}_2 \tag{1.65}$$

and

$$\left(\frac{\partial}{\partial t} + \mathbf{v}_1 \cdot \frac{\partial}{\partial \mathbf{r}_1} + \mathbf{v}_2 \cdot \frac{\partial}{\partial \mathbf{r}_2}\right) g_{12}$$

$$= \frac{1}{m} \frac{\partial \phi_{12}}{\partial \mathbf{r}_1} \cdot \left(\frac{\partial}{\partial \mathbf{v}_1} - \frac{\partial}{\partial \mathbf{v}_2}\right) (f_1(\mathbf{r}_1, \mathbf{v}_1 ; t) f_1(\mathbf{r}_2, \mathbf{v}_2 ; t) + g_{12})$$

$$+ \frac{N}{m} \int \left(\frac{\partial f_1}{\partial \mathbf{v}_1} \cdot \frac{\partial \phi_{13}}{\partial \mathbf{r}_1} g_{23} + \frac{\partial f_1}{\partial \mathbf{v}_2} \cdot \frac{\partial \phi_{23}}{\partial \mathbf{r}_2} g_{13}\right) d\mathbf{r}_3 \, d\mathbf{v}_3 \tag{1.66}$$

where we again omitted external forces. We further neglected the influence of the self-consistent field on the flux in the velocity space. This neglection cancels the third term on the left-hand side of (1.20) and the fourth and sixth term on the left-hand side of (1.21). Trivially this procedure is exact for a homogeneous system. Inhomogeneity is admissible to the extent that the corresponding self-consistent field is small in comparison to the average correlation field.

We emphasize the fact that space dependence of f_1 in (1.65), (1.66), and the following is to be understood small in the sense specified above.

As in the case of the Boltzmann equation, we aim to eliminate g_{12} from (1.65) with the help of (1.66). Now let us recall that in the dimensionless representation (1.4) the first term on the right-hand side of (1.66) has the factor Π_{co}, and the second has the factor $\Pi_{\text{co}}\Pi_{\text{d}}$. Consequently, we omit this latter term, which is small of higher order (see (1.64)).

[†] Although we are using specific distribution functions, only one set of field particle coordinates $(\mathbf{r}_2, \mathbf{v}_2)$ represents together with N the effect of all field particles on the right-hand side due to the basic assumption given on p. 86.

Physically speaking, this means the neglect of the interaction of the field particle with a third particle in the environment of the test particle. At this point we loose the screening mechanism.

Remembering (1.23), we can omit g_{12} in the first term on the right-hand side of (1.66), too. Then we have

$$\left(\frac{\partial}{\partial t} + \mathbf{v}_1 \cdot \frac{\partial}{\partial \mathbf{r}_1} + \mathbf{v}_2 \cdot \frac{\partial}{\partial \mathbf{r}_2}\right) g_{12} = \frac{1}{m} \frac{\partial \phi_{12}}{\partial \mathbf{r}_1} \cdot \left(\frac{\partial}{\partial \mathbf{v}_1} - \frac{\partial}{\partial \mathbf{v}_2}\right) f_1(\mathbf{r}_1, \mathbf{v}_1; t) f_1(\mathbf{r}_2, \mathbf{v}_2; t) \tag{1.67}$$

which is a partial differential equation for the function g_{12}.

Equation (1.67) may be rewritten in the form

$$\frac{d}{dt} g_{12} = \frac{1}{m} \frac{\partial \phi_{12}}{\partial \mathbf{r}_1} \cdot \left(\frac{\partial}{\partial \mathbf{v}_1} - \frac{\partial}{\partial \mathbf{v}_2}\right) f_1(\mathbf{r}_1, \mathbf{v}_1; t) f_1(\mathbf{r}_2, \mathbf{v}_2; t) + \left(\dot{\mathbf{v}}_1 \cdot \frac{\partial}{\partial \mathbf{v}_1} + \dot{\mathbf{v}}_2 \cdot \frac{\partial}{\partial \mathbf{v}_2}\right) g_{12} \tag{1.68}$$

Integrating along straight paths

$$\dot{\mathbf{v}}_1 = \dot{\mathbf{v}}_2 = 0, \qquad \mathbf{r}_{12}(t') = \mathbf{r}_{12}(t) + \mathbf{v}_{12}(t' - t) \tag{1.69}$$

the second term on the right-hand side yields zero. However, applying the straight path approximation, we have to remember that the variables $\mathbf{r}_i$ and $\mathbf{v}_i$ are coupled by Newton's law. From the estimate

$$\dot{v}_1 = O\left(\frac{1}{m} \int \frac{\partial \phi_{12}}{\partial r_1} f_1(\mathbf{r}_2, \mathbf{v}_2; t)\, d\mathbf{r}_2\, d\mathbf{v}_2\right) = O\left(\frac{\phi_c}{m r_c}\right) \tag{1.70}$$

it can be shown that the relative error of this straight path approximation is of order Π_{co}. With the abbreviation

$$A(\mathbf{r}_1, \mathbf{r}_2, \mathbf{v}_1, \mathbf{v}_2; t) = \frac{1}{m} \frac{\partial \phi_{12}}{\partial \mathbf{r}_1} \cdot \left(\frac{\partial}{\partial \mathbf{v}_1} - \frac{\partial}{\partial \mathbf{v}_2}\right) f_1(\mathbf{r}_1, \mathbf{v}_1; t) f_1(\mathbf{r}_2, \mathbf{v}_2; t) \tag{1.71}$$

for the remaining term on the right-hand side of (1.68) we find after integration

$$g_{12}(\ldots; t) = \int_{-\infty}^{t} A(\ldots; t')\, dt' + g_{12}(\ldots; -\infty) \tag{1.72}$$

We approach our correlated state at time t from an uncorrelated state, $g_{12}(\ldots, -\infty) = 0$.

Under the assumption that the variation of f_1 may be neglected over an interaction distance $r_c (L \gg r_c)$, the integrand depends only on the relative distance. With the substitution $\tau := t - t'$ we get

$$g_{12}(|\mathbf{r}_1 - \mathbf{r}_2|, \mathbf{v}_1, \mathbf{v}_2, t) = \int_0^\infty A(\mathbf{r}_1(t) - \mathbf{v}_1\tau, \mathbf{r}_2(t) - \mathbf{v}_2\tau, \mathbf{v}_1, \mathbf{v}_2; t - \tau)\, d\tau \tag{1.73}$$

Introducing a flux $\mathbf{J}$ in velocity space by

$$\left(\frac{\delta f_1}{\delta t}\right)_{\text{coll}} = -\frac{\partial}{\partial \mathbf{v}_1} \cdot \mathbf{J} \tag{1.74}$$

we find from (1.65), (1.71), and (1.73)

$$\mathbf{J} = -\frac{N}{m^2} \int d\mathbf{r}_2 \int d\mathbf{v}_2 \int_0^\infty d\tau \frac{\partial \phi_{12}}{\partial \mathbf{r}_1} \cdot \left[\frac{\partial \phi_{12}}{\partial \mathbf{r}_1} \cdot \left[\frac{\partial}{\partial \mathbf{v}_1} - \frac{\partial}{\partial \mathbf{v}_2}\right] f_1(\mathbf{r}_1, \mathbf{v}_1; t) f_1(\mathbf{r}_1, \mathbf{v}_2; t)\right]_{\substack{t-\tau \\ \mathbf{r}_{12} - \mathbf{v}_{12}\tau}} \tag{1.75}$$

Using the expansion

$$(f_1(\mathbf{r}_1, \mathbf{v}_1; t) f_1(\mathbf{r}_1, \mathbf{v}_2; t))_{t-\tau} = (f_1 f_1)_t - (t - \tau)\left(\frac{\partial}{\partial t}(f_1 f_1)\right)_t + \cdots \tag{1.76}$$

in (1.75) we get the **Landau equation with memory.**

If the distribution function does not change appreciably during the interaction time $(T \gg t_c)$, we may restrict ourselves to the first term of this expansion thus losing the "memory" of the system. Then we arrive at

$$\mathbf{J} = -N(e/m)^2 \int d\mathbf{v}_2 \langle \mathbf{E})(\mathbf{E}\rangle_{\mathbf{r}_2} \cdot \left\{ f_1(\mathbf{r}_1, \mathbf{v}_2; t) \frac{\partial f_1(\mathbf{r}_1, \mathbf{v}_1; t)}{\partial \mathbf{v}_1} - f_1(\mathbf{r}_1, \mathbf{v}_1; t) \frac{\partial f_1(\mathbf{r}_1, \mathbf{v}_2; t)}{\partial \mathbf{v}_2} \right\} \tag{1.77}$$

where the coefficient $\langle \mathbf{E})(\mathbf{E}\rangle_{\mathbf{r}_2}$ is given by

$$\langle \mathbf{E})(\mathbf{E}\rangle_{\mathbf{r}_2} = \int d\mathbf{r}_2 \int_0^\infty d\tau\, \mathbf{E}[\mathbf{r}_{12}])(\mathbf{E}[\mathbf{r}_{12} - \mathbf{v}_{12}\tau] \tag{1.78}$$

and $\mathbf{E}$ is the Coulomb field.

$\langle \mathbf{E})(\mathbf{E}\rangle_{\mathbf{r}_2}$ depending on $\mathbf{v}_1, \mathbf{v}_2$ is the autocorrelation tensor of the pair field in the straight-path approximation. The index $\mathbf{r}_2$ indicates averaging over $\mathbf{r}_2$ only.

Equations (1.74) and (1.77) represent the **Landau equation**, the explicit evaluation of the field autocorrelation coefficient still pending. In the form given here, the Landau equation presents itself as the Boltzmann equation for weak interactions [see (1.8)]. It is therefore interesting to note that we were obliged to use the same basic assumptions in its derivation. The requirement $g_{12}(\cdots; -\infty) = 0$ corresponds to Boltzmann's initial chaos assumption. The conditions (1.53) for the space and time dependence of the distribution function recur here in the requirements $L \gg r_c$ and $T \gg t_c$.

Evaluation of the Field Correlation Coefficient

This evaluation is a geometrical task. It is advisable to choose a cylindrical coordinate system with the z axis in the direction of the relative velocity $\mathbf{g} = \mathbf{v}_2 - \mathbf{v}_1$. We again designate the radial coordinate by b (see p. 152).

In this coordinate system, the field is expressed by

$$\mathbf{E}[\mathbf{r}_1 - \mathbf{r}_2 + \mathbf{g}\tau] = \frac{e}{\{b^2 + (z - g\tau)^2\}^{3/2}} \begin{pmatrix} b\cos\varphi \\ b\sin\varphi \\ z - g\tau \end{pmatrix} \tag{1.79}$$

The correlation tensor is then given by

$$\mathbf{E}\,|_0)(\mathbf{E}\,|_\tau = e^2\{[b^2 + z^2][b^2 + (z - g\tau)^2]\}^{-3/2} \times \begin{pmatrix} b^2\cos^2\varphi & b^2\sin\varphi\cos\varphi & (z - g\tau)b\cos\varphi \\ b^2\sin\varphi\cos\varphi & b^2\sin^2\varphi & (z - g\tau)b\sin\varphi \\ bz\cos\varphi & bz\sin\varphi & z(z - g\tau) \end{pmatrix} \tag{1.80}$$

and the volume element by

$$d\mathbf{r}_2 = dz\, b\, db\, d\varphi \tag{1.81}$$

Concerning the cutoffs for the integration over the impact parameter b, we refer to the discussion of the range of validity of the Landau–Fokker–Planck procedure yielding $b_{\min} = r_w$ and $b_{\max} = \lambda_D$, respectively. (The rigorous justification of this procedure follows in the next section on the Bogolubov–Lenard–Balescu equation.) The integration with respect to the angle φ in (1.80) trivially cancels all off-diagonal terms. Integration with respect to z renders the zz element A_{zz} zero due to the symmetry relation

$$\int_0^\infty A_{zz}(-z, g\tau)\, d\tau = -\int_0^\infty A_{zz}(z, g\tau)\, d\tau \tag{1.82}$$

The terms A_{xx} and A_{yy} are elementary integrable and produce for the correlation coefficient

$$\langle \mathbf{E})(\mathbf{E}\rangle_{\mathbf{r}_2} = 2\pi e^2 \frac{\ln \Lambda}{g} \begin{pmatrix} 1 & 0 & 0 \\ 0 & 1 & 0 \\ 0 & 0 & 0 \end{pmatrix} \tag{1.83}$$

or

$$\langle \mathbf{E})(\mathbf{E}\rangle_{\mathbf{r}_2} = 2\pi e^2 (\ln \Lambda / g^3)(g\underset{2}{\mathbf{I}} - \mathbf{g})(\mathbf{g}) \tag{1.84}$$

where $\underset{2}{\mathbf{I}}$ is the identity tensor. Introducing this result together with (1.77) into (1.74), we have the final form

$$\left(\frac{\delta f_1}{\delta t}\right)_{\text{coll}} = 2\pi N \left(\frac{e^2}{m}\right)^2 \ln \Lambda \frac{\partial}{\partial \mathbf{v}_1} \cdot \int d\mathbf{v}_2 \frac{g\underset{2}{\mathbf{I}} - \mathbf{g})(\mathbf{g}}{g^3} \times \left\{ f_1(\mathbf{r}_1, \mathbf{v}_2; t) \frac{\partial f_1(\mathbf{r}_1, \mathbf{v}_1; t)}{\partial \mathbf{v}_1} - f_1(\mathbf{r}_1, \mathbf{v}_1; t) \frac{\partial f_1(\mathbf{r}_1, \mathbf{v}_2; t)}{\partial \mathbf{v}_2} \right\} \tag{1.85}$$

of the **Landau equation** (Landau, 1936).

The Landau equation is one form of the kinetic equation for weak interactions without screening. It is of interest to see how the equivalent Fokker–Planck form follows from the BBGKY hierarchy. This can be seen by starting from (1.77) and relating the field correlation coefficient to the first two moments of velocity scattering.

Relation of the Field Correlation to the Velocity Scattering Moments

In the following, we use the first two moments of velocity scattering per unit time, the coefficient of **dynamic friction**

$$\langle \Delta \mathbf{v}_1 \rangle = (1/\tau_c)(e/m) \left\langle \int_0^{\tau_c} \mathbf{E} \,|_{t'}\, dt' \right\rangle \tag{1.86}$$

and the coefficient of **momentum diffusivity**

$$\langle \Delta \mathbf{v}_1)(\Delta \mathbf{v}_1 \rangle = (1/\tau_c)(e/m)^2 \left\langle \int_0^{\tau_c} dt' \int_0^{\tau_c} dt'' \, \mathbf{E} \,|_{t'})(\mathbf{E}_{t''} \right\rangle \tag{1.87}$$

E is the total field acting on particle 1 under consideration; the brackets $\langle\ \rangle$ denote averaging with respect to the phase coordinates of all field particles; τ_c is the mean time of free flight which is proportional to $\tau_p \Lambda / \ln \Lambda$ [see (III.1.7)].

It is convenient to consider the coefficient of momentum diffusivity first. A simple transformation $t'' = t' + \tau$ in (1.87) yields

$$\langle \Delta \mathbf{v}_1)(\Delta \mathbf{v}_1 \rangle = (1/\tau_c)(e/m)^2 \int_0^{\tau_c} dt' \int_{-t'}^{\tau_c - t'} d\tau \, \langle \mathbf{E} |_{t'})(\mathbf{E} |_{t'+\tau} \rangle \tag{1.88}$$

where the integrand is given by the field autocorrelation coefficient. We evaluate this expression using two approximations we have already applied before.

(a). Interactions beyond a certain distance r_c do not contribute, so that there is a maximum interaction time $t_c = r_c/v_c$ with

$$\langle \mathbf{E} |_{t'})(\mathbf{E} |_{t'+\tau} \rangle = 0 \qquad \text{for} \quad \tau > t_c \tag{1.89}$$

(b). The variation of f_1 is slow in the sense already applied above when we used the "no memory-approximation" $f_1(t - \tau) \approx f_1(t)$. More precisely, if T is the characteristic time for the variation of f_1, then we request $T \gg t_c$ so that $\langle \mathbf{E} |_{t'})(\mathbf{E} |_{t'+\tau} \rangle$ is independent of t' for $t' < T$.

It follows that

$$\begin{aligned} \langle \Delta \mathbf{v}_1)(\Delta \mathbf{v}_1 \rangle &= (e/m)^2 (1/\tau_c) \int_0^{\tau_c} dt' \int_{|\tau| < t_c} d\tau \, \langle \mathbf{E} |_0)(\mathbf{E} |_\tau \rangle \\ &= 2(e/m)^2 \int_0^{\infty} d\tau \, \langle \mathbf{E} |_0)(\mathbf{E} |_\tau \rangle \end{aligned} \tag{1.90}$$

In this expression, the field $\mathbf{E}$ is given by the linear superposition of the Coulomb contributions of all field particles. Since Landau's treatment is based on the assumption of uncorrelated identical field particles, the momentum diffusivity (1.90) can be written as N times the average pair contribution in the form

$$\langle \Delta \mathbf{v}_1)(\Delta \mathbf{v}_1 \rangle = 2N(e/m)^2 \int d\mathbf{v}_2 \, \langle \mathbf{E})(\mathbf{E} \rangle_{\mathbf{r}_2} f_1(\mathbf{r}_1, \mathbf{v}_2 ; t) \tag{1.91}$$

where we have used (1.78).

Now we consider the coefficient of dynamic friction

$$\langle \Delta \mathbf{v}_1 \rangle = (e/m)(1/\tau_c) \int_0^{\tau_c} \langle \mathbf{E}[\mathbf{r}_{12}(t')] \rangle \, dt' \tag{1.92}$$

where the integrand may be written as

$$\mathbf{E}[\mathbf{r}_{12}(t')] = \mathbf{E}\left[\mathbf{r}_{12} |_0 + \mathbf{g} |_0 \, t' - 2(e/m) \int_0^{t'} dt'' \int_0^{t''} dt''' \, \mathbf{E}[\mathbf{r}_{12}(t''')]\right] \tag{1.93}$$

Since we are restricted to the case of weak interactions with only small deflections from straight paths, we have

$$\mathbf{E}[\mathbf{r}_{12}(t')] = \mathbf{E}[\mathbf{r}_{12} + \mathbf{g}t'] - \frac{\partial \mathbf{E}}{\partial \mathbf{r}_{12}}\bigg|_{t'} \cdot 2\,\frac{e}{m} \int_0^{t'} dt'' \int_0^{t''} dt''' \,\mathbf{E}\,|_{t'''} \tag{1.94}$$

or integrating by parts

$$\mathbf{E}[\mathbf{r}_{12}(t')] = \mathbf{E}[\mathbf{r}_{12} + \mathbf{g}t'] - \frac{\partial \mathbf{E}}{\partial \mathbf{r}_{12}}\bigg|_{t'} \cdot 2\,\frac{e}{m} \int_0^{t'} dt''\,(t' - t'')\,\mathbf{E}\,|_{t''} \tag{1.95}$$

The first term on the right-hand side (resulting from the straight-path approximation) does not contribute to the average (1.92). Therefore we find

$$\langle \Delta \mathbf{v}_1 \rangle = -\left(\frac{e}{m}\right)^2 \frac{2}{\tau_c} \int_0^{\tau_c} dt' \int_0^{t'} dt''(t' - t'') \left\langle \frac{\partial \mathbf{E}}{\partial \mathbf{r}_{12}}\bigg|_{t'} \cdot \mathbf{E}\,|_{t''} \right\rangle \tag{1.96}$$

With the substitution $\tau = t' - t''$, we find

$$\langle \Delta \mathbf{v}_1 \rangle = -\left(\frac{e}{m}\right)^2 \frac{2}{\tau_c} \int_0^{\tau_c} dt' \int_0^{t'} d\tau\,\tau \left\langle \frac{\partial \mathbf{E}[\mathbf{r}_{12} + \mathbf{g}t']}{\partial \mathbf{r}_{12}} \cdot \mathbf{E}\,|_{t'-\tau} \right\rangle \tag{1.97}$$

or shifting the average about $\tau - t'$

$$\begin{aligned} \langle \Delta \mathbf{v}_1 \rangle &= 2\left(\frac{e}{m}\right)^2 \frac{\partial}{\partial \mathbf{v}_1} \cdot \frac{1}{\tau_c} \int_0^{\tau_c} dt' \int_0^{t'} d\tau\,\langle \mathbf{E}\,|_\tau\,)(\,\mathbf{E}\,|_0 \rangle \\ &= 2\left(\frac{e}{m}\right)^2 \frac{\partial}{\partial \mathbf{v}_1} \cdot \int_0^{\tau_c} \left(1 - \frac{\tau}{t_c}\right) \langle \mathbf{E}\,|_\tau\,)(\,\mathbf{E}\,|_0 \rangle\, d\tau \end{aligned} \tag{1.98}$$

Since $\tau_c/t_c \approx \Lambda/\ln \Lambda \gg 1$ according to our general supposition, we find

$$\begin{aligned} \langle \Delta \mathbf{v}_1 \rangle &= 2\left(\frac{e}{m}\right)^2 \frac{\partial}{\partial \mathbf{v}_1} \cdot \int_0^{\infty} \langle \mathbf{E}\,|_\tau\,)(\,\mathbf{E}\,|_0 \rangle\, d\tau \\ &= 2N\left(\frac{e}{m}\right)^2 \frac{\partial}{\partial \mathbf{v}_1} \cdot \int d\mathbf{v}_2\, \langle \mathbf{E}\,)(\,\mathbf{E} \rangle_{\mathbf{r}_2} f_1(\mathbf{r}_1\,,\,\mathbf{v}_2\,;\,t) \end{aligned} \tag{1.99}$$

and recognize the important relation between the coefficient of dynamic friction and the coefficient of momentum diffusivity

$$\langle \Delta \mathbf{v}_1 \rangle = \frac{\partial}{\partial \mathbf{v}_1} \cdot \langle \Delta \mathbf{v}_1\,)(\,\Delta \mathbf{v}_1 \rangle \tag{1.100}$$

Fokker–Planck Form of the Weak Interaction Equation

With the information contained in (1.91) and (1.100), it is now easy to transform the Landau form of the weak interaction equation into the Fokker–Planck form.

To this end, we start from (1.74) and (1.77) and introduce the result (1.91) arriving at

$$\left(\frac{\delta f_1}{\delta t}\right)_{\text{coll}} = \frac{\partial}{\partial \mathbf{v}_1} \cdot \left\{ \frac{1}{2} \langle \Delta \mathbf{v}_1)(\Delta \mathbf{v}_1 \rangle \cdot \frac{\partial f_1}{\partial \mathbf{v}_1} - N \left(\frac{e}{m}\right)^2 \int d\mathbf{v}_2 \, \langle \mathbf{E})(\mathbf{E} \rangle_{\mathbf{r}_2} f_1(\mathbf{r}_1, \mathbf{v}_1; t) \cdot \frac{\partial f_1(\mathbf{r}_1, \mathbf{v}_2; t)}{\partial \mathbf{v}_2} \right\} \tag{1.101}$$

Integration by parts and use of (1.91) gives

$$\left(\frac{\delta f_1}{\delta t}\right)_{\text{coll}} = \frac{\partial}{\partial \mathbf{v}_1} \cdot \left\{ \frac{1}{2} \langle \Delta \mathbf{v}_1)(\Delta \mathbf{v}_1 \rangle \cdot \frac{\partial f_1}{\partial \mathbf{v}_1} - \frac{1}{2} f_1 \frac{\partial}{\partial \mathbf{v}_1} \cdot \langle \Delta \mathbf{v}_1)(\Delta \mathbf{v}_1 \rangle \right\} \tag{1.102}$$

which yields, together with (1.100),

$$\left(\frac{\delta f_1}{\delta t}\right)_{\text{coll}} = -\frac{\partial}{\partial \mathbf{v}_1} \cdot \{ \langle \Delta \mathbf{v}_1 \rangle f_1 \} + \frac{1}{2} \frac{\partial}{\partial \mathbf{v}_1} \cdot \left\{ \frac{\partial}{\partial \mathbf{v}_1} \cdot [\langle \Delta \mathbf{v}_1)(\Delta \mathbf{v}_1 \rangle f_1] \right\} \tag{1.103}$$

This is the **Fokker–Planck** form of the weak interaction equation. It describes generally the motion of particles undergoing numerous weak deflections and was first deduced for the case of the Brownian motion of large molecules (Planck, 1917; Fokker, 1914).

The derivation given above is not identical with that presented by Fokker and Planck. Following the general trend of our investigation, we considered the problem on the basis of the BBGKY hierarchy for the specific parameter values $\Pi_{\text{co}} = O(\varepsilon)$ and $\Pi_{\text{d}} = O(\varepsilon)$. We used "the two-particle interaction straight-path approximation without memory."

For comparison, we present in the following another derivation (Chandrasekhar, 1943) which is brief and does not require the "two-particle interaction straight path, no-memory approximation." But then—as we will see—it does not contain all the information which we derived above. We start from the basic Chapman–Kolmogorov equation for Markovian processes in the form

$$f(\mathbf{v}; t + \Delta t) = \int f(\mathbf{v} - \Delta \mathbf{v}; t) \, P_{\Delta t}(\mathbf{v} - \Delta \mathbf{v}, \Delta \mathbf{v}) \, d\Delta \mathbf{v} \tag{1.104}$$

Here we identify f with the one-particle distribution function in a homogeneous system without external forces. $P_{\Delta t}(\mathbf{v}, \Delta\mathbf{v})$ designates the probability that a particle with the velocity $\mathbf{v}$ at the time t will experience a velocity change $\Delta\mathbf{v}$ during the time interval Δt. Expanding the integrand of (1.104) in the form

$$f(\mathbf{v} - \Delta\mathbf{v}; t)\, P_{\Delta t}(\mathbf{v} - \Delta\mathbf{v}, \Delta\mathbf{v}) \\ = f(\mathbf{v}; t)\, P_{\Delta t}(\mathbf{v}, \Delta\mathbf{v}) - \Delta\mathbf{v} \cdot \partial/\partial\mathbf{v}\,(f\, P_{\Delta t}) + \tfrac{1}{2}(\Delta\mathbf{v} \cdot \partial/\partial\mathbf{v})^2 (f\, P_{\Delta t}) - + \cdots \tag{1.105}$$

we get from (1.104)

$$\frac{f(\mathbf{v}; t + \Delta t) - f(\mathbf{v}; t)}{\Delta t} \\ = -\frac{\partial}{\partial\mathbf{v}} \cdot [\langle\Delta\mathbf{v}\rangle f] + \frac{1}{2}\frac{\partial}{\partial\mathbf{v}} \cdot \left\{\frac{\partial}{\partial\mathbf{v}} \cdot [\langle\Delta\mathbf{v}\,)(\,\Delta\mathbf{v}\rangle f]\right\} - + \cdots \tag{1.106}$$

where we have used

$$\int \begin{Bmatrix} 1 \\ \Delta\mathbf{v} \\ \Delta\mathbf{v}\,)(\,\Delta\mathbf{v} \end{Bmatrix} P_{\Delta t}\, d\Delta\mathbf{v} = \begin{Bmatrix} 1 \\ \Delta t\langle\Delta\mathbf{v}\rangle \\ \Delta t\langle\Delta\mathbf{v}\,)(\,\Delta\mathbf{v}\rangle \end{Bmatrix} \tag{1.107}$$

In the limit $\Delta t \to 0$, (1.106) is the Fokker–Planck equation (1.103). We have used in this derivation only the two assumptions of Markovian behavior and weak deflections. We did not use the concepts of "straight path," "two-particle interaction," etc.

However, this formally greater generality of the derivation based on the Chapman–Kolmogorov equation was possible only since we did not care to evaluate the scattering coefficients. If we intended such an evaluation, we would need the same simplifications as above. Comparing (1.85) with the Fokker–Planck equation, we can write these scattering coefficients in the form

$$\langle\Delta\mathbf{v}_1\,)(\,\Delta\mathbf{v}_1\rangle = 4\pi N\left(\frac{e^2}{m}\right)^2 \ln\Lambda \int \left(\frac{\mathbf{l}}{g} - \frac{\mathbf{g}\,)(\,\mathbf{g}}{g^3}\right) f_1(\mathbf{r}_1, \mathbf{v}_2; t)\, d\mathbf{v}_2 \\ = 4\pi N\left(\frac{e^2}{m}\right)^2 \ln\Lambda \frac{\partial}{\partial\mathbf{v}_1}\,)(\,\frac{\partial}{\partial\mathbf{v}_1} \int g\, f_1(\mathbf{r}_1, \mathbf{v}_2; t)\, d\mathbf{v}_2 \tag{1.108}$$

and

$$\langle\Delta\mathbf{v}_1\rangle = \frac{\partial}{\partial\mathbf{v}_1} \cdot \langle\Delta\mathbf{v}_1\,)(\,\Delta\mathbf{v}_1\rangle = 8\pi N(e^2/m)^2 \ln\Lambda \frac{\partial}{\partial\mathbf{v}_1} \int [f_1(\mathbf{r}_1, \mathbf{v}_2; t)/g]\, d\mathbf{v}_2 \tag{1.109}$$

The results (1.108) and (1.109) were first derived by Rosenbluth, McDonald, and Judd (1957).

1.4. The Bogolubov–Lenard–Balescu Equation

As in the preceding sections, we start from the specification (1.8) of the kinetic parameters

$$\Pi_{\mathrm{co}} = O(\varepsilon), \qquad \Pi_{\mathrm{d}} \geqslant O(1) \tag{1.110}$$

This means that we have weak coupling as in the case of the Landau–Fokker–Planck equation but cannot restrict our consideration to binary collisions. Consequently, we use the formulation of the previous section for our present investigation up to the point where we omitted the second term on the right-hand side of (1.66) as small of higher order. With our present value $\Pi_{\mathrm{d}} \geqslant O(1)$, this argument is not valid anymore, and we have to take into account the effect of the pair interaction of a third particle with the test particle $(\mathbf{r}_1, \mathbf{v}_1)$ and the field particle $(\mathbf{r}_2, \mathbf{v}_2)$. Our system of equations then reads

$$\left(\frac{\delta f_1}{\delta t}\right)_{\mathrm{coll}} = \frac{N}{m}\frac{\partial}{\partial \mathbf{v}_1}\cdot\int d\mathbf{r}_2 \int d\mathbf{v}_2 \frac{\partial \phi_{12}}{\partial \mathbf{r}_1} g_{12} \tag{1.111}$$

and

$$g_{12}(t) = \int_0^\infty A(\mathbf{r}_1 - \mathbf{v}_1\tau, \mathbf{r}_2 - \mathbf{v}_2\tau, \mathbf{v}_1, \mathbf{v}_2 ; t - \tau)\, d\tau + \int_0^\infty B_1(\mathbf{r}_1 - \mathbf{v}_1\tau, \mathbf{v}_1 ; t - \tau)\, d\tau + \int_0^\infty B_2(\mathbf{r}_2 - \mathbf{v}_2\tau, \mathbf{v}_2 ; t - \tau)\, d\tau \tag{1.112}$$

where A is again given by

$$A(\mathbf{r}_1, \mathbf{r}_2, \mathbf{v}_1, \mathbf{v}_2 ; t) = \frac{1}{m}\frac{\partial \phi_{12}}{\partial \mathbf{r}_1}\cdot\left[\frac{\partial}{\partial \mathbf{v}_1} - \frac{\partial}{\partial \mathbf{v}_2}\right] f_1(\mathbf{r}_1, \mathbf{v}_1 ; t) f_1(\mathbf{r}_2, \mathbf{v}_2 ; t) \tag{1.113}$$

and B_ν is given by

$$B_\nu(\mathbf{r}_\nu, \mathbf{v}_\nu, t) = \frac{N}{m}\int d\mathbf{r}_3 \int d\mathbf{v}_3 \frac{\partial f_1(\mathbf{r}_\nu, \mathbf{v}_\nu ; t)}{\partial \mathbf{v}_\nu}\cdot\frac{\partial \phi_{\nu 3}}{\partial \mathbf{r}_\nu} g_{3(3-\nu)} \tag{1.114}$$

for $\nu = 1,2$.

Equation (1.112) has to be solved for g_{12}. The problem, of course, is that we now have an integral equation describing the modification of the effect of the field-particle–test-particle interaction represented by the function A through a third particle represented by the functions B_1 and B_2.

As in the preceding, we assume

$$t_c = r_c/v_c \ll T, \qquad r_c \ll L \tag{1.115}$$

so that we may use

$$f(t-\tau) \approx f(t), \qquad g_{ij}(\mathbf{r}_i, \mathbf{r}_j, ...) = g_{ij}(\mathbf{r}_i - \mathbf{r}_j, ...) = g_{ij}(\mathbf{r}_{ij}, ...) \tag{1.116}$$

Further, we assume (following Bogolubov, 1946) that in the time scale of f_1 the function g_{12} can be treated quasistationary, so that g_{12} depends on time only functionally through f_1. We will elaborate in detail on this point in the section on the Bogolubov multiple time-scale theory. With this assumption, we write (1.112) in the form

$$\begin{aligned} g_{12} = {} & \mathbf{K}(\mathbf{r}_{12}, \mathbf{v}_{12}) \cdot \left[\frac{\partial}{\partial \mathbf{v}_1} - \frac{\partial}{\partial \mathbf{v}_2}\right] f_1(\mathbf{r}_1, \mathbf{v}_1; t) f_1(\mathbf{r}_2, \mathbf{v}_2; t) \\ & + N \int d\mathbf{v}_3 \int d\boldsymbol{\rho}\, \mathbf{K}(\mathbf{r}_{12} - \boldsymbol{\rho}, \mathbf{v}_{12}) \\ & \cdot \left\{ \frac{\partial f_1(\mathbf{r}_1, \mathbf{v}_1; t)}{\partial \mathbf{v}_1} g_{23}(-\boldsymbol{\rho}, \mathbf{v}_2, \mathbf{v}_3) - \frac{\partial f_1(\mathbf{r}_2, \mathbf{v}_2; t)}{\partial \mathbf{v}_2} g_{13}(\boldsymbol{\rho}, \mathbf{v}_1, \mathbf{v}_3) \right\} \end{aligned} \tag{1.117}$$

with the kernel

$$\mathbf{K}(\mathbf{r}_{ij}, \mathbf{v}_{ij}) = \frac{1}{m} \int_0^\infty \frac{\partial}{\partial \mathbf{r}_i} \phi(\mathbf{r}_{ij} - \mathbf{v}_{ij}\tau)\, d\tau, \qquad \mathbf{v}_{ij} = \mathbf{v}_i - \mathbf{v}_j \tag{1.118}$$

Here we have used the coordinate transformation

$$\mathbf{r}_{23} = -\boldsymbol{\rho} \quad \text{and} \quad \mathbf{r}_{13} = +\boldsymbol{\rho} \tag{1.119}$$

for the integration variable in the first and second terms, B_1 and B_2, respectively.

Spatial Fourier transformation recommends itself for the solution of (1.117). The procedure is straightforward if we apply the convolution theorem and the symmetry relation $\tilde{g}_{12}(-\mathbf{k}) = \tilde{g}_{12}(\mathbf{k})$ which follows from the fact that $g_{12}(\boldsymbol{\rho})$ is a real quantity. The Fourier transform of the kernel is given by

$$\tilde{\mathbf{K}}(\mathbf{k}, \mathbf{v}_{12}) = -(i/m) \int_0^\infty \mathbf{k}\tilde{\phi}(\mathbf{k}) \exp(-i\mathbf{k} \cdot \mathbf{v}_{12}\tau)\, d\tau \tag{1.120}$$

Engaging a convergence factor ε, we find

$$\tilde{\mathbf{K}}(\mathbf{k}, \mathbf{v}_{12}) = -\frac{i}{m}\mathbf{k}\tilde{\phi}(\mathbf{k}) \lim_{\varepsilon\to 0} \int_0^\infty \exp\left[-i(\mathbf{k}\cdot\mathbf{v}_{12} - i\varepsilon)\tau\right] d\tau \qquad (1.121)$$

$$= \frac{i}{m}\frac{\mathbf{k}\tilde{\phi}(\mathbf{k})}{\mathbf{k}\cdot\mathbf{v}_{12} - i0}$$

Introducing this into (1.117) the Fourier transformation results in[†]

$$\tilde{g}_{12}(\mathbf{k}, \mathbf{v}_1, \mathbf{v}_2) = \frac{1}{m}\frac{\mathbf{k}\tilde{\phi}(\mathbf{k})}{\mathbf{k}\cdot\mathbf{v}_{12} - i0} \cdot \Big\{\Big[\frac{\partial}{\partial\mathbf{v}_1} - \frac{\partial}{\partial\mathbf{v}_2}\Big] f_1(\mathbf{r}_1, \mathbf{v}_1; t) f_1(\mathbf{r}_2, \mathbf{v}_2; t)$$

$$+ N(2\pi)^{3/2}\frac{\partial f_1(\mathbf{r}_1, \mathbf{v}_1; t)}{\partial\mathbf{v}_1} \int d\mathbf{v}_3\, \tilde{g}_{23}(-\mathbf{k}, \mathbf{v}_2, \mathbf{v}_3)$$

$$- N(2\pi)^{3/2}\frac{\partial f_1(\mathbf{r}_2, \mathbf{v}_2; t)}{\partial\mathbf{v}_2} \int d\mathbf{v}_3\, \tilde{g}_{13}(\mathbf{k}, \mathbf{v}_1, \mathbf{v}_3)\Big\} \qquad (1.122)$$

On the other hand, the Fourier representation of (1.111) yields, using Parseval's theorem,

$$\left(\frac{\delta f_1}{\delta t}\right)_{\text{coll}} = -i\frac{N}{m}\frac{\partial}{\partial\mathbf{v}_1} \cdot \int \mathbf{k}\tilde{\phi}(\mathbf{k})\tilde{g}(\mathbf{k}, \mathbf{v}_1)\, d\mathbf{k} \qquad (1.123)$$

with

$$\tilde{g}(\mathbf{k}, \mathbf{v}_1) := \int \tilde{g}_{12}(\mathbf{k}, \mathbf{v}_1, \mathbf{v}_2)\, d\mathbf{v}_2 \qquad (1.124)$$

Observe that we here, as in the preceding, do not distinguish the functions by their symbols but by their arguments.

Since the real function $\phi(\mathbf{r}_{12})$ depends only on $r_{12} = |\mathbf{r}_{12}|$, $\tilde{\phi}(\mathbf{k})$ is real. For $\tilde{g}(\mathbf{k}, \mathbf{v}_1)$, we have the obvious symmetry relation

$$\tilde{g}(-\mathbf{k}, \mathbf{v}_1) = \tilde{g}^*(\mathbf{k}, \mathbf{v}_1) \qquad (1.125)$$

From this, we conclude that only the imaginary part of $\tilde{g}(\mathbf{k}, \mathbf{v}_1)$ contributes to our kinetic equation

$$\left(\frac{\delta f_1}{\delta t}\right)_{\text{coll}} = \frac{N}{m}\frac{\partial}{\partial\mathbf{v}_1} \cdot \int \mathbf{k}\tilde{\phi}(\mathbf{k})\, \mathrm{Im}[\tilde{g}(\mathbf{k}, \mathbf{v}_1)]\, d\mathbf{k} \qquad (1.126)$$

[†] Here it is essential to note that due to the basic assumption introduced on p. 157 following (1.65) and (1.66) the space variation of the distribution function f_1 may be neglected in comparison to the space variation of the pair correlation function.

To find this quantity $\mathrm{Im}[\tilde{g}(\mathbf{k}, \mathbf{v}_1)]$, we integrate (1.122) with respect to $\mathbf{v}_2$ and find

$$\tilde{g}(\mathbf{k}, \mathbf{v}_1) = \frac{1}{m} \int \frac{\mathbf{k}\phi(\mathbf{k})}{\mathbf{k} \cdot \mathbf{v}_{12} - i0} \cdot \Big\{ f_1(\mathbf{r}_2, \mathbf{v}_2) \frac{\partial f_1(\mathbf{r}_1, \mathbf{v}_1)}{\partial \mathbf{v}_1} - f_1(\mathbf{r}_1, \mathbf{v}_1) \frac{\partial f_1(\mathbf{r}_2, \mathbf{v}_2)}{\partial \mathbf{v}_2}$$
$$+ (2\pi)^{3/2} N \Big[\frac{\partial f_1(\mathbf{r}_1, \mathbf{v}_1)}{\partial \mathbf{v}_1} \tilde{g}^*(\mathbf{k}, \mathbf{v}_2) - \frac{\partial f_1(\mathbf{r}_2, \mathbf{v}_2)}{\partial \mathbf{v}_2} \tilde{g}(\mathbf{k}, \mathbf{v}_1) \Big] \Big\} \, d\mathbf{v}_2 \tag{1.127}$$

respectively

$$\Big[1 + \frac{1}{m} \tilde{\phi}(\mathbf{k}) \, \chi(\mathbf{k}, \mathbf{v}_1) \Big] \tilde{g}(\mathbf{k}, \mathbf{v}_1)$$
$$= \frac{1}{m} \int \frac{\mathbf{k}\phi(\mathbf{k})}{\mathbf{k} \cdot \mathbf{v}_{12} - i0} \cdot \Big\{ f_1(\mathbf{r}_2, \mathbf{v}_2) \frac{\partial f_1(\mathbf{r}_1, \mathbf{v}_1)}{\partial \mathbf{v}_1}$$
$$- f_1(\mathbf{r}_1, \mathbf{v}_1) \frac{\partial f_1(\mathbf{r}_2, \mathbf{v}_2)}{\partial \mathbf{v}_2} + (2\pi)^{3/2} N \frac{\partial f_1(\mathbf{r}_1, \mathbf{v}_1)}{\partial \mathbf{v}_1} \tilde{g}^*(\mathbf{k}, \mathbf{v}_2) \Big\} \, d\mathbf{v}_2 \tag{1.128}$$

with the abbreviation

$$\chi(\mathbf{k}, \mathbf{v}_1) = (2\pi)^{3/2} N \int \frac{\mathbf{k} \cdot [\partial f_1(\mathbf{r}_2, \mathbf{v}_2)/\partial \mathbf{v}_2]}{\mathbf{k} \cdot \mathbf{v}_{12} - i0} \, d\mathbf{v}_2 \tag{1.129}$$

We want to eliminate the last term on the right-hand side of (1.128). Therefore we integrate this equation with respect to the velocity components ($\mathbf{v}_{1\perp}$) perpendicular to $\mathbf{k}$ and find[†]

$$\Big[1 + \frac{1}{m} \tilde{\phi}(\mathbf{k}) \, \chi(\mathbf{k}, u_1) \Big] \tilde{g}(\mathbf{k}, u_1)$$
$$= \frac{1}{m} \int \frac{k\tilde{\phi}(\mathbf{k})}{k u_{12} - i0} \Big\{ f_1(\mathbf{r}_2, \mathbf{v}_2) \frac{\partial f(\mathbf{r}_1, \mathbf{u}_1)}{\partial u_1}$$
$$- f(\mathbf{r}_1, u_1) \frac{\partial f_1(\mathbf{r}_2, \mathbf{v}_2)}{\partial u_2} + (2\pi)^{3/2} N \frac{\partial f(\mathbf{r}_1, \mathbf{u}_1)}{\partial u_1} \tilde{g}^*(\mathbf{k}, \mathbf{v}_2) \Big\} \, d\mathbf{v}_2 \tag{1.130}$$

Here we have introduced the following abbreviations:

$$u_1 = (\mathbf{v}_1 \cdot \mathbf{k})/k, \qquad u_2 = (\mathbf{v}_2 \cdot \mathbf{k})/k$$
$$f(\mathbf{r}_1, u_1) = \int f_1(\mathbf{r}_1, \mathbf{v}_1) \, d\mathbf{v}_{1\perp}, \qquad \tilde{g}(\mathbf{k}, u_1) = \int \tilde{g}(\mathbf{k}, \mathbf{v}_1) \, d\mathbf{v}_{1\perp} \tag{1.131}$$
$$u_{12} = u_1 - u_2$$

[†] Note that χ depends only on the velocity component u_1.

Multiplication of (1.128) by $\partial f(\mathbf{r}_1, u_1)/\partial u_1$, and of (1.130) by $\partial f_1(\mathbf{r}_1, \mathbf{v}_1)/\partial u_1$, and subtraction results in

$$\left[1 + \frac{1}{m}\tilde{\phi}(\mathbf{k})\,\chi(\mathbf{k}, u_1)\right]\left(\tilde{g}(\mathbf{k}, \mathbf{v}_1)\frac{\partial f}{\partial u_1} - \tilde{g}(\mathbf{k}, u_1)\frac{\partial f_1}{\partial u_1}\right)$$

$$= \frac{(2\pi)^{-3/2}}{mN}\chi(\mathbf{k}, u_1)\tilde{\phi}(\mathbf{k})\left[f(\mathbf{r}_1, u_1)\frac{\partial f_1}{\partial u_1} - f_1(\mathbf{r}_1, \mathbf{v}_1)\frac{\partial f}{\partial u_1}\right] \quad (1.132)$$

Taking the imaginary part of this equation, we have the desired quantity $\mathrm{Im}[\tilde{g}(\mathbf{k}, \mathbf{v}_1)]$ if we use $\mathrm{Im}[\tilde{g}(k, u_1)] = 0$. For a proof of this fact—which is not difficult but lengthy—we refer to the appendix of the original paper by Lenard (1960).

Taking the imaginary part of (1.132), we get

$$\frac{\partial f}{\partial u_1}\mathrm{Im}[\tilde{g}(\mathbf{k}, \mathbf{v}_1)] = \frac{(2\pi)^{-3/2}}{mN}\left[f(\mathbf{r}_1, u_1)\frac{\partial f_1}{\partial u_1} - f_1(\mathbf{r}_1, \mathbf{v}_1)\frac{\partial f}{\partial u_1}\right]\frac{\tilde{\phi}(\mathbf{k})\,\mathrm{Im}\,\chi}{|\,1 + (1/m)\tilde{\phi}\chi\,|^2} \quad (1.133)$$

We read from Plemelj's formula for χ

$$\chi(\mathbf{k}, u_1) = (2\pi)^{3/2}N\,\mathcal{P}\int\frac{\partial f(\mathbf{r}_2, u_2)/\partial u_2}{u_1 - u_2}\,du_2 - i\pi(2\pi)^{3/2}N\left.\frac{\partial f(\mathbf{r}_2, u_2)}{\partial u_2}\right|_{u_2=u_1}$$

$$\mathrm{Im}[\chi(\mathbf{k}, u_1)] = -(2\pi)^{3/2}\pi N\left.\frac{\partial f(\mathbf{r}_2, u_2)}{\partial u_2}\right|_{u_2=u_1} \quad (1.134)$$

$$= -\pi N\frac{\partial f(\mathbf{r}_2, u_1)}{\partial u_1}(2\pi)^{3/2}$$

Introducing (1.133) and (1.134) into (1.126), we find

$$\left(\frac{\delta f_1}{\delta t}\right)_{\mathrm{coll}} = -\frac{\pi N}{m^2}\frac{\partial}{\partial \mathbf{v}_1}\cdot\int\mathbf{k}\frac{[\tilde{\phi}(k)]^2}{|\,1 + (1/m)\tilde{\phi}\chi\,|2}$$

$$\times\left[f(\mathbf{r}_1, u_1)\frac{\partial f_1}{\partial u_1} - f_1(\mathbf{r}_1, \mathbf{v}_1)\frac{\partial f}{\partial u_1}\right]d\mathbf{k} \quad (1.135)$$

or

$$\left(\frac{\delta f_1}{\delta t}\right)_{\mathrm{coll}} = N\frac{\partial}{\partial \mathbf{v}_1}\cdot\int\underset{2}{\mathbf{Q}}(\mathbf{v}_1, \mathbf{v}_2)\cdot\left[\frac{\partial}{\partial \mathbf{v}_1} - \frac{\partial}{\partial \mathbf{v}_2}\right]f_1(\mathbf{r}_1, \mathbf{v}_1)f_1(\mathbf{r}_2, \mathbf{v}_2)\,d\mathbf{v}_2 \quad (1.136)$$

with the second-rank tensor $\underset{2}{\mathbf{Q}}$

$$\underset{2}{\mathbf{Q}}(\mathbf{v}_1, \mathbf{v}_2) = -\frac{\pi}{m^2}\int d\mathbf{k}\,\frac{[\tilde{\phi}(\mathbf{k})]^2\,\mathbf{k})(\mathbf{k}}{|\,1 + (1/m)\,\tilde{\phi}\chi\,|^2}\,\delta[\mathbf{k}\cdot(\mathbf{v}_1 - \mathbf{v}_2)] \tag{1.137}$$

This is the **Lenard–Balescu equation.**

In the preceding derivation, we essentially followed the approach of Lenard on the basis of Bogolubov's equations. Simultaneously with Lenard's investigation, a calculation by Balescu (1960) produced the same result. Balescu, however, did not start from Bogolubov's equations but derived his result on the basis of diagram techniques. At least this is true up to (1.17). From there on, his evaluation is similar to the method given by Lenard.

Landau Equation with Screening

We remember that due to the logarithmic divergence of the tensor $\underset{2}{\mathbf{Q}}$ for large impact parameters Landau had to introduce an artificial cutoff. The derivation of the kinetic equation by Lenard and Balescu accounts for interactions with a third particle and therewith includes screening effects. Consequently, we may hope to recover the Landau equation from (1.136) and (1.137) without introducing the maximum value cutoff. To prove this, we choose a cylindrical coordinate system with the z axis in the direction of the relative velocity $\mathbf{g} = \mathbf{v}_2 - \mathbf{v}_1$ so that we have

$$Q_{ij} = -\frac{\pi}{m^2}\int dk_\perp\, d\varphi\, dk_z\, k_\perp \delta(k_z g)\,\frac{[\tilde{\phi}(\mathbf{k})]^2\, k_i k_j}{|\,1 + (1/m)\,\tilde{\phi}\chi\,|^2} \tag{1.138}$$

where $\mathbf{k}_\perp$ is the projection of the vector $\mathbf{k}$ into the plane perpendicular to $\mathbf{g}$. φ specifies its orientation in this plane. Integration with respect to k_z yields

$$Q_{ij} = \begin{cases} -\dfrac{\pi}{gm^2}\displaystyle\int dk_\perp\, d\varphi\, k_\perp \dfrac{[\tilde{\phi}(\mathbf{k}_\perp)]^2\, k_i k_j}{|\,1 + (1/m)\,\tilde{\phi}(\mathbf{k}_\perp)\,\chi(\varphi, \mathbf{v}_{1\perp})|^2} & i,j \neq z \\ 0 \qquad \text{if} \quad (i = z) \vee (j = z) \end{cases} \tag{1.139}$$

Now we use the decomposition

$$\begin{aligned} &\frac{1}{|\,1 + (1/m)\,\tilde{\phi}(\mathbf{k}_\perp)\,\chi(\varphi, \mathbf{v}_{1\perp})|^2} \\ &\quad = \frac{i}{2\,\mathrm{Im}[(1/m)\,\tilde{\phi}(\mathbf{k})\,\chi(\varphi, \mathbf{v}_{1\perp})]} \\ &\qquad \times \left[\frac{1}{1 + (1/m)\,\tilde{\phi}(\mathbf{k}_\perp)\,\chi(\varphi, \mathbf{v}_{1\perp})} - \frac{1}{1 + (1/m)\,\tilde{\phi}(\mathbf{k}_\perp)\,\chi^*(\varphi, \mathbf{v}_{1\perp})}\right] \end{aligned} \tag{1.140}$$

and the Coulomb spectral function

$$\tilde{\phi}(\mathbf{k}_\perp) = \tilde{\phi}(k_\perp) = (2/\pi)^{1/2} (e^2/k_\perp{}^2) \tag{1.141}$$

and find through elementary integration

$$Q_{ij} = -\frac{e^4}{m^2 g} \int d\varphi\, \hat{k}_{\perp i} \hat{k}_{\perp j} \frac{\operatorname{Im}[\chi(\varphi, \mathbf{v}_{1\perp}) \ln (1 + (\pi/2)^{1/2} (\Theta^2 m/\chi e^6))]}{\operatorname{Im}[\chi(\varphi, \mathbf{v}_{1\perp})]} \tag{1.142}$$

Actually the integral (1.139) diverges logarithmically for $k_\perp \to \infty$, a problem which we avoided by introducing the cutoff

$$k_{\perp \max} = 1/r_w = \Theta/e^2 \tag{1.143}$$

This divergence was to be expected, since it has its origin in the range of small impact parameters not accessible to a weak interaction theory and not removable by the "third-particle screening" taken into account in the Bogolubov–Lenard–Balescu equation.

On the other hand, the divergence for small values of $k_\perp$ which corresponds to Landau's large range divergence has indeed been removed by the Bogolubov–Lenard–Balescu approach. This can be seen from the fact that the denominator in (1.139) increases rapidly with small values of k in such a fashion that it overcomes the increase of the numerator which would cause the usual Landau divergence.

Under the logarithm of (1.142), it seems sufficient to use an approximation for the function χ. It follows from (1.129) that we have

$$\chi(\varphi, \mathbf{v}_{1\perp}) = O(mn/\Theta) \tag{1.144}$$

so that we find

$$Q_{ij} = \begin{cases} -(e^4\pi/m^2 g) \ln(1 + \Theta^3/ne^6), & i, j \neq z, \quad i = j \\ 0, & \text{otherwise} \end{cases} \tag{1.145}$$

Remembering that

$$\Theta^3/ne^6 = O(\Lambda^2) \gg 1 \tag{1.146}$$

and moving back to an arbitrary coordinate system, we find the general representation

$$Q_{ij} = -2\pi e^4/m^2 \ln \Lambda (g^2 \delta_{ij} - g_i g_j)/g^3 \tag{1.147}$$

Comparison with (1.85) demonstrates that, in fact, the Bogolubov–Lenard–Balescu equation removes the long-range divergence of the Landau equation without making an "ad hoc cutoff."

The relative error comitted above by the approximation of χ in the logarithm is of the order of the ratio

$$\left| \frac{\ln \chi - \ln(r_0/r_w)^3}{\ln(r_0/r_w)^3} \right| = O\left(\left| \frac{\ln(\Theta/mv^2)}{\ln(r_0/r_w)} \right| \right) \tag{1.148}$$

This becomes substantial only for values $v \gg \langle v \rangle$, that is, in the region where $f_1(v)$ will be very small and therefore does not contribute much to $(\delta f_1/\delta t)_{\text{coll}}$.

Physical Interpretation of the Bogolubov–Lenard–Balescu Equation

From the relations (1.85), (1.136), and (1.137), it is clear that the Landau–Fokker–Planck and the Bogolubov–Lenard–Balescu equations are distinguished through their scattering coefficients Q_{ij} only, which read, respectively,

$$\underset{2}{\mathbf{Q}}_{\mathbf{LFP}} = -(e^2/m^2) \int d\mathbf{r}_2 \int_0^\infty d\tau\, \mathbf{E}[\mathbf{r}_1 - \mathbf{r}_2]\,)(\,\mathbf{E}[\mathbf{r}_1 - \mathbf{r}_2 - \mathbf{v}_1\tau + \mathbf{v}_2\tau] \tag{1.149}$$

and

$$\underset{2}{\mathbf{Q}}_{\mathbf{BLB}} = -\frac{\pi}{m^2} \int d\mathbf{k}\, \frac{[\tilde{\phi}(k)]^2\, \mathbf{k}\,)(\,\mathbf{k}}{|\,1 + (1/m)\tilde{\phi}\chi\,|^2}\, \delta(\mathbf{k} \cdot \mathbf{v}_{12}) \tag{1.150}$$

If the fields $\mathbf{E}$ in (1.149) are given as the gradient of a general potential ψ, then we have

$$e\mathbf{E}(\mathbf{r}) = -[i/(2\pi)^{3/2}] \int \mathbf{k} e^{i\mathbf{k}\cdot\mathbf{r}} \tilde{\psi}(\mathbf{k})\, d\mathbf{k} \tag{1.151}$$

Equation (1.149) can be written in the form

$$\underset{2}{\mathbf{Q}}_{\mathbf{LFP}} = (1/m^2) \int_0^\infty d\tau \int d\mathbf{k}\, \mathbf{k}\,)(\,\mathbf{k} \exp(-i\mathbf{k} \cdot \mathbf{v}_{12}\tau)\, |\tilde{\psi}(\mathbf{k})|^2 \tag{1.152}$$

or

$$\underset{2}{\mathbf{Q}}_{\mathbf{LFP}} = (\pi/m^2) \int d\mathbf{k}\, \mathbf{k}\,)(\,\mathbf{k}\, \delta(\mathbf{k} \cdot \mathbf{v}_{12})\, |\tilde{\psi}(\mathbf{k})|^2 \tag{1.153}$$

Comparison of (1.153) with (1.150) shows that the Bogolubov–Lenard–Balescu equation may be interpreted as the "unscreened" Landau–Fokker–Planck equation for field particles acting upon the test particle with an effective potential

$$\tilde{\psi}(\mathbf{k}) = \tilde{\phi}(\mathbf{k})/[1 + (1/m)\tilde{\phi}\chi] \tag{1.154}$$

which, with the Fourier transform of the Coulomb potential,

$$\tilde{\phi}(\mathbf{k}) = (2/\pi)^{1/2} e^2/k^2 \tag{1.155}$$

reads

$$\tilde{\psi}(k) = \left(\frac{2}{\pi}\right)^{1/2} \frac{e^2}{k^2 + (2/\pi)^{1/2} e^2\chi(\mathbf{k}, \mathbf{v}_1)/m} \tag{1.156}$$

To provide a crude interpretation of this effective potential (1.156) we consider the forces acting on the test particle as the superposition of binary interactions accounting for the screening of the Coulomb interaction by the average dielectric reaction of the plasma on the field particle. To this end we refer to (III.3.12), which relates the time–space Fourier transform of the effective potential in a plasma to the corresponding spectral functions of the charge density $\tilde{n}$ and the dielectric constant† ε. This relation is

$$\tilde{\tilde{\psi}}(\mathbf{k}, \omega) = 4\pi e^2\tilde{\tilde{n}}(\mathbf{k}, \omega)/k^2\varepsilon(\mathbf{k}, \omega) \tag{1.157}$$

where ε in the quasistationary approximation according to (III.3.10) field is given by

$$\varepsilon(\mathbf{k}, \omega) = 1 - \frac{\omega_{\mathrm{p}}^2}{k^2} \int_{-\infty}^{+\infty} \frac{\mathbf{k} \cdot (\partial f/\partial \mathbf{v})}{\mathbf{k} \cdot \mathbf{v} - \omega + i0}\, d\mathbf{v} \tag{1.158}$$

In the straight-path approximation, the charge distribution of the field electron is given by

$$n(\mathbf{r}, t) = \delta(\mathbf{r} - \mathbf{r}_2 - \mathbf{v}_2 t) \tag{1.159}$$

which yields, for the corresponding spectral function,

$$\tilde{\tilde{n}}(\mathbf{k}, \omega) = (1/2\pi)\, \delta(\omega - \mathbf{k} \cdot \mathbf{v}_2) \tag{1.160}$$

so that the plasma potential distribution caused by this field particle can be written in the form

$$\tilde{\tilde{\psi}}(\mathbf{k}, \omega) = 2e^2\delta(\omega - \mathbf{k} \cdot \mathbf{v}_2)/k^2\varepsilon(\mathbf{k}, \omega) \tag{1.161}$$

Obviously the index 2 designates the field particle. For reasons of convenience, we have choosen the origin of our coordinate system in the initial position of the field particle ($\mathbf{r}_{10}$).

† In this respect, see p. 116.

Inverting the time transformation yields

$$\tilde{\psi}(\mathbf{k}, t) = \left(\frac{2}{\pi}\right)^{1/2} \frac{e^2}{k^2} e^{-i\mathbf{k}\cdot\mathbf{v}_2 t} \frac{1}{\varepsilon(\mathbf{k}, \mathbf{k}\cdot\mathbf{v}_1)} \tag{1.162}$$

Consequently, the potential of the field particle including screening by the plasma as observed at the moving test particle is given by

$$\tilde{\tilde{\psi}}(\mathbf{k}, t) = (2/\pi)^{1/2} (e^2/k^2)[1/\varepsilon(\mathbf{k}, \mathbf{k}\cdot\mathbf{v}_2)]\ e^{i\mathbf{k}\cdot\mathbf{v}_{12} t} \tag{1.163}$$

where the additional factor $\exp(i\mathbf{k}\cdot\mathbf{v}_1 t)$ accounts for the fact that we consider the time variation of the plasma potential at the point of the moving particle. Note that due to the Dirac function $\delta(\mathbf{k}\cdot\mathbf{v}_{12})$ in (1.153) we are allowed to omit—as far as application in (1.153) is concerned—the exponential factor in (1.163) and to replace $\mathbf{v}_1$ by $\mathbf{v}_2$ in the dielectric constant $\varepsilon(\mathbf{k}, \mathbf{k}\cdot\mathbf{v}_2)$. That means we may write

$$\tilde{\tilde{\psi}}(\mathbf{k}, t) = \left(\frac{2}{\pi}\right)^{1/2} \frac{e^2}{k^2} \frac{1}{\varepsilon(\mathbf{k}, \mathbf{k}\cdot\mathbf{v}_1)} \tag{1.164}$$

Introducing (1.158), we arrive at

$$\tilde{\tilde{\psi}}(\mathbf{k}, t) = \left(\frac{2}{\pi}\right)^{1/2} e^2 \Big/ \left[k^2 + \omega_p{}^2 \int \frac{\mathbf{k}\cdot\partial f/\partial\mathbf{v}}{\mathbf{k}\cdot(\mathbf{v}_1 - \mathbf{v}) - i0}\, d\mathbf{v}\right] \tag{1.165}$$

which is identical with (1.156).

For a more detailed and sophisticated discussion of the coefficients of the "Screened Fokker–Planck–Equation" we refer the reader to the work of Thompson and Hubbard (1960, 1961).

Limitation of the Bogolubov–Lenard–Balescu Equation

It is obvious from (1.137) and (1.154) that the Bogolubov–Lenard–Balescu equation can be valid only if $\tilde{\tilde{\psi}}(\mathbf{k}, t)$ exists. This is identical with the requirement that $\varepsilon(\mathbf{k}, \mathbf{k}\cdot\mathbf{v}_1)$ does not vanish. This means that the image curve

$$G(\hat{\mathbf{k}}, \mathbf{v}_1) = \omega_p{}^2 \int \frac{\hat{\mathbf{k}}\cdot\partial f/\partial\mathbf{v}}{\hat{\mathbf{k}}\cdot\mathbf{v} - \hat{\mathbf{k}}\cdot\mathbf{v}_1 + i0}\, d\mathbf{v} \tag{1.166}$$

of the real $\hat{\mathbf{k}}\cdot\mathbf{v}_1$ axis does not cross the real positive axis in the G-plane. The conditions for this crossing were given by Penrose and discussed in detail in Chapter III, Section 2.

We conclude that the stability of the system in the sense of Penrose is a necessary condition for the validity of the Bogolubov–Lenard–Balescu equation. It is essential to note that this must hold for all times and that the stability of f_1 is not necessarily conserved by the Bogolubov–Lenard–Balescu equation. If, however, $f_1(\mathbf{v})$ is isotropic, then it remains isotropic and is always single humped and therefore stable.

2. Macroscopic Equations

2.1. Moment Equations from the Boltzmann Equation: The Grad Method

In the preceding section, we derived the Boltzmann equation from the BBGKY hierarchy. Generalized to systems with several components including external forces, the Boltzmann equations for the general distribution functions $f_\mu^{(1)} = f_\mu$ read

$$\frac{\partial f_\mu}{\partial t} + \mathbf{v}_\mu \cdot \frac{\partial f_\mu}{\partial \mathbf{r}} + \frac{\partial}{\partial \mathbf{v}_\mu} \cdot \dot{\mathbf{v}}_\mu f_\mu = \sum_\nu \int d\mathbf{v}_\nu \, {}^{\mu\nu}g \int dQ_{\mu\nu}(f_\mu' f_\nu' - f_\mu f_\nu) \tag{2.1}$$

with

$$\begin{aligned} {}^{\mu\nu}g &= |\mathbf{v}_\mu - \mathbf{v}_\nu| \qquad \text{for} \quad \nu \neq \mu \\ {}^{\mu\mu}g &= |\mathbf{v}_\mu - \mathbf{v}_{2\mu}| \qquad \text{for} \quad \nu = \mu \end{aligned} \tag{2.2}$$

where $dQ_{\mu\nu}$ designates the differential scattering cross section $b\, db\, d\varphi$. Note that we distinguish the velocity coordinates but not the configuration space coordinates of the various components. This is possible only because Boltzmann's collision concept limits interactions to distances where field and test particle are practically in the same position.

To derive the **Maxwell–Boltzmann transport equations** from the Boltzmann equation, let us consider any function $\Psi(\mathbf{r}, \mathbf{v}_\mu; t)$ and define the corresponding macroscopic transport quantity by

$$\langle \Psi \rangle_{\mathbf{v}_\mu}(\mathbf{r}, t) = [1/n_\mu(\mathbf{r}, t)] \int d\mathbf{v}_\mu \, \Psi(\mathbf{r}, \mathbf{v}_\mu; t) f_\mu(\mathbf{r}, \mathbf{v}_\mu; t) \tag{2.3}$$

with the particle number density

$$n_\mu(\mathbf{r}, t) = \int d\mathbf{v}_\mu f_\mu(\mathbf{r}, \mathbf{v}_\mu; t) \tag{2.4}$$

We multiply (2.1) by $\Psi(\mathbf{r}, \mathbf{v}_\mu; t)$ and integrate over the velocity space $\mathbf{v}_\mu$. Straightforward partial integrations of the left-hand side under the sensible expectation that the quantity Ψf_μ vanishes sufficiently for $v_\mu \to \infty$ yield for the convection terms

$$\frac{\partial}{\partial t} n_\mu \langle \Psi \rangle_{\mathbf{v}_\mu} + \frac{\partial}{\partial \mathbf{r}} \cdot n_\mu \langle \mathbf{v}_\mu \Psi \rangle_{\mathbf{v}_\mu} - n_\mu \left\{ \left\langle \frac{\partial}{\partial t} \Psi \right\rangle_{\mathbf{v}_\mu} + \left\langle \mathbf{v}_\mu \cdot \frac{\partial}{\partial \mathbf{r}} \Psi \right\rangle_{\mathbf{v}_\mu} + \left\langle \dot{\mathbf{v}}_\mu \cdot \frac{\partial}{\partial \mathbf{v}_\mu} \Psi \right\rangle_{\mathbf{v}_\mu} \right\} \tag{2.5}$$

Exchanging primed and unprimed variables in the first term of the collision integral applying Liouville's law and using the invariance of ${}^{\mu\nu}g$ and $dQ_{\mu\nu}$ with respect to this procedure we find for this first term

$$\begin{aligned} \int d\mathbf{v}_\mu \int d\mathbf{v}_\nu \, {}^{\mu\nu}g \int dQ_{\mu\nu} \Psi(\mathbf{v}_\mu) f_\mu(\mathbf{v}_\mu') f_\nu(\mathbf{v}_\nu') &= \int d\mathbf{v}_\mu' \int d\mathbf{v}_\nu' \, {}^{\mu\nu}g \int dQ_{\mu\nu} \Psi(\mathbf{v}_\mu') f_\mu(\mathbf{v}_\mu) f_\nu(\mathbf{v}_\nu) \\ &= \int d\mathbf{v}_\mu \int d\mathbf{v}_\nu \, {}^{\mu\nu}g \int dQ_{\mu\nu} \Psi(\mathbf{v}_\mu') f_\mu(\mathbf{v}_\mu) f_\nu(\mathbf{v}_\nu) \end{aligned} \tag{2.6}$$

which then results in

$$\begin{aligned} &\sum_\nu \int d\mathbf{v}_\mu \int d\mathbf{v}_\nu \, {}^{\mu\nu}g \int dQ_{\mu\nu} \, \Psi(\mathbf{v}_\mu)(f_\mu' f_\nu' - f_\mu f_\nu) \\ &\quad = \sum_\nu \int d\mathbf{v}_\mu \int d\mathbf{v}_\nu \, {}^{\mu\nu}g \int dQ_{\mu\nu} (\Psi(\mathbf{v}_\mu') - \Psi(\mathbf{v}_\mu)) f_\mu f_\nu \\ &\quad = n_\mu \sum_\nu n_\nu \left\langle \left\langle {}^{\mu\nu}g \int dQ_{\mu\nu} (\Psi' - \Psi) \right\rangle_{\mathbf{v}_\mu} \right\rangle_{\mathbf{v}_\nu} \end{aligned} \tag{2.7}$$

for the collision contribution.

Relations (2.5) and (2.7) together present the **Maxwell–Boltzmann transport equation**

$$\begin{aligned} &\frac{\partial}{\partial t} n_\mu \langle \Psi \rangle_{\mathbf{v}_\mu} + \frac{\partial}{\partial \mathbf{r}} \cdot n_\mu \langle \mathbf{v}_\mu \Psi \rangle_{\mathbf{v}_\mu} \\ &- n_\mu \left\{ \left\langle \frac{\partial}{\partial t} \Psi \right\rangle_{\mathbf{v}_\mu} + \left\langle \mathbf{v}_\mu \cdot \frac{\partial}{\partial \mathbf{r}} \Psi \right\rangle_{\mathbf{v}_\mu} + \left\langle \mathbf{v}_\mu \cdot \frac{\partial}{\partial \mathbf{v}_\mu} \Psi \right\rangle_{\mathbf{v}_\mu} \right\} \\ &= n_\mu \sum_\nu n_\nu \left\langle \left\langle {}^{\mu\nu}g \int dQ_{\mu\nu} (\Psi' - \Psi) \right\rangle_{\mathbf{v}_\mu} \right\rangle_{\mathbf{v}_\nu} \end{aligned} \tag{2.8}$$

Macroscopic transport quantities which are open to an immediate physical interpretation result for the following velocity moments:

The **zero-order moment** $\Psi = m_\mu$ yields

$$\langle \Psi \rangle_{\mathbf{v}_\mu} = m_\mu \tag{2.9}$$

which provides the transport equation for the **mass densities**

$$\rho_\mu = n_\mu m_\mu\,, \qquad \rho = \sum_\nu \rho_\nu \tag{2.10}$$

ρ being defined as the total mass density.

The **first-order moment** $\mathbf{\Psi} = m_\mu \mathbf{g}_\mu$ with

$$\mathbf{g}_\mu = \mathbf{v}_\mu - \mathbf{u} = \mathbf{v}_\mu - (1/\rho) \sum_\nu \rho_\nu \langle \mathbf{v}_\nu \rangle_{\mathbf{v}_\nu} \tag{2.11}$$

yields the transport equation for the **relative momentum** $\rho_\mu \langle \mathbf{g}_\mu \rangle_{\mathbf{v}_\mu}$.

The **second-order moment** $\underset{2}{\mathbf{\Psi}} = m_\mu \mathbf{g}_\mu \mathbf{g}_\mu$ yields the transport equation for the **pressure tensor**

$$\underset{2}{\mathbf{p}}_\mu = \rho_\mu \langle \mathbf{g}_\mu \mathbf{g}_\mu \rangle_{\mathbf{v}_\mu} \tag{2.12}$$

which is related to the **hydrostatic partial pressure** p_μ and the **temperature** Θ_μ by

$$p_\mu = \tfrac{1}{3}\,\mathrm{tr}(\underset{2}{\mathbf{p}}_\mu) = n_\mu \Theta_\mu \tag{2.13}$$

and the **stress tensor** by

$$\underset{2}{\mathring{\mathbf{p}}}_\mu = \underset{2}{\mathbf{p}}_\mu - p_\mu \underset{2}{\mathbf{I}} \tag{2.14}$$

The trace-free tensor $((\mathbf{a})^n)^\circ$ for $n = 2, 3$ is defined by

$$\begin{aligned} (\mathbf{aa})^\circ &= ((\mathbf{a})^2)^\circ = \mathbf{aa} - \tfrac{1}{3}\,\mathrm{tr}(\mathbf{aa})\underset{2}{\mathbf{I}} \\ (\mathbf{aaa})^\circ &= ((\mathbf{a})^3)^\circ = \mathbf{aaa} - \tfrac{3}{5}\,\mathrm{sym}(\mathbf{a}\,\mathrm{tr}(\mathbf{aa})\underset{2}{\mathbf{I}}) \end{aligned} \tag{2.15}$$

The **third-order moment**

$$\underset{3}{\mathbf{\Psi}} = m_\mu \mathbf{g}_\mu \mathbf{g}_\mu \mathbf{g}_\mu$$

yields the transport equation for the **energy current tensor**

$$\underset{3}{\mathbf{q}}_\mu = \rho_\mu \langle \mathbf{g}_\mu \mathbf{g}_\mu \mathbf{g}_\mu \rangle_{\mathbf{v}_\mu} \tag{2.16}$$

which is related to the **energy current** $\mathbf{q}_\mu$ by

$$\mathbf{q}_\mu = \tfrac{1}{2} \rho_\mu \langle \mathbf{g}_\mu \mathbf{g}_\mu \cdot \mathbf{g}_\mu \rangle_{\mathbf{v}_\mu} = \tfrac{1}{2}\,\mathrm{tr}(\underset{3}{\mathbf{q}}_\mu) \tag{2.17}$$

and to the **heat current tensor** by

$$\underset{3}{\mathbf{h}}_\mu = \underset{3}{\mathbf{q}}_\mu - 3p_\mu\langle \mathrm{sym}(\mathbf{g}_\mu \underset{2}{\mathbf{I}})\rangle_{\mathbf{v}_\mu} \tag{2.18}$$

with the **heat current** given by

$$\mathbf{h}_\mu = \mathbf{q}_\mu - \tfrac{5}{2}n_\mu\Theta_\mu\langle \mathbf{g}_\mu\rangle \tag{2.19}$$

Systematic Solution of the Transport Equation Hierarchy by Expansion of the Distribution Function

It is not possible to solve the Maxwell–Boltzmann equations for a certain transport quantity $\langle \Psi\rangle_{\mathbf{v}_\mu}$ separately. The reason is that this differential equation contains on its left-hand side other transport quantities, e.g., $\langle \mathbf{v}_\mu \Psi\rangle_{\mathbf{v}_\mu}$, and its collision term cannot be reduced to a functional of the transport quantity $\langle \Psi\rangle_{\mathbf{v}_\mu}$ only.

A systematic approach to tackle this problem is the following: We write down the set of transport equations for the quantities $\underset{n}{\boldsymbol{\Psi}} = (\mathbf{v}_\mu)^n$. Then on the left-hand side of the equations the moments $\langle \underset{n}{\boldsymbol{\Psi}}\rangle_{\mathbf{v}_\mu}$ occur, whereas the collision terms on the right-hand side do not readily present themselves as functions of these moments $\langle \underset{n}{\boldsymbol{\Psi}}\rangle_{\mathbf{v}_\mu} = \langle(\mathbf{v}_\mu)^n\rangle_{\mathbf{v}_\mu}$. Therefore one expands the distribution function f_μ and relates the coefficients $a_{k\mu}$ of the expansion to the moments $\langle \underset{n}{\boldsymbol{\Psi}}\rangle_{\mathbf{v}_\mu}$ via (2.3), which yields the collision terms expressed through the moments.

The procedure described results in a hierarchy for the moments $\langle(\mathbf{v}_\mu)^n\rangle_{\mathbf{v}_\mu}$ which one will aim to truncate. It depends on the choice of the expansion insofar as the complexity of the relations between the expansion coefficients and the physical moments or the collisional terms is influenced.

Bopp and Meixner have suggested the general expansion (Suchy, 1964)

$$f_\mu(\mathbf{r}, \mathbf{v}_\mu; t) = \left\{1 - \mathbf{a}_\mu \cdot \frac{\partial}{\partial \mathbf{g}_\mu} + \frac{1}{2!}\underset{2}{\mathbf{a}}_\mu : \frac{\partial}{\partial \mathbf{g}_\mu}\frac{\partial}{\partial \mathbf{g}_\mu} - \frac{1}{3!}\underset{3}{\mathbf{a}}_\mu \vdots \frac{\partial}{\partial \mathbf{g}_\mu}\frac{\partial}{\partial \mathbf{g}_\mu}\frac{\partial}{\partial \mathbf{g}_\mu} + \cdots\right\}{}^{(0)}f_\mu(g_\mu{}^2) \tag{2.20}$$

where ${}^{(0)}f_\mu$ is the zero-order distribution function which should be chosen close to reality.

If one applies the Maxwell distribution

$${}^{(0)}f_\mu(g_\mu{}^2) = \frac{n_\mu}{(2\pi\Theta_\mu/m_\mu)^{1/2}}\exp\left(-\frac{m_\mu g_\mu{}^2}{2\Theta_\mu}\right) \tag{2.21}$$

then the development (2.20) corresponds to the **Grad expansion** in

Hermitian polynomials, which is of central importance in the transport theory (Grad, 1958).

The coefficients of the Grad expansion are related to the physical moments in a simple way. This can be seen by introducing the expansion (2.20) into the definition (2.3).

Straightforward integration yields

$$\begin{aligned}
\mathbf{a}_\mu &= \langle \mathbf{g}_\mu \rangle_{\mathbf{v}_\mu} \\
\underset{2}{\mathbf{a}}_\mu &= (1/\rho_\mu)\, \underset{2}{\mathring{\mathbf{p}}}_\mu = \langle \mathbf{g}_\mu \mathbf{g}_\mu \rangle_{\mathbf{v}_\mu} - \tfrac{1}{3} \langle \mathrm{tr}(\mathbf{g}_\mu \mathbf{g}_\mu) \rangle_{\mathbf{v}_\mu} \\
\underset{3}{\mathbf{a}}_\mu &= (1/\rho_\mu)\, \underset{3}{\mathbf{h}}_\mu = \langle \mathbf{g}_\mu \mathbf{g}_\mu \mathbf{g}_\mu \rangle_{\mathbf{v}_\mu} - (3\Theta_\mu/m_\mu) \langle \mathrm{sym}(\mathbf{g}_\mu \underset{2}{\mathbf{I}}) \rangle_{\mathbf{v}_\mu} \\
\underset{4}{\mathbf{a}}_\mu &= \langle \mathbf{g}_\mu \mathbf{g}_\mu \mathbf{g}_\mu \mathbf{g}_\mu \rangle_{\mathbf{v}_\mu} - (6\Theta_\mu/m_\mu \rho_\mu)\, \mathrm{sym}(\underset{2}{\mathring{\mathbf{p}}}_\mu \underset{2}{\mathbf{I}}) - 3(\Theta_\mu/m_\mu)^2\, \underset{4}{\mathbf{I}}
\end{aligned} \tag{2.22}$$

Following our concept, we now express the collision terms by the Grad coefficients and [via (2.22)] through the moments. This yields the desired hierarchy for the moments.

Here a cautionary remark is called for. Although we may evaluate the moments, we cannot calculate the distribution function with accuracy, not even if we restrict ourselves to systems which satisfy the condition

$$|\langle \mathbf{g}_\mu \rangle_{\mathbf{v}_\mu}| \ll (\Theta_\mu/m_\mu)^{1/2} \tag{2.23}$$

In Grad's case, the convergence of the Bopp–Meixner expansion (2.20) is poor. One can show that the coefficients are all small in comparison to one, but they are not of decreasing magnitude.

Evaluation of the Collision Terms

We first integrate with respect to the collision parameters. To this end, we transform the velocities relative to the total mass motion $\mathbf{u}$, into the center of mass and relative particle velocities ${}^{\mu\nu}\mathbf{g}_G$, ${}^{\mu\nu}\mathbf{g}$, which yields

$$m_\mu[(\mathbf{g}_\mu')^n - (\mathbf{g}_\mu)^n] = M_{\mu\nu} \sum_{k=1}^{n} \binom{n}{k} \left(\frac{M_{\mu\nu}}{m_\mu}\right)^k ({}^{\mu\nu}\mathbf{g}_G)^{n-k} (({}^{\mu\nu}\mathbf{g}')^k - ({}^{\mu\nu}\mathbf{g})^k) \tag{2.24}$$

with

$${}^{\mu\nu}\mathbf{g}_G = \frac{m_\mu \mathbf{g}_\mu + m_\nu \mathbf{g}_\nu}{m_\mu + m_\nu}, \qquad {}^{\mu\nu}\mathbf{g} = \mathbf{v}_\mu - \mathbf{v}_\nu \tag{2.25}$$

where $M_{\mu\nu}$ denotes the reduced masses.

We did not present the details of this transformation. Quite generally we restrain ourselves in the following from the presentation of straightforward but cumbersome algebraic transformations which are not the focus of interest here.

The elements of the $({}^{\mu\nu}\mathbf{g}')^k$ tensor can be considered as a function of the absolute value ${}^{\mu\nu}g$ and the scattering angle χ. An integration of the tensor $({}^{\mu\nu}\mathbf{g}')^k$ with respect to the azimuth angle φ yields

$$\begin{aligned}\int dQ_{\mu\nu}({}^{\mu\nu}\mathbf{g}')^k &= \int d\varphi\, b\, db({}^{\mu\nu}\mathbf{g}')^k \\ &= 2\pi \sum_{j=0}^{k} b_{kj}\, \mathrm{sym}[(({}^{\mu\nu}\mathbf{g})^j)^\circ\, \mathbf{I}_2^{(k-j)/2}] b\, db\, P_j(\cos\chi)\end{aligned} \tag{2.26}$$

where the coefficients b_{kj} are characterized by

$$\begin{aligned} b_{kj} &= 0 \qquad \text{for} \quad k + j = 2m + 1 \\ b_{kk} &= 1, \qquad b_{20} = \tfrac{1}{3}, \qquad b_{31} = \tfrac{3}{5}\end{aligned} \tag{2.27}$$

We are now in a position to carry out the integration

$$\begin{aligned} m_\mu \int dQ_{\mu\nu}[(\mathbf{g}_\mu')^n - (\mathbf{g}_\mu)^n] &= -M_{\mu\nu} \sum_{k=1}^{n} \sum_{j=0}^{k} \binom{n}{k} \left(\frac{M_{\mu\nu}}{m_\mu}\right)^k b_{kj} \\ &\quad \times \mathrm{sym}[({}^{\mu\nu}\mathbf{g}_G)^{n-k}\, (({}^{\mu\nu}\mathbf{g})^j)^\circ\, \mathbf{I}_2^{(k-j)/2}]\, Q^j({}^{\mu\nu}g)\end{aligned} \tag{2.28}$$

where we have used the cross sections

$$Q^j({}^{\mu\nu}g) = 2\pi \int b\, db\, (1 - P_j(\cos\chi)) \tag{2.29}$$

The evaluation of the cross sections (2.29) is a purely geometrical task and offers no principal difficulty, provided the interaction law is known. For the simple case that the particle interaction can be presented by a power law of the form

$$\Phi(r) = \pm C/r^s \tag{2.30}$$

the cross sections can be approximated by

$$Q^j({}^{\mu\nu}g) = (2C/M_{\mu\nu})^{2/s}\, (A_s^j({}^{\mu\nu}g))^{-4/s} \tag{2.31}$$

where A_s^j is independent of ${}^{\mu\nu}g$ for $s > 1$. For $s = 1$, we have

$$A_1^j(\gamma_c) = \tfrac{1}{4}\gamma_c^2 + \left(\frac{1+\gamma_c}{2+\gamma_c}\right)^2 P_j^{(1)}\left(\frac{2/\gamma_c + 2 + \gamma_c}{2+\gamma_c}\right) \ln|1+\gamma_c|$$

$$-\sum_{l=0}^{j} \frac{1}{4l!(1-l)}\left[\frac{-2/\gamma_c - 4 - 2\gamma_c}{2+\gamma_c}\right]^l$$

$$\times \frac{(1+\gamma_c)^{2l-1} - 1}{2+\gamma_c} P_j^{(l)}\left\{\frac{2/\gamma_c + 2 + \gamma_c}{2+\gamma_c}\right\} \tag{2.32}$$

with

$$\gamma_c = r_c M_{\mu\nu}{}^{\mu\nu}g^2/C \tag{2.33}$$

Carrying out the double average, we find the collision integral in the form

$$\int d\mathbf{g}_\mu \int d\mathbf{g}_\nu \exp\left(-\frac{m_\mu g_\mu{}^2}{2\Theta_\mu} - \frac{m_\nu g_\nu{}^2}{2\Theta_\nu}\right)\left\{1 - \mathbf{a}_\mu \cdot \frac{\partial}{\partial \mathbf{g}_\mu} + \frac{1}{2!}\underset{2}{\mathbf{a}}_\mu : \frac{\partial}{\partial \mathbf{g}_\mu}\frac{\partial}{\partial \mathbf{g}_\mu} - \cdots\right\}$$

$$\times \left\{1 - \mathbf{a}_\nu \cdot \frac{\partial}{\partial \mathbf{g}_\nu} + \frac{1}{2!}\underset{2}{\mathbf{a}}_\nu : \frac{\partial}{\partial \mathbf{g}_\nu}\frac{\partial}{\partial \mathbf{g}_\nu} - \cdots\right\}$$

$$\times \sum_{k=1}^{n}\sum_{j=0}^{k}\binom{n}{k}(M_{\mu\nu}/m_\mu)^k\, b_{kj}\, \mathrm{sym}[({}^{\mu\nu}\mathbf{g}_G)^{n-k}\,(({}^{\mu\nu}\mathbf{g})^j)^\circ\, \underset{2}{\mathbf{I}}^{[(k-j)/2]}\, Q^j({}^{\mu\nu}g) \tag{2.34}$$

for the case of Grad's distributions.

The evaluation of (2.34) proceeds in the following way: The nth-order term of the collision integral is defined as the collection of all those terms with a differential operator of order n, each of them being multiplied by the product of two Grad coefficients. To prepare the application of these differential operators, we transform operator and argument to the relative velocities ${}^{\mu\nu}\mathbf{g}$ and the velocity ${}^{\mu\nu}\mathbf{g}_G$

$${}^{\mu\nu}\mathbf{g}_G = ((m_\mu \mathbf{g}_\mu/\Theta_\mu) + (m_\nu \mathbf{g}_\nu/\Theta_\nu))/((m_\mu/\Theta_\mu) + (m_\nu/\Theta_\nu)) \tag{2.35}$$

Then the integration with respect to ${}^{\mu\nu}\mathbf{g}_G$ and the angular integration can be performed.

The result presents itself as a sum over Grad coefficients and products of Grad coefficients. These terms are multiplied with factors which can be expressed as linear combinations of the so-called **thermodynamic transport coefficients**

$$q^{l|m} = \frac{2}{(m+\frac{1}{2})!}\int \gamma_{\mu\nu}^2\, d\gamma_{\mu\nu}\, (\gamma_{\mu\nu}^2)^{m+1/2}\, Q^l({}^{\mu\nu}g)\exp(-\gamma_{\mu\nu}^2) \tag{2.36}$$

where we have used the abbreviation

$$\gamma_{\mu\nu} = {}^{\mu\nu}g/\gamma_0 , \qquad \gamma_0 = [(2\Theta_\mu/m_\mu) + (2\Theta_\nu/m_\nu)]^{1/2} \tag{2.37}$$

The Hierarchy of the Transport Equations in the Grad Approximation

In the preceding section we have shown how the collision terms can be expressed through the Grad coefficients and the thermodynamic transport coefficients. We may use the relations (2.22) to substitute the physical moments for the Grad coefficients. Together with the trivial formulation of the left-hand side in terms of the moments, we have then the hierarchy which, apart from the moments, involves the thermodynamic transport coefficients.

As far as the left-hand side of the hierarchy is concerned, only the next order Grad coefficient occurs in addition to the Grad coefficient under consideration. As far as these complicated terms are concerned, a truncation of the hierarchy is possible including all equations up to odd moments larger than three, provided that the condition

$$a_{k\mu}/(\Theta_\mu/m_\mu)^{k/2} \ll 1 \tag{2.38}$$

is fulfilled.

In the collisional terms on the right-hand side of the hierarchy equations, the main contribution comes from the Grad coefficient of the order under consideration. The additional contributions, nonlinear in the Grad coefficients, can be neglected if condition (2.38) holds. The additional linear contributions are small to the extent that the condition

$$\mathring{q}^{l|m} := q^{l|m+1} - q^{l|m} \ll q^{l|m} \tag{2.39}$$

is met. Relation (2.39) is the basic requirement for the truncation of the right-hand side of the hierarchy.

The condition (2.38) is fulfilled if the average velocities are small in comparison to the average thermal velocities which we require for the following.

Condition (2.39) depends decisively on the particle interaction law. Information about its implications may be derived from the formula

$$\frac{\mathring{q}^{l|m}}{q^{l|m}} = \frac{1 - 4/s}{2m + 3} \tag{2.40}$$

which is deduced for an interaction law of the simple form

$$\Phi(r) = \pm C/r^s \tag{2.41}$$

For power laws, $s > 2$ and $s < -2$, condition (2.39) is reasonably satisfied.

Twenty-Moments Approximation

If one limits the Grad approximation to the third-order contributions, one has the 20-moments approximation. For this approximation, the truncation of the left-hand side is correct provided condition (2.38) is satisfied. Whether the right-hand side allows the limitation to the third-order terms depends, as discussed, on the interaction law of the particles.

The twenty moments are: zero order, the number density; first order, three components of the momentum; second order, six components of the pressure tensor; and third order, ten components of the energy current tensor.

Thirteen-Moments Approximation

The 13-moments approximation follows from the 20-moments approximation if we neglect all components of the energy current tensor except for the components of the heat current density. For this approximation, the truncation of the left-hand side of the hierarchy does not anymore follow conclusively from condition (2.38) but must be justified by additional physical reasoning.

Eight-Moments Approximation

The 13-moments approach still yields a rather complicated system of equations. It can be reduced to the 8-moments approximation if we replace the pressure tensor by the hydrostatic pressure. This reduction is consequent within the range of our condition (2.38), provided that the interaction law of the particles has an appropriate form.

We present in the following the equations of the 8-moments approach. This system is still very involved. To reach further simplification, we choose our coordinate system so that the mass velocity has the value zero and neglect all terms small of second order according to (2.38). Under these circumstances, we have:

Continuity equation

$$\frac{\partial \rho_\mu}{\partial t} + \frac{\partial}{\partial \mathbf{r}} \cdot \langle \mathbf{g}_\mu \rangle = 0 \tag{2.42}$$

Momentum equation

$$\frac{\partial}{\partial t}\rho_\mu\langle\mathbf{g}_\mu\rangle+\frac{\partial}{\partial\mathbf{r}}\cdot p_\mu\mathbf{I}_2-\rho_\mu\langle\dot{\mathbf{g}}_\mu\rangle=\rho_\mu\sum_\nu n_\nu M_{\mu\nu}\gamma_0$$
$$\times\left\{q^{1|1}(\langle\mathbf{g}_\nu\rangle-\langle\mathbf{g}_\mu\rangle)+\frac{\mathring{q}^{1|1}}{\Theta_\mu/m_\mu+\Theta_\nu/m_\nu}(\mathbf{h}_\nu/\rho_\nu-\mathbf{h}_\mu/\rho_\mu)\right\} \tag{2.43}$$

Energy equation

$$\frac{3}{2}\frac{\partial}{\partial t}p_\mu+\frac{\partial}{\partial\mathbf{r}}\cdot\mathbf{h}_\mu+\frac{5}{2}p_\mu\langle\mathbf{g}_\mu\rangle-\dot{\mathbf{g}}_\mu\cdot\rho_\mu\langle\mathbf{g}_\mu\rangle$$
$$=\rho_\mu\sum_\nu n_\nu M_{\mu\nu}\gamma_0\frac{q^{1|1}}{m_\mu+m_\nu}(\Theta_\nu-\Theta_\mu) \tag{2.44}$$

Heat-flow equation

$$\frac{\partial}{\partial t}\mathbf{h}_\mu+\frac{5}{2}p_\mu\frac{\partial}{\partial\mathbf{r}}\cdot\frac{\Theta_\mu}{m_\mu}\mathbf{I}_2$$
$$=\rho_\mu\sum_\nu n_\nu M_{\mu\nu}\gamma_0\left\{\left(\frac{M_{\mu\nu}}{m_\mu}\right)^2\mathring{q}^g\left(\frac{3\Theta_\mu}{m_\mu}+\frac{3\Theta_\nu}{m_\nu}\right)(\langle\mathbf{g}_\nu\rangle-\langle\mathbf{g}_\mu\rangle)\right.$$
$$\left.-\frac{1}{m_\mu+m_\nu}\left(\frac{4}{3}\frac{M_{\mu\nu}}{m_\mu}q^h(\mathbf{h}_\mu/n_\mu+\mathbf{h}_\nu/n_\nu)+3\frac{M_{\mu\nu}}{m_\nu}q^{1|1}(\mathbf{h}_\mu/n_\mu-\mathbf{h}_\nu/n_\nu)\right)\right\} \tag{2.45}$$

with

$$\mathring{q}^g=\frac{2m_\mu{}^2}{3M_{\mu\nu}^2}\left(\frac{3M_{\mu\nu}^2}{2m_\mu{}^2}\mathring{q}^{1|1}-2\frac{M_{\mu\nu}}{m_\mu}\bar{\tau}(3q^{1|1}-q^{2|2})+\frac{9}{2}\bar{\tau}^2q^{1|1}\right)$$
$$q^h=\frac{3m_\mu}{4M_{\mu\nu}}\left(\frac{4M_{\mu\nu}}{3m_\mu}q^{2|2}-\bar{\tau}(6q^{1|1}-\frac{8}{3}q^{1|2}-5\mathring{q}^{1|1}\right) \tag{2.46}$$
$$\bar{\tau}=\frac{M_{\mu\nu}\Theta_\mu-M_{\mu\nu}\Theta_\nu}{m_\nu\Theta_\mu+m_\mu\Theta_\nu}$$

Applicability to the Plasma

So far our presentations have been of general character, since they have not specified the interaction law of the individuals which enters the coefficients $q^{l|m}$, $\mathring{q}^{l|m}$. We now consider the case of the Coulomb system for which the interaction is given through the Coulomb potential with $s = 1$. Of course, the average velocities due to the external fields and the inhomogeneities must be small in comparison to the thermal velocity. The evaluation of the thermodynamic constants $q^{l|m}$ for $s = 1$ shows that their magnitude decreases only slowly with increasing order. As a consequence, the expansion of the collision terms in a series of moments does not converge rapidly. Therefore, the justification of the approximations described above for the case of the Coulomb system is not obvious.

If one nevertheless assumes that the 8-moments approach is applicable to the fully ionized plasma, then it is interesting to compare the corresponding Grad equations with relations which have been used to describe the fully ionized plasma (Maecker, Peters 1956; Schlüter 1950, 1951).

Restricting our comparison to the case $\Theta_+ = \Theta_-$ and using the relation

$$M_{+-} \approx m_-/m_+ \ll 1 \tag{2.47}$$

we find from the general Grad equation for the electron and ion component the **continuity equations**

$$\frac{\partial}{\partial t} \rho_\pm + \frac{\partial}{\partial \mathbf{r}} \cdot \rho_\pm \langle \mathbf{g}_\pm \rangle = 0 \tag{2.48}$$

Comparison with the literature quoted above shows that these relations are identical.

For the total pressure or, respectively, the temperature, we have according to Grad's approach the **energy balance**

$$-\frac{\partial}{\partial t} p + \frac{\partial}{\partial \mathbf{r}} \cdot \mathbf{h} - \dot{\mathbf{g}}_- \cdot \rho_- \langle \mathbf{g}_- \rangle - \dot{\mathbf{g}}_+ \cdot \rho_+ \langle \mathbf{g}_+ \rangle = 0 \tag{2.49}$$

where the total heat current density $\mathbf{h}$ is given by

$$\mathbf{h} = \mathbf{h}_+ + \mathbf{h}_- \tag{2.50}$$

Equation (2.49) again is identical with the corresponding relation used in the literature cited.

For the **momentum balances** of the two components, we find from the Grad expansion with (2.47) and (2.23)

$$\begin{aligned} &\frac{\partial}{\partial t} \rho_- \langle \mathbf{g}_- \rangle + \frac{\partial}{\partial \mathbf{r}} p_- - \rho_- \dot{\mathbf{g}}_- \\ &= -\frac{\partial}{\partial t} \rho_+ \langle \mathbf{g}_+ \rangle - \frac{\partial}{\partial \mathbf{r}} p_+ + \rho_+ \dot{\mathbf{g}}_+ \\ &= -m_- \gamma_0 q^{1|1} n_- n_+ \{\langle \mathbf{g}_- \rangle - \langle \mathbf{g}_+ \rangle\} - \frac{2m_-}{\gamma_0} \mathring{q}^{1|1} n_- n_+ \{\mathbf{h}_-/\rho_- - \mathbf{h}_+/\rho_+\} \end{aligned} \tag{2.51}$$

with

$$\{\mathbf{h}_-/\rho_- - \mathbf{h}_+/\rho_+\} = -\frac{5}{4}\frac{\gamma_0}{m_- R^2}\left[q_{2-}\, n_+ + M_{+-}\left\{3q_{+-}^{1|1} - \frac{2^{1/2}}{3} M_{+-}\, q_{--}^{2|2}\right\} n_-\right] \frac{\partial}{\partial \mathbf{r}}\Theta$$

$$- \frac{3\Theta \mathring{q}^{1|1}}{m_-\, R^2}\left[\frac{2^{1/2}}{3} M_{+-}^3\, q_{--}^{2|2} n_-{}^2 + \frac{(2M_{+-})^{1/2}}{3}\, q_{++}^{2|2}\, n_{+2}\right.$$

$$\left. + 3M_{+-}\, q_{+-}^{1|1}\, n_- n_+\right] \{\langle \mathbf{g}_- \rangle - \langle \mathbf{g}_+ \rangle\} \tag{2.52}$$

and the abbreviations

$$\begin{aligned} q_{+-} &= 3M_{+-}\, q_{+-}^{1|1} + \tfrac{4}{3} q_{+-}^{2|2} \\ q_{2-} &= \frac{\sqrt{2}}{3}\, q_{--}^{2|2} + M_{+-}\{3q_{+-}^{1|1} - \tfrac{4}{3} q_{+-}^{2|2}\} \\ q_{2+} &= \frac{(2M_{+-})^{1/2}}{3}\, q_{++}^{2|2} + M_{+-}\{3q_{+-}^{1|1} - \tfrac{4}{3} q_{+-}^{2|2}\} \\ R^2 &= (2M_{+-})^{1/2}\, q_{--}^{2|2}\, q_{+-}^{1|1}\, n_-{}^2 + \frac{(2M_{+-})^{1/2}}{3}\, q_{++}^{2|2}\, q_{+-}\, n_+{}^2 \\ &\quad + \{4M_{+-}^2\, q_{+-}^{1|1}\, q_{+-}^{2|2} + \tfrac{2}{9}(M_{+-})^{1/2}\, q_{++}^{2|2}\, q_{--}^{2|2}\}\, n_+ n_- \end{aligned} \tag{2.53}$$

On the other hand, the corresponding relations given by Maecker and Peters (1956) are

$$\frac{\partial}{\partial t}\rho_-\langle \mathbf{g}_- \rangle + \frac{\partial}{\partial \mathbf{r}} p_- - \rho_- \dot{\mathbf{g}}_- = -\frac{\partial}{\partial t}\rho_+\langle \mathbf{g}_+ \rangle - \frac{\partial}{\partial \mathbf{r}} p_+ + \rho_+ \dot{\mathbf{g}}_+$$

$$= -\nu'_{+-}\, n_+ n_-\{\langle \mathbf{g}_- \rangle - \langle \mathbf{g}_+ \rangle\} - \frac{\alpha n_+ n_-}{n_+ + n_-}\frac{\partial}{\partial \mathbf{r}}\Theta \tag{2.54}$$

Obviously, the same principal terms occur in both sets of equations. However, the coefficients are quite different even in their dependence on the ratio of the densities.

For the total **heat current density**, the Grad development—in the quasistationary approximation (Grad, 1958)—yields

$$\mathbf{h} = -\frac{5}{4}\frac{\gamma_0}{R^2}\{3M_{+-}\, q_{+-}^{1|1}\, n_-{}^2 + M_{+-}^2\, q_{+-}\, n_+{}^2 + (q_{2-} + q_{2+})\, n_+ n_-\}\frac{\partial}{\partial \mathbf{r}}\Theta$$

$$+ \frac{3\Theta \mathring{q}^{1|1}}{R^2}\left[-M_{+-}\left\{3q_{+-}^{1|1} - \frac{\sqrt{2}}{3} M_{+-}\, q_{--}^{2|2}\right\} n_- + q_{2+}\, n_+\right]\{\langle \mathbf{g}_- \rangle - \langle \mathbf{g}_+ \rangle\} \tag{2.55}$$

whereas the corresponding relation has the form

$$\mathbf{h} = -\frac{\kappa}{\kappa_B}\frac{\partial}{\partial \mathbf{r}}\Theta + \alpha\Theta\frac{n_+ n_-}{n_+ + n_-}\{\langle \mathbf{g}_- \rangle - \langle \mathbf{g}_+ \rangle\} \tag{2.56}$$

In this case, again, the coefficients of the terms are different. Also the restriction to the quasistationary form is incisive. It is also noteworthy that—if one takes into account magnetic fields—additional terms appear in (2.55).

2.2. Moment Equations from the Boltzmann Equations: The Chapman–Enskog Method

Another approach which yields a set of macroscopic equations from Boltzmann's fundamental equation has been given by Chapman (1916, 1917) and Enskog (1917, 1921). For the sake of formal simplicity, we first discuss their procedure for the case of a single-component system. Chapman and Enskog base their calculation on the following three equations for the first five moments of the distribution function:

$$\frac{\partial \rho}{\partial t} + \frac{\partial}{\partial \mathbf{r}} \cdot (\rho \mathbf{u}) = 0 \tag{2.57}$$

$$\rho\left[\frac{\partial}{\partial t} + \mathbf{u} \cdot \frac{\partial}{\partial \mathbf{r}}\right]\mathbf{u} = \frac{\rho}{m}\mathbf{F} - \frac{\partial}{\partial \mathbf{r}} \cdot \underset{2}{\mathbf{p}} \tag{2.58}$$

$$\rho\left[\frac{\partial}{\partial t} + \mathbf{u} \cdot \frac{\partial}{\partial \mathbf{r}}\right]\Theta = -\frac{2}{3}m\frac{\partial}{\partial \mathbf{r}} \cdot \mathbf{q} - \frac{2}{3}m\underset{2}{\mathbf{p}} : \mathrm{sym}\,\frac{\partial \mathbf{u}}{\partial \mathbf{r}} \tag{2.59}$$

(With respect to the terminology and the derivation of these equations, we refer the reader to the previous section.) Chapman and Enskog then consider cases where the distribution function can be expressed in the form

$$f(\mathbf{r}, \mathbf{v}; t) = f(\mathbf{r}, \mathbf{v}, \rho, \mathbf{u}, \Theta) \tag{2.60}$$

and use the expansion

$$f = \sum_{\mu=0}^{\infty} {}^{(\mu)}f \tag{2.61}$$

which they specify through the systematic procedure for the calculation of ${}^{(\mu)}f$ from the distribution functions ${}^{(\nu)}f$, with $\nu < \mu$.

If one has derived the distribution function f up to the order (s), one is in a position to calculate the pressure tensor $\underset{2}{\mathbf{p}}$ and the energy current $\mathbf{q}$ to the same order (s). Introducing these qualities into (2.57), (2.58), and (2.59), we have the macroscopic equations with the accuracy limited by ${}^{(s+1)}f$.

Single-Component Solutions

To specify the expansion (2.61), we write the time differential operator in the form

$$\partial/\partial t = \sum_{\nu=1}^{\infty} \partial_\nu/\partial t \tag{2.62}$$

where the operators $\partial_\nu/\partial t$ in their application to ρ, $\mathbf{u}$, and Θ are defined by

$$\begin{aligned}
&\frac{\partial_0}{\partial t}\rho := -\frac{\partial}{\partial \mathbf{r}}\cdot(\rho\mathbf{u}) \\
&\frac{\partial_\nu}{\partial t}\rho := 0 \qquad (\nu > 0) \\
&\frac{\partial_0}{\partial t}\mathbf{u} := -\mathbf{u}\cdot\frac{\partial}{\partial \mathbf{r}}\mathbf{u} + \frac{1}{m}\mathbf{F} - \frac{1}{\rho}\frac{\partial}{\partial \mathbf{r}}\cdot{}^{(0)}\underset{2}{\mathbf{p}} \\
&\frac{\partial_\nu}{\partial t}\mathbf{u} := -\frac{1}{\rho}\frac{\partial}{\partial \mathbf{r}}\cdot{}^{(\nu)}\underset{2}{\mathbf{p}} \qquad (\nu > 0) \\
&\frac{\partial_0}{\partial t}\Theta := -\mathbf{u}\cdot\frac{\partial}{\partial \mathbf{r}}\Theta - \frac{2}{3}\frac{m}{\rho}\frac{\partial}{\partial \mathbf{r}}\cdot{}^{(0)}\mathbf{q} - \frac{2}{3}\frac{m}{\rho}{}^{(0)}\underset{2}{\mathbf{p}} : \operatorname{sym}\frac{\partial \mathbf{u}}{\partial \mathbf{r}} \\
&\frac{\partial_\nu}{\partial t}\Theta := -\frac{2}{3}\frac{m}{\rho}\frac{\partial}{\partial \mathbf{r}}\cdot{}^{(\nu)}\mathbf{q} - \frac{2}{3}\frac{m}{\rho}{}^{(\nu)}\underset{2}{\mathbf{p}} : \operatorname{sym}\frac{\partial \mathbf{u}}{\partial \mathbf{r}} \qquad (\nu > 0)
\end{aligned} \tag{2.63}$$

Due to the fact that f is assumed to depend on time only through the moments ρ, $\mathbf{u}$, Θ [see (2.60)], we have therewith also defined $\partial_\nu/\partial t$ in its application to f:

$$\frac{\partial_\nu}{\partial t}{}^{(\mu)}f = \left(\frac{\partial}{\partial \rho}{}^{(\mu)}f\right)\frac{\partial_\nu}{\partial t}\rho + \left(\frac{\partial}{\partial \mathbf{u}}{}^{(\mu)}f\right)\cdot\frac{\partial_\nu}{\partial t}\mathbf{u} + \left(\frac{\partial}{\partial \Theta}{}^{(\mu)}f\right)\frac{\partial_\nu}{\partial t}\Theta \tag{2.64}$$

Introducing (2.61) and (2.62) into the Boltzmann equation (2.1), we find

$$\begin{aligned}
&\left(\frac{\partial_0}{\partial t} + \frac{\partial_1}{\partial t} + \cdots\right)({}^{(0)}f + {}^{(1)}f + \cdots) \\
&\quad + \left(\mathbf{v}\cdot\frac{\partial}{\partial \mathbf{r}} + \frac{1}{m}\mathbf{F}\cdot\frac{\partial}{\partial \mathbf{v}}\right)({}^{(0)}f + {}^{(1)}f + \cdots) \\
&\quad = I({}^{(0)}f + {}^{(1)}f + \cdots \mid {}^{(0)}f + {}^{(1)}f + \cdots)
\end{aligned} \tag{2.65}$$

where we used the abbreviation

$$I({}^{(r)}f \mid {}^{(s)}f) = \int d\mathbf{v}_1 \int dQ \mid \mathbf{v} - \mathbf{v}_1 \mid \\ \times \{{}^{(r)}f(\mathbf{v}')\, {}^{(s)}f(\mathbf{v}_1') - {}^{(r)}f(\mathbf{v})\, {}^{(s)}f(\mathbf{v}_1)\} \tag{2.66}$$

The approximations ${}^{(\mu)}f$ are further specified through the decomposition of (2.65) which balances all terms on the left-hand side which have the same sum over indices with all terms on the right-hand side which have a sum over indices one higher than that of the left-hand side. This procedure yields the following set of equations:

$$\begin{gathered} I({}^{(0)}f \mid {}^{(0)}f) = 0 \\ \frac{\partial_0}{\partial t}\, {}^{(0)}f + \left(\mathbf{v} \cdot \frac{\partial}{\partial \mathbf{r}} + \frac{1}{m}\mathbf{F} \cdot \frac{\partial}{\partial \mathbf{v}}\right) {}^{(0)}f = \sum_{r+s=1} I({}^{(r)}f \mid {}^{(s)}f) \\ \sum_{\mu+\nu=n-1} \frac{\partial_\nu}{\partial t}\, {}^{(\mu)}f + \left(\mathbf{v} \cdot \frac{\partial}{\partial \mathbf{r}} + \frac{1}{m}\mathbf{F} \cdot \frac{\partial}{\partial \mathbf{v}}\right) {}^{(n-1)}f = \sum_{r+s=n} I({}^{(s)}f \mid {}^{(r)}f) \end{gathered} \tag{2.67}$$

The solutions of this set of integro-differential equations are not unique. Uniqueness follows through the conditions

$$\int d\mathbf{v}\, {}^{(s)}f \begin{Bmatrix} 1 \\ \mathbf{v} \\ v^2 \end{Bmatrix} = 0, \qquad s > 0 \tag{2.68}$$

It should be noted that these conditions (2.68) do not imply that the first five moments are independent of the degree of approximation s to which the evaluation is progressed. Rather, the distribution function ${}^{(0)}f$ depends on the values of the first five moments as calculated from (2.57), (2.58), and (2.59), which in turn depend on the order of approximation (s) via $\underset{2}{\mathbf{p}}$ and $\mathbf{q}$.

Zero-Order Solution

In this approximation, the distribution function is determined by the relation

$$I({}^{(0)}f \mid {}^{(0)}f) = 0 \tag{2.69}$$

which trivially has the Maxwell solution

$${}^{(0)}f = n[m/2\pi\Theta]^{3/2} \exp\{-(m/2\Theta)(\mathbf{v} - \mathbf{u})^2\} \tag{2.70}$$

Calculating the pressure tensor and the energy current using (2.70) and introducing these quantities into (2.57), (2.58), and (2.59), we arrive at the Euler relations

$$\frac{\partial \rho}{\partial t} + \frac{\partial}{\partial \mathbf{r}} \cdot (\rho \mathbf{u}) = 0$$

$$\left[\frac{\partial}{\partial t} + \mathbf{u} \cdot \frac{\partial}{\partial \mathbf{r}}\right] \mathbf{u} + \frac{1}{\rho} \frac{\partial}{\partial \mathbf{r}} \frac{\rho \Theta}{m} = \frac{1}{m} \mathbf{F} \tag{2.71}$$

$$\left[\frac{\partial}{\partial t} + \mathbf{u} \cdot \frac{\partial}{\partial \mathbf{r}}\right] \Theta + \frac{2}{3} \Theta \frac{\partial}{\partial \mathbf{r}} \cdot \mathbf{u} = 0$$

which are the hydrodynamical equations of zero order.

First-Order Solution

For the first-order solution, we have to solve

$$\left(\frac{\partial_0}{\partial t} + \mathbf{v} \cdot \frac{\partial}{\partial \mathbf{r}} + \frac{1}{m} \mathbf{F} \cdot \frac{\partial}{\partial \mathbf{v}}\right) {}^{(0)}f = I({}^{(0)}f \mid {}^{(1)}f) + I({}^{(1)}f \mid {}^{(0)}f) \tag{2.72}$$

using for ${}^{(0)}f$ the result of (2.70) and applying the conditions (2.68).

A number of straightforward but cumbersome transformations are now required. As in the preceding section, we do not present these details here. The result for the left-hand side of (2.72) is

$${}^{(0)}f \left\{ \frac{1}{\Theta} \frac{\partial \Theta}{\partial \mathbf{r}} \cdot \mathbf{v}_r \left(\frac{m}{2\Theta} \mathbf{v}_r^2 - \frac{5}{2} \right) + \frac{1}{\Theta} m \operatorname{sym} \frac{\partial \mathbf{u}}{\partial \mathbf{r}} : (\mathbf{v}_r \mathbf{v}_r)^{\circ} \right\} \tag{2.73}$$

where we have used the abbreviation $\mathbf{v}_r = \mathbf{v} - \mathbf{u}$.

The transformation of the right-hand side of (2.72) yields

$${}^{(0)}f \, C\left(\frac{{}^{(1)}f}{{}^{(0)}f}\right) \tag{2.74}$$

where C is defined by

$$C(z) := \int d\mathbf{v}_1 \int dQ \; |\mathbf{v} - \mathbf{v}_1| \; {}^{(0)}f(\mathbf{v}_1) \times \{z(\mathbf{v}_1') + z(\mathbf{v}') - z(\mathbf{v}_1) - z(\mathbf{v})\} \tag{2.75}$$

The results (2.73) and (2.74) together with (2.72) provide an integral equation for ${}^{(1)}f$.

This equation has the homogeneous solution

$${}^{(1)}f_{\mathrm{H}} = -{}^{(0)}f\{\alpha + \boldsymbol{\beta} \cdot \mathbf{v}_r + \gamma \mathbf{v}_r^2\} \tag{2.76}$$

which together with the structure of the left-hand side of (2.72) suggests an ansatz of the form

$$ {}^{(1)}f = -\,{}^{(0)}f \left\{ \alpha + \boldsymbol{\beta} \cdot \mathbf{v}_r + \gamma \mathbf{v}_r^2 + \left(\frac{\partial}{\partial \mathbf{r}} \ln \Theta \right) \cdot \mathbf{v}_r x \right. $$
$$ \left. + \frac{1}{\Theta} m \operatorname{sym} \frac{\partial \mathbf{u}}{\partial \mathbf{r}} : (\mathbf{v}_r \mathbf{v}_r)^\circ y \right\} \tag{2.77} $$

The constants α, $\boldsymbol{\beta}$, and γ follow from the conditions (2.68). Since $\alpha = 0$ and $\gamma = 0$, we can rewrite (2.77) as

$$ {}^{(1)}f = -{}^{(0)}f \left\{ \left(\frac{\partial}{\partial \mathbf{r}} \ln \Theta \right) \cdot \mathbf{v}_r\, x + \frac{1}{\Theta} m \operatorname{sym} \frac{\partial \mathbf{u}}{\partial \mathbf{r}} : (\mathbf{v}_r \mathbf{v}_r)^\circ y \right\} \tag{2.78} $$

For the functions $x = x(\mathbf{v}_r^2, \rho, \Theta)$ and $y = y(\mathbf{v}_r^2, \rho, \Theta)$, one has the integral equations

$$ \begin{aligned} C(\mathbf{v}_r\, x) &= -((m/2\Theta)\, \mathbf{v}_r^2 - \tfrac{5}{2})\, \mathbf{v}_r \\ C((\mathbf{v}_r \mathbf{v}_r)^\circ\, y) &= -(\mathbf{v}_r \mathbf{v}_r)^\circ \end{aligned} \tag{2.79} $$

which can be evaluated if one has information about the particle interaction law.

We now use the distribution function $f = {}^{(0)}f + {}^{(1)}f$ to calculate the pressure tensor and the energy current density. For the pressure tensor, we find

$$ {}^{(0)}\underset{2}{\mathbf{p}} + {}^{(1)}\underset{2}{\mathbf{p}} = n\Theta \underset{2}{\mathbf{I}} - 2\eta(\operatorname{sym}(\partial \mathbf{u}/\partial \mathbf{r}))^\circ \tag{2.80} $$

where the coefficient η is given by

$$ \eta = (m^2/15\Theta) \int d\mathbf{v}_r \mathbf{v}_r^4\, y(\rho, \Theta, \mathbf{v}_r^2)\, {}^{(0)}f \tag{2.81} $$

and designates the **friction coefficient**. Note that the friction coefficient depends only on the function y.

For the energy current density, we find

$$ {}^{(1)}\mathbf{q} = -\kappa\, \partial\Theta/\partial\mathbf{r} \tag{2.82} $$

where the **heat conduction coefficient** is given by

$$ \kappa = (m/6\Theta) \int d\mathbf{v}_r\, \mathbf{v}_r^4\, x(\rho, \Theta, \mathbf{v}_r^2)\, {}^{(0)}f \tag{2.83} $$

Observe that the heat conduction coefficient depends only on the function x.

Introducing the pressure tensor (2.80) and the heat conduction current (2.82) into (2.57), (2.58), and (2.59), we find

$$\frac{\partial \rho}{\partial t} + \frac{\partial}{\partial \mathbf{r}} \cdot (\rho \mathbf{u}) = 0$$

$$\rho \left(\frac{\partial}{\partial t} + \mathbf{u} \cdot \frac{\partial}{\partial \mathbf{r}} \right) \mathbf{u} = \frac{\rho}{m} \mathbf{F} - \frac{\partial}{\partial \mathbf{r}} \left(\frac{\rho}{m} \Theta - \frac{\eta}{3} \frac{\partial}{\partial \mathbf{r}} \cdot \mathbf{u} \right)$$

$$+ \eta \frac{\partial}{\partial \mathbf{r}} \cdot \frac{\partial}{\partial \mathbf{r}} \mathbf{u} + 2 \frac{\partial \eta}{\partial \mathbf{r}} \cdot \left(\operatorname{sym} \frac{\partial \mathbf{u}}{\partial \mathbf{r}} \right)^{\circ}$$

$$\rho \left(\frac{\partial}{\partial t} + \mathbf{u} \cdot \frac{\partial}{\partial \mathbf{r}} \right) \Theta = m \left(\frac{2}{3} \kappa \frac{\partial}{\partial \mathbf{r}} \cdot \frac{\partial}{\partial \mathbf{r}} \Theta + \frac{2}{3} \frac{\partial}{\partial \mathbf{r}} \kappa \cdot \frac{\partial}{\partial \mathbf{r}} \Theta \right)$$

$$+ \frac{4}{3} \eta m \left(\operatorname{sym} \frac{\partial \mathbf{u}}{\partial \mathbf{r}} \right)^{\circ} : \left(\operatorname{sym} \frac{\partial \mathbf{u}}{\partial \mathbf{r}} \right) - \frac{2}{3} \rho \Theta \frac{\partial}{\partial \mathbf{r}} \cdot \mathbf{u} \tag{2.84}$$

This is a form of generalized Navier–Stokes equations. The usual Navier–Stokes equations result from the relation (2.84) if we neglect all derivatives of κ and μ:

$$\frac{\partial \rho}{\partial t} + \frac{\partial}{\partial \mathbf{r}} \cdot (\rho \mathbf{u}) = 0$$

$$\rho \left(\frac{\partial}{\partial t} + \mathbf{u} \cdot \frac{\partial}{\partial \mathbf{r}} \right) \mathbf{u} = \frac{\rho}{m} \mathbf{F} - \frac{\partial}{\partial \mathbf{r}} \left(\frac{\rho}{m} \Theta - \frac{\eta}{3} \frac{\partial}{\partial \mathbf{r}} \cdot \mathbf{u} \right) + \eta \frac{\partial}{\partial \mathbf{r}} \cdot \frac{\partial}{\partial \mathbf{r}} \mathbf{u}$$

$$\rho \left(\frac{\partial}{\partial t} + \mathbf{u} \cdot \frac{\partial}{\partial \mathbf{r}} \right) \Theta = \frac{2}{3} \kappa m \frac{\partial}{\partial \mathbf{r}} \cdot \frac{\partial}{\partial \mathbf{r}} \Theta - \frac{2}{3} \rho \Theta \frac{\partial}{\partial \mathbf{r}} \cdot \mathbf{u} - \frac{4}{3} \eta m \left(\frac{\partial}{\partial \mathbf{r}} \cdot \mathbf{u} \right)^2 \tag{2.85}$$

$$+ \frac{2}{3} \eta m \left(\frac{\partial}{\partial \mathbf{r}} \cdot \frac{\partial}{\partial \mathbf{r}} u^2 - 2\mathbf{u} \cdot \left(\frac{\partial}{\partial \mathbf{r}} \cdot \frac{\partial}{\partial \mathbf{r}} \right) \mathbf{u} - \left| \frac{\partial}{\partial \mathbf{r}} \times \mathbf{u} \right|^2 \right)$$

Second-Order Solution

Second-order solutions of this procedure have been developed. They are extremely involved and result in correction terms to the pressure tensor and energy current density. There are hardly any features of general importance which can be deduced from these solutions (Chapman and Cowling, 1952).

Multiple Component Solutions

The principle ideas described in the case of the single component apply in full analogy to the multiple-component system. Instead of (2.57), (2.58), and (2.59), one has for a system of σ components a set of $3\sigma + 4$ equations: 3σ continuity equations, three equations for the average mass velocity, and one for the mean temperature. The relations are

$$\frac{\partial}{\partial t}\rho_k + \frac{\partial}{\partial \mathbf{r}} \cdot (\rho_k \mathbf{u}_k) = 0$$

$$\rho\left(\frac{\partial}{\partial t} + \mathbf{u} \cdot \frac{\partial}{\partial \mathbf{r}}\right)\mathbf{u} = \sum_{\mu=1}^{\sigma} \frac{\rho_\mu}{m_\mu}\mathbf{F}_\mu - \frac{\partial}{\partial \mathbf{r}} \cdot \underset{2}{\mathbf{p}}$$

$$\sum_{\mu=1}^{\sigma} \frac{\rho_\mu}{m_\mu}\left(\frac{\partial}{\partial t} + \mathbf{u} \cdot \frac{\partial}{\partial \mathbf{r}}\right)\Theta = -\frac{2}{3}\frac{\partial}{\partial \mathbf{r}} \cdot \mathbf{q} - \frac{2}{3}\underset{2}{\mathbf{p}} : \frac{\partial \mathbf{u}}{\partial \mathbf{r}} + \Theta \sum_{\mu=1}^{\sigma} \frac{\partial}{\partial \mathbf{r}} \cdot \left(\frac{\rho_\mu}{m_\mu}(\mathbf{u}_\mu - \mathbf{u})\right) + \frac{2}{3}\sum_{\mu=1}^{\sigma}\frac{\rho_\mu}{m_\mu}\mathbf{F}_\mu \cdot (\mathbf{u}_\mu - \mathbf{u}) \tag{2.86}$$

with the definitions

$$n = \sum_{\mu=1}^{\sigma} n_\mu, \qquad \rho = \sum_{\mu=1}^{\sigma} \rho_\mu, \qquad \mathbf{u} = (1/\rho)\sum_{\mu=1}^{\sigma}\rho_\mu \mathbf{u}_\mu$$
$$\Theta = (1/n)\sum_{\mu=1}^{\sigma} n_\mu \Theta_\mu, \qquad \underset{2}{\mathbf{p}} = \sum_{\mu=1}^{\sigma} \underset{2}{\mathbf{p}_\mu}, \qquad \mathbf{q} = \sum_{\mu=1}^{\sigma} \mathbf{q}_\mu \tag{2.87}$$

In the case of the multiple-component system, the basic assumption is the dependence of f on time only via ρ_μ and the average quantities $\mathbf{u}$ and Θ. The expansion of the distribution functions f_μ and the differential operators are used in full analogy to (2.61) and (2.62). We have an equation of the type (2.65) for each component, with the modification that the collision term on the right-hand side is now of the form

$$\sum_{\nu=1}^{\sigma} I(f_\mu \mid f_\nu) \tag{2.88}$$

With respect to the definition of the approximation ${}^{(r)}f_\mu$, we refer to (2.67) and the conditions (2.68).

Zero-Order Solution for Multiple Components

Again we find displaced Maxwell distributions as the solution for the zero-order distribution

$$ {}^{(0)}f_\mu = n_\mu [m_\mu / 2\pi\Theta]^{3/2} \exp[-(m_\mu/2\Theta)\, v_{r\mu}^2] \tag{2.89} $$

Observe that these Maxwell distributions are all displaced by the same average velocity $\mathbf{u}$, which means that in this approximation the average velocities of the components $\mathbf{u}_\mu$ are identical with the mass velocity $\mathbf{u}$. The same statement holds for the temperatures.

If we now calculate the pressure tensor and the heat current density with (2.89) and introduce the results into Equation (2.86), summing at the same time over all continuity equations for ρ_μ, then we arrive at the set of Euler equations for ρ, $\mathbf{u}$, and Θ corresponding to (2.71).

First-Order Calculation for Multiple Components

The equations to determine the first-order distributions are given by

$$ \left(\frac{\partial_0}{\partial t} + \mathbf{v}_\mu \cdot \frac{\partial}{\partial \mathbf{r}} + \frac{1}{m_\mu} \mathbf{F}_\mu\right) {}^{(0)}f_\mu = \sum_{\nu=1}^{\sigma} (I\{{}^{(0)}f_\mu \mid {}^{(1)}f_\nu\} + I\{{}^{(1)}f_\mu \mid {}^{(0)}f_\nu\}) \tag{2.90} $$

and, following the procedure used for the single component system, the first-order distributions present themselves in the form

$$ {}^{(1)}f_\mu = -{}^{(0)}f_\mu \left(A_\mu \mathbf{v}_{r\mu} \cdot \frac{\partial}{\partial \mathbf{r}} \ln \Theta + B_\mu (\mathbf{v}_{r\mu}\mathbf{v}_{r\mu})^\circ : \frac{\partial \mathbf{u}}{\partial \mathbf{r}} \right. $$

$$ \left. + \mathbf{v}_{r\mu} \cdot \sum_{\nu=1}^{\sigma} C_\mu^{\ \nu}\, \mathbf{d}_\nu \right) \tag{2.91} $$

where we have used the abbreviation

$$ \mathbf{d}_\nu = \frac{\partial}{\partial \mathbf{r}} \left[\frac{n_\nu}{n}\right] + \left[\frac{n_\nu}{n} - \frac{\rho_\nu}{\rho}\right] \frac{\partial}{\partial \mathbf{r}} \ln(n\Theta) - \frac{n_\nu}{n\Theta} \mathbf{F}_\nu + \frac{\rho_\nu}{n\Theta\rho} \sum_{\lambda=1}^{\sigma} n_\lambda \mathbf{F}_\lambda \tag{2.92} $$

The coefficients A_μ, B_μ, and $C_\mu^{\ \nu}$ are defined through complicated integral equations which result from insertion of (2.91) into (2.90). We refer the reader in this respect to the literature (Kogan, 1969).

With the first-order distribution function, we calculate the mean velocity, the pressure tensor, and the energy current density. We find for the mean velocity

$$ {}^{(1)}\mathbf{u}_\mu = (n^2/n_\mu \rho) \sum_{\nu=1}^{\sigma} m_\nu D_\mu^{\ \nu}\, \mathbf{d}_\nu - \frac{1}{\rho_\mu} D_\mu \frac{\partial}{\partial \mathbf{r}} \ln \Theta \tag{2.93} $$

with the abbreviations

$$D_\mu{}^\nu = (\rho/3nm_\nu)[m_\mu/2\Theta]^{1/2} \int d\mathbf{v}_\mu C_\mu{}^\nu(\mathbf{v}_{r\mu})\, \mathbf{v}_{r\mu}^2\, {}^{(0)}f_\mu$$

$$D_\mu = (m_\mu/3)[m_\mu/2\Theta]^{1/2} \int d\mathbf{v}_\mu A_\mu(\mathbf{v}_{r\mu})\, \mathbf{v}_{r\mu}^2\, {}^{(0)}f_\mu \tag{2.94}$$

These equations determine the coefficients of **mobility, diffusion**, and **thermodiffusion.**

For the pressure tensor, we find

$${}^{(1)}\underset{2}{\mathbf{p}} = -2\eta \left(\text{sym}\left(\frac{\partial \mathbf{u}}{\partial \mathbf{r}}\right)\right)^{\circ} \tag{2.95}$$

with the friction coefficient

$$\eta = \frac{1}{15} \sum_{\nu=1}^{\sigma} m_\nu \int d\mathbf{v}_\nu\, B_\nu(\mathbf{v}_{r\nu})\, \mathbf{v}_{r\nu}^4\, {}^{(0)}f_\nu \tag{2.96}$$

Finally, for the energy current density, we find

$${}^{(1)}\mathbf{q} = -\kappa_\Theta\, \partial \ln \Theta/\partial \mathbf{r} + \sum_{\nu=1}^{\sigma} \kappa_\nu \mathbf{d}_\nu \tag{2.97}$$

with the coefficients of heat conduction

$$\kappa_\Theta = \sum_{\lambda=1}^{\sigma} \tfrac{1}{6} m_\lambda \int d\mathbf{v}_\lambda\, A_\lambda(\mathbf{v}_{r\lambda})\, \mathbf{v}_{r\lambda}^4\, {}^{(0)}f_\lambda$$

$$\kappa_\mu = \sum_{\lambda=1}^{\sigma} \tfrac{1}{6} m_\lambda \int d\mathbf{v}_\lambda\, C_\mu{}^\lambda(\mathbf{v}_{r\lambda})\, \mathbf{v}_{r\lambda}^4\, {}^{(0)}f_\lambda \tag{2.98}$$

Note that the coefficients of physical interest depend each only on one of the coefficients A, B, and C, respectively. The latter are decisively influenced by the particle interaction law.

Comparison with Grad's 13-*Moments Approach*

To evaluate the functions x and y for a one-component system from the integral equations (2.79), it is common practice to expand these quantities in terms of Sonine polynomials. If one limits the expansion of the x functions to the first two terms and the expansion of the y functions to a single term, then the distribution function as calculated from the first-order Chapman–Enskog expansion is identical with the result of Grad's 13-moments approach.

Relation to the Hilbert Expansion

In the section on Grad's method, the basic concept was the systematic Bopp–Meixner expansion. The Chapman–Enskog method as described above cannot be readily associated with such a systematic expansion procedure. However, additional insight into this problem can be gained from consideration of the **Hilbert expansion** (Hilbert, 1912, 1924). To this end we measure the space coordinate $\mathbf{r}$ in units of L (L being the characteristic length of the distribution function f), the velocity $\mathbf{v}$ in units of v_{th}, the time t in units of L/v_{th}, the collision parameter b in units r_c (r_c is range of interaction), and normalize the distribution function by the factor $L^3 v_{\text{th}}^3$.

Then the Boltzmann equation for the general distribution function reads

$$\frac{\partial f}{\partial t} + \mathbf{v} \cdot \frac{\partial f}{\partial \mathbf{r}} = (1/\varepsilon) \int [f(\mathbf{r}, \mathbf{v}'; t) f(\mathbf{r}, \mathbf{v}_2'; t) - f(\mathbf{r}, \mathbf{v}; t) f(\mathbf{r}, \mathbf{v}_2; t)] \times g\, b\, db\, d\varphi\, d\mathbf{v}_2 = (1/\varepsilon) I(f \mid f) \tag{2.99}$$

We draw attention to the fact that all quantities in this equation are dimensionless. The parameter ε is called the Knudsen number which is given by

$$\varepsilon = (\pi r_c^2\, nL)^{-1} = \lambda_c / L \tag{2.100}$$

and λ_c is the mean-free path.

Hilbert considers the case of small Knudsen numbers and uses ε as the expansion parameter for the distribution function f in the form

$$f = \sum_{\nu=0}^{\infty} \varepsilon^{\nu}\, {}^{(\nu)}f \tag{2.101}$$

Introducing this into (2.99) and collecting all terms of same order in ε, we find the following set of equations:

$$\frac{\partial}{\partial t}\, {}^{(n-1)}f + \mathbf{v} \cdot \frac{\partial}{\partial \mathbf{r}}\, {}^{(n-1)}f - \sum_{m=1}^{n-1} I\{{}^{(m)}f \mid {}^{(n-m)}f\} = 2\, I\{{}^{(0)}f \mid {}^{(n)}f\} \tag{2.102}$$

The solution of these equations may be composed of a particular solution of the inhomogeneous equation ${}^{(n)}f_{\text{inh}}$ and the general solution of the homogeneous part of the equation which presents itself as a linear combination of the collision invariants:

$$
\begin{gathered}
{}^{(n)}f = {}^{(n)}f_{\text{inh}} + \sum_{s=0}^{4} {}^{(n)}\gamma_s(\mathbf{r}, t)\, \psi_s(\mathbf{v})\, {}^{(0)}f \\
\psi_0 = m; \qquad \psi_i = v_i\,, \qquad i = 1, 2, 3; \qquad \psi_4 = v^2
\end{gathered}
\tag{2.103}
$$

Obviously we need initial conditions to determine the coefficients ${}^{(n)}\gamma_s(\mathbf{r}, t)$. Since we have five such parameters in each equation, we need five initial values for each function ${}^{(n)}f$. We may, for instance, prescribe the first five moments at $t = 0$:

$$
\left\langle \begin{matrix} m \\ \mathbf{v} \\ v^2 \end{matrix} \right\rangle_{t=0} = \int d\mathbf{v}\, f(t = 0) \begin{pmatrix} m \\ \mathbf{v} \\ v^2 \end{pmatrix} = \sum_{\nu=0}^{\infty} \varepsilon^{\nu}\; {}^{(\nu)}\!\left\langle \begin{matrix} m \\ \mathbf{v} \\ v^2 \end{matrix} \right\rangle_{t=0}
\tag{2.104}
$$

If we were not concerned with convergence problems, we could choose an arbitrary decomposition of the given quantities $\langle\ \rangle_{t=0}$ into ${}^{(\nu)}\langle\ \rangle_{t=0}$. In fact, for complete summation, Hilbert has shown that the distribution at time t is independent of the initial decomposition at time $t = 0$. In addition, Hilbert gave the necessary proofs of existence and uniqueness of the distribution function (Hilbert, 1912, 1924).

Actually, however, the question of convergence of the expansion (2.101) is of interest. The results of Hilbert's extended investigations showed that (2.101) belongs to the group of asymptotic convergent series. Consequently, in deciding how far we may carry the Hilbert expansion, it is always important to estimate the magnitude of the remainder term. Grad has given a formulation of this term in his investigations (Grad, 1963).

Comparing the Hilbert expansion with the Chapman–Enskog method, one cannot readily recognize the intercorrelations. Grad (1963) shows how Hilbert's ideas may be used in the consideration of the convergence of the Chapman–Enskog expansion. He comes to the conclusion that the Chapman–Enskog expansion is also asymptotically convergent.

2.3. Moment Equations from the Bhatnagar–Gross–Krook Equation

The central problem of the derivation of macroscopic equations originates from the collision term of the kinetic equations. In this connection, a model is of interest which has been proposed by Bhatnagar, Gross, and Krook (1954) and which replaces the collision integral by a single "relaxation" term. The basic idea is that small deviations from the local equilibrium distribution always tend to decay towards this local equilibrium distribution with a constant relaxation time of the order of the collision time.

Applying this concept the Bhatnagar–Gross–Krook equation reads

$$\frac{\partial f}{\partial t} + \mathbf{v} \cdot \frac{\partial f}{\partial \mathbf{r}} + \dot{\mathbf{v}} \cdot \frac{\partial f}{\partial \mathbf{v}} = \frac{{}^{(0)}f - f}{\tau_0} \tag{2.105}$$

where τ_0 denotes the decay constant and

$${}^{(0)}f = [n(m/2\pi\Theta)^{3/2}] \exp(-m\mathbf{g}^2/2\Theta) \tag{2.106}$$

is the local equilibrium distribution. It is essential to the idea of Bhatnagar, Gross, and Krook that the above equilibrium distribution contains as parameters—similar to the Chapman–Enskog zero-order function—the correct values of the first five moments of the distribution function. As a consequence, the Bhatnagar–Gross–Krook equation is not a linear equation as one might think at first sight.

Since the Bhatnagar–Gross–Krook equation is a simplified version of the Boltzmann equation, one might expect that it is open to the application of the Grad expansion and the Chapman–Enskog method described in the preceding sections. We present here only the corresponding results for a two-component system applying Grad's expansion:

Continuity equation

$$\frac{\partial}{\partial t} \rho_\mu + \frac{\partial}{\partial \mathbf{r}} \cdot \rho_\mu \langle \mathbf{g}_\mu \rangle = 0 \tag{2.107}$$

Momentum equation

$$\frac{\partial}{\partial t} \rho_\mu \langle \mathbf{g}_\mu \rangle + \frac{\partial}{\partial \mathbf{r}} \cdot \underset{2}{\mathbf{I}} p_\mu - \rho_\mu \langle \dot{\mathbf{g}}_\mu \rangle = \frac{\rho_\nu}{\tau_{\mu\nu}} \langle {}^{\mu\nu}\mathbf{g} \rangle \tag{2.108}$$

Energy equation

$$\frac{3}{2} \frac{\partial}{\partial t} p_\mu + \frac{\partial}{\partial \mathbf{r}} \cdot \left(\mathbf{h}_\mu + \frac{5}{2} p_\mu \langle \mathbf{g}_\mu \rangle \right) - \rho_\mu \langle \mathbf{g}_\mu \rangle \langle \dot{\mathbf{g}}_\mu \rangle$$

$$= \frac{\rho_\mu}{\tau_\mu} \left(\langle \mathbf{g}_\mu \rangle^2 - \frac{p_\mu}{\rho_\mu} + \frac{\Theta_\mu}{m_\mu} \right) + \frac{\rho_\nu}{\tau_{\mu\nu}} \left(\langle {}^{\mu\nu}\mathbf{g} \rangle^2 + \frac{\Theta_{\mu\nu}}{m_\mu} \right) \tag{2.109}$$

Heat flow equation

$$\frac{\partial}{\partial t}\mathbf{h}_\mu + \frac{5}{2}p_\mu \frac{\partial}{\partial \mathbf{r}} \cdot \underset{2}{\mathbf{I}}\, \frac{\Theta_\mu}{m_\mu} = \frac{\rho_\mu}{\tau_\mu}\left(\langle \mathbf{g}_\mu \rangle^3 + \frac{\Theta_\mu}{m_\mu}\langle \mathbf{g}_\mu \rangle - \frac{\mathbf{h}_\mu}{\rho_\mu}\right)$$

$$+ \frac{\rho_\nu}{\tau_{\mu\nu}}\left(\langle {}^{\mu\nu}\mathbf{g} \rangle^3 + \frac{3\Theta_{\mu\nu}}{m_\mu}\langle {}^{\mu\nu}\mathbf{g} \rangle\right) + 3p\langle \dot{\mathbf{g}}_\mu \rangle \tag{2.110}$$

Here we have used the abbreviations

$$\begin{aligned}
\langle {}^{\mu\nu}\mathbf{g} \rangle &= a_{\mu\mu}\langle \mathbf{g}_\mu \rangle + a_{\mu\nu}\langle \mathbf{g}_\nu \rangle \\
\Theta_{\mu\nu} &= b_{\mu\mu}\Theta_\mu + b_{\mu\nu}\Theta_\nu + D_{\mu\nu}\langle \mathbf{g}_\mu \rangle^2 + E_{\mu\nu}\langle \mathbf{g}_\mu \rangle \cdot \langle \mathbf{g}_\nu \rangle + F_{\mu\nu}\langle \mathbf{g}_\nu \rangle^2 \\
a_{\mu\mu} &= 1 - a_{\mu\nu} = 0.113\,\frac{m_\mu m_\nu}{m_\mu + m_\nu} \\
D_{\mu\nu} &= -\tfrac{1}{2}E_{\mu\nu} = F_{\mu\nu} = \frac{1}{30}\,\frac{m_\mu m_\nu}{m_\mu + m_\nu} \\
b_{\mu\mu} &= 1 - b_{\mu\nu} = B_{\mu\nu}
\end{aligned} \tag{2.111}$$

where τ_{jk} is the mean time between collisions and B_{jk} is the coefficient of direct heat transfer of the two components.

The friction coefficient and the heat conduction coefficient are therefore given by

$$\eta_\mu = 2\Theta_\mu\, \tau_\mu / m_\mu\,, \qquad \nu_\mu = -5\Theta_\mu\, \tau_\mu / {m_\mu}^2 \tag{2.112}$$

The above equations are the 8-moments reduction of the 35-moments results derived by Devanathan, Bhatnagar, and Uberoi (1965).

The application of the Chapman–Enskog procedure yields for the first-order contribution to the distribution function ${}^{(1)}f$, the result

$${}^{(1)}f = {}^{(0)}f \cdot \tau_0 \left\{ \mathbf{g} \cdot \left(\frac{m\mathbf{g}^2}{2\Theta} - \frac{5}{2}\right)\frac{\partial}{\partial \mathbf{r}} \ln \Theta + \frac{m}{\Theta}(\mathbf{g}\mathbf{g})^\circ : \frac{\partial}{\partial \mathbf{r}}\langle \mathbf{v} \rangle \right\} \tag{2.113}$$

which, of course, is the expression with x and y evaluated for the Bhatnagar–Gross–Krook collision term. With the help of this distribution function, we may again formulate transport equations and find the Navier–Stokes equations presented in the relations (2.84) and (2.85). Note, however, that the coefficients of these equations due to the basic model assumption depend on the phenomenological constant τ_0 and consequently may be used to gather information about this quality.

The use of a constant relaxation time τ_0 might suggest the idea that the results derived from the Bhatnagar–Gross–Krook equation should be identical with those deduced from the Boltzmann equation for Maxwell molecules. ($s = 4$, in (2.30)). This conclusion would be incorrect, since the Bhatnagar–Gross–Krook equation uses in addition a very simplified model neglecting all details of the collision process.

The relaxation term of the Bhatnagar–Gross–Krook equation was only heuristically justified. Several authors tried to prove its validity with mathematical rigor (Gross, Jackson 1959; Liepmann, Narusimha, and Chahine 1962). With a linearized theory, they succeeded in specifying the condition for relaxation terms of the Bhatnagar–Gross–Krook type.

3. Review of Systematic Methods

So far our approach to the BBGKY hierarchy has been to derive kinetic equations by truncation and transformation. This is the content of Sections 1.2, 1.3, and 1.4. We now intend to study more general methods which—at least in principle—do not restrict themselves to the first- and second-order equations of the hierarchy. Each of these procedures opens up a wide field of problems which cannot possibly be treated thoroughly within the frame of this investigation. We consequently restrict ourselves from the beginning—as in Chapter I, Section 1—to a review of the basic ideas and results without expounding the details of the evaluation formalisms.

3.1. Bogolubov's Theory

We describe in the following the features of the Bogolubov theory since this theory, although primarily designed for dilute systems with dominating binary interaction, provides the basis for several modern approaches to the description of plasmas.

Characteristic Constants of Time Variation

We start again from the general hierarchy equations

$$\frac{\partial f_s}{\partial t} + \sum_{i=1}^{s} \mathbf{v}_i \cdot \frac{\partial f_s}{\partial \mathbf{r}_i} - \frac{1}{m} \sum_{i,k}^{s}{}' \frac{\partial}{\partial \mathbf{r}_i} \phi(\mathbf{r}_i, \mathbf{r}_k) \cdot \frac{\partial}{\partial \mathbf{v}_i} f_s$$

$$= \frac{N}{m} \int d\mathbf{r}' \int d\mathbf{v}' \sum_{i=1}^{s} \frac{\partial}{\partial \mathbf{r}_i} \phi(\mathbf{r}_i, \mathbf{r}') \cdot \frac{\partial}{\partial \mathbf{v}_i} f_{s+1}(\mathbf{r}', \mathbf{v}'; t) \tag{3.1}$$

As on p. 142 we bring these equations into a dimensionless form through the transformations

$$\mathring{\mathbf{r}} = \mathbf{r}/r_c\,, \qquad \mathring{\mathbf{v}} = \mathbf{v}/v_c\,, \qquad \mathring{t} = tv_c/r_c\,, \qquad\qquad (3.2)$$
$$\mathring{\phi} = \phi/\phi_c\,, \qquad \mathring{f}_s = v_c^{3s}\, V^s\, f_s$$

resulting in

$$\frac{\partial \mathring{f}_s}{\partial \mathring{t}} + \sum_{i=1}^{s} \mathring{\mathbf{v}}_i \cdot \frac{\partial \mathring{f}_s}{\partial \mathring{\mathbf{r}}_i} - \Pi_{co} \sum_{i,k}{}' \frac{\partial}{\partial \mathring{\mathbf{r}}_i} \mathring{\phi} \cdot \frac{\partial \mathring{f}_s}{\partial \mathring{\mathbf{v}}_i}$$
$$= \Pi_{co}\Pi_d \int d\mathring{\mathbf{r}}' \int d\mathring{\mathbf{v}}' \sum_{i=1}^{s} \frac{\partial}{\partial \mathring{\mathbf{r}}_i} \mathring{\phi}(\mathring{\mathbf{r}}_i\,, \mathring{\mathbf{r}}') \cdot \frac{\partial}{\partial \mathring{\mathbf{v}}_i} \mathring{f}_{s+1} \qquad (3.3)$$

with the coupling and density parameters

$$\Pi_{co} = \phi_c/mv_c^2, \qquad \Pi_d = nr_c^3 \qquad (3.4)$$

We are interested in an order of magnitude estimate of the terms in (3.3). To this end we remark that the quantities $\mathring{\phi}$ and $\mathring{v}$ are of order one

$$\mathring{\phi} = O(1), \qquad \mathring{v} = O(1) \qquad (3.5)$$

and that we may use the relation

$$\mathring{f}_{s+1} = O(\mathring{f}_s \mathring{f}_1) = O(\mathring{f}_s) \qquad (3.6)$$

Further we assume that

$$\partial \mathring{f}_s/\partial \mathring{\mathbf{v}}_i = O(\mathring{f}_s), \qquad \partial \mathring{f}_s/\partial \mathring{\mathbf{r}}_i = \frac{r_c}{L}\, O(\mathring{f}_s) \qquad (3.7)$$

holds where L is the characteristic length for macroscopic variations. We then find from (3.3) for the dilute system with $\Pi_{co} = O(1)$

$$\frac{\partial \mathring{f}_s}{\partial t} = \frac{v_c}{r_c} \frac{\partial \mathring{f}_s}{\partial \mathring{t}} = \sum_{l=1}^{3} \left(\frac{\partial \mathring{f}_s}{\partial t} \right)_l$$
$$\left(\frac{\partial \mathring{f}_s}{\partial t} \right)_1 = O\left(\frac{v_c}{L} \mathring{f}_s \right)$$
$$\left(\frac{\partial \mathring{f}_s}{\partial t} \right)_2 = (1 - \delta_{1,s})\, O\left(\frac{v_c}{r_c} \mathring{f}_s \right) \qquad (3.8)$$
$$\left(\frac{\partial \mathring{f}_s}{\partial t} \right)_3 = O\left(\frac{v_c}{r_c} \Pi_d \mathring{f}_s \right)$$

where, respectively,

$$\left(\frac{\partial \mathring{f}_s}{\partial t}\right)_1 = O\left(\frac{\mathring{f}_s}{\tau_h}\right)$$

$$\left(\frac{\partial \mathring{f}_s}{\partial t}\right)_2 = (1 - \delta_{1,s})\, O\left(\frac{\mathring{f}_s}{\tau_{in}}\right) \tag{3.9}$$

$$\left(\frac{\partial \mathring{f}_s}{\partial t}\right)_3 = O\left(\frac{\mathring{f}_s}{\tau_c}\right)$$

The three characteristic time constants which for the dilute system are given by

$$\tau_h = L/v_c\,, \qquad \tau_{in} = r_c/v_c\,, \qquad \tau_c = r_c/v_c \Pi_d \tag{3.10}$$

have in general the following meaning:

τ_{in} characterizes the average duration of the internal interaction between the particles of the s-configuration;

τ_c marks the average time between collisions of the s-particles with the other particles of the system;

τ_h is the characteristic time constant of the macroscopic hydrodynamic flow in our system.

The fact that these times are in general of different magnitude constitutes the basis of Bogolubov's theory. To estimate their magnitudes we recall

$$v_c \approx v_{th} \tag{3.11}$$

and

$$\Pi_d = n r_c^3 \approx n q r_c = r_c/\lambda \tag{3.12}$$

where q designates the collision cross section and λ the corresponding mean free path.

This gives

$$\tau_c \approx \lambda/v_{th}\,, \qquad \tau_{in} \approx r_c/v_{th} \tag{3.13}$$

We see that the three time constants τ_h, τ_c, τ_{in} governing the evolution of f_s obey the relation

$$\tau_h \gg \tau_c \gg \tau_{in} \tag{3.14}$$

provided that the conditions

$$L \gg \lambda \gg r_c \tag{3.15}$$

are fulfilled. Note that these conditions are violated in the theory of plasma oscillations where $L \approx r_c \approx \lambda_D$ holds.

Time Evolution of f_s

The three time constants discussed above provide a suitable means to specify the time evolution of f_s. At the time $t = 0$ we start with an arbitrary state described through a set of reduced distribution functions $f_s(0)$.

f_1 practically stays constant during the time $t \lesssim \tau_{\text{in}}$ since its fastest time variation operates on the scale τ_c, and we know that $\tau_c \gg \tau_{\text{in}}$ holds. On the other hand, for $s > 1$ all functions f_s relax to an average value $\langle f_s \rangle$ within the time τ_{in} and continue with rapid fluctuations around the value $\langle f_s \rangle$. This phenomenon is due to the direct interaction of the s specified particles. The average quantity $\langle f_s \rangle$ depends on f_1. After a time $t = O(\tau_{\text{in}})$ when the average value has established itself one may advocate to present the functions f_s in a functional form

$$f_s \simeq \langle f_s \rangle = f_s(\mathbf{r}_1, ..., \mathbf{r}_s, \mathbf{v}_1, ..., \mathbf{v}_s \mid f_1) \tag{3.16}$$

However, clearly this is not fully correct, since the quantity f_1 cannot describe the individual fluctuations of f_s around the average value $\langle f_s \rangle$.

In the time scale τ_c all f_s—including $s = 1$—show variations due to the average interaction of the s-group with the rest system. This interaction causes f_1 to relax to a distribution varying only through its dependence on a certain set of moments. These moments at the same time stay constant within intervals of $O(\tau_c)$ due to our assumption $\tau_h \gg \tau_c$.

The time interval $0 < t \lesssim \tau_{\text{in}}$ is referred to as "synchronization phase," whereas the interval $\tau_{\text{in}} < t \lesssim \tau_c$ is called "kinetic phase."

The System of Functional Equations

With the arguments given above, Bogolubov starts with his postulate of functional dependence

$$f_s(\mathbf{r}_1, ..., \mathbf{v}_s; t) = f_s(\mathbf{r}_1, ..., \mathbf{v}_s \mid f_1(t)) \tag{3.17}$$

This is the key assumption of Bogolubov's theory. As discussed above it introduces a certain loss of information.

Of course, the assumption (3.17) by itself does not truncate the hierarchy. However, in the case of a dilute gas a successive procedure of solution can be derived if one uses an expansion in the density parameter Π_d equivalent to the virial expansion applied to the equation of state. The fact that the density expansion will effectively truncate the hierarchy to a given order in Π_d —independent of the assumption of functional dependence—is obvious since the higher order correlation function f_{s+1}—which is the only link to the higher part of the hierarchy—is multiplied with Π_d. Let us therefore use for all distribution functions f_s $(s > 1)$ the expansion

$$f_s = {}^{(0)}f_s + \Pi_d\, {}^{(1)}f_s + \Pi_d^2\, {}^{(2)}f_s + \cdots \tag{3.18}$$

Observe particularly that f_1 itself is not expanded but—in analogy to the hydrodynamic quantities in the Chapman–Enskog procedure—is used as a functional parameter.

To solve the hierarchy we also need the time derivatives of the functions f_s, including the case $s = 1$. We develop them also in the form

$$\partial f_s/\partial t = A_s^{(0)} + \Pi_d A_s^{(1)} + \Pi_d^2\, A_s^{(2)} \tag{3.19}$$

Due to Bogolubov's postulate $f_s = f_s(\cdots \mid f_1)$ we must expect a relation between the coefficients $A_s^{(\nu)}$ for $s \geqslant 2$ and $A_1^{(\nu)}$ which follows from†

$$\sum A_s^{(\nu)} \Pi_d{}^\nu = \frac{\partial f_s}{\partial t} = \frac{\delta f_s}{\delta f_1} \circ \frac{\partial f_1}{\partial t} = \sum \frac{\delta\, {}^{(i)}f_s}{\delta f_1} \circ A_1^{(j)} \Pi_d^{i+j} \tag{3.20}$$

yielding

$$A_s^{(\nu)} = \sum_{i=0}^{\nu} \frac{\delta\, {}^{(i)}f_s}{\delta f_1} \circ A_1^{(\nu-i)} \tag{3.21}$$

Let us now write the BBGKY hierarchy in the abbreviated form

$$\partial f_s/\partial t + i\mathscr{L}_s\, f_s = \Pi_d\, \mathscr{C}_s\, f_{s+1} \tag{3.22}$$

†Note that the symbol $(\delta f_s/\delta f_1) \circ$ represents an operator acting on $\partial f_1/\partial t$. If, for instance,

$$f_s(x_1, \ldots, x_s \mid f_1) = \int \prod_{i=1}^{m} d\xi_i\, f_1(\xi_i\,;t)\, K(\xi_1, \ldots, \xi_m; x_1, \ldots, x_s)$$

then

$$\partial f_s/\partial t = \int d\xi \left[\sum_{\nu=1}^{m} \int \prod_{i\neq\nu} d\xi_i\, f_1(\xi_i\,;t)\;\; K(\xi_i, \ldots, \xi_{\nu-1}, \xi, \xi_{\nu+1}, \ldots, \xi_m; x_1, \ldots, x_s)\right] \partial f_1(\xi\,;t)/\partial t$$

holds. We therefore have in this case

$$\frac{\delta f_s}{\delta f_1} = \sum_{\nu=1}^{m} \int \prod_{i\neq\nu} d\xi_i\, f_1(\xi_i\,;t)\, K$$

and the $\circ$ symbolizes an integration with respect to ξ. $K = \prod_i \delta(\xi_i - x_i)$ covers the product ansatz used as initial condition in the following.

where $\mathscr{C}_s$ represents the correlation operator and $\mathscr{L}_s = i\{H_s, ...\}$ is the s-particle Liouville operator. Then introducing (3.18), (3.19), and (3.21) we arrive at

$$A_1^{(0)} = -i\mathscr{L}_1 f_1, \qquad A_1^{(\nu)} = \mathscr{C}_1 {}^{(\nu-1)}f_2 \tag{3.23}$$

for the case $s = 1$, and

$$\begin{aligned} &\frac{\delta\, {}^{(0)}f_s}{\delta f_1} \circ A_1^{(0)} + i\mathscr{L}_s {}^{(0)}f_s = 0 \\ &\sum_{i=0}^{\nu} \frac{\delta\, {}^{(i)}f_s}{\delta f_1} \circ A_1^{(\nu-i)} + i\mathscr{L}_s {}^{(\nu)}f_s = \mathscr{C}_s {}^{(\nu-1)}f_{s+1} \end{aligned} \tag{3.24}$$

for $s > 1$.

The solution of the system (3.23) and (3.24) to a given order ν in Π_{d} proceeds—in principle—as follows:

In zero order ($\nu = 0$) we have the equations

$$\begin{aligned} A_1^{(0)} &= -i\mathscr{L}_1 f_1 \\ \frac{\delta\, {}^{(0)}f_s}{\delta f_1} \circ A_1^{(0)} &= -i\mathscr{L}_s {}^{(0)}f_s \qquad (s > 1) \end{aligned} \tag{3.25}$$

which express $A_1^{(0)}$ and ${}^{(0)}f_s$ ($s = 2, ..., n+1$) functionally through f_1. In the order $(\nu + 1)$, correspondingly the relations

$$\begin{aligned} &A_1^{(\nu+1)} = \mathscr{C}_1 {}^{(\nu)}f_2 \\ &\frac{\delta\, {}^{(\nu+1)}f_s}{\delta f_1} \circ A_1^{(0)} + i\mathscr{L}_s {}^{(\nu+1)}f_s = \mathscr{C}_s {}^{(\nu)}f_{s+1} - \sum_{i=0}^{\nu} \frac{\delta\, {}^{(i)}f_s}{\delta f_1} \circ A_1^{(\nu+1-i)} \end{aligned} \tag{3.26}$$

hold. These determine the functional dependence of $A_1^{(\nu+1)}$ and ${}^{(\nu+1)}f_s$ ($s = 2, ..., n-\nu$) on f_1 assuming that all $A_1^{(i)}$ and ${}^{(i)}f_s$ for $i \leqslant \nu$ and $s = 1, ..., n+1-i$ are already known from the analogous lower-order equations.

In this way we may, in particular, calculate ${}^{(n-1)}f_2$, and from that

$$A_1^{(n)} = \mathscr{C}_1 {}^{(n-1)}f_2 \tag{3.27}$$

With that f_1 is determined through a kinetic equation

$$\partial f_1/\partial t = \sum_{\nu=0}^{n} A_1^{(\nu)}(f_1) = -i\mathscr{L}_1 f_1 + \mathscr{C}_1 \sum_{\nu=0}^{n-1} {}^{(\nu)}f_2(f_1) \tag{3.28}$$

correct to the order $\Pi_d{}^n$. Of course, the higher distributions ${}^{(\nu)}f_s$ $(\nu < n)$ not used in the procedure may be calculated from (3.26) too.

Initial Conditions

Of course, an exact solution of the BBGKY hierarchy would require a set of initial conditions prescribing the values of f_s at the time $t = 0$. These initial distributions $f_s(0)$ can be chosen arbitrarily except that the conditions

$$f_s = a_s \int f_{s+1}\, d\mathbf{r}_{s+1}\, d\mathbf{v}_{s+1} \tag{3.29}$$

are to be fulfilled, where a_s is a constant depending on the normalization of f_s.

The functional dependence postulated in the Bogolubov ansatz transforms the system of differential equations which we have in the BBGKY hierarchy into a system of functional differential equations. The loss of generality of possible solutions due to this approximative procedure is, of course, also reflected in the initial conditions.

We can certainly choose an initial value f_{1a}. If we choose further $f_{sa} = f_{sa}(f_{1a})$, then with the equations of functional dependence we can calculate the general functional dependence $f_s = f_s(f_1)$ on $f_1(t)$ from our system of equations. However, due to the limitations imposed by the relaxation process, an arbitrary choice of the initial functional dependencies $f_{sa} = f_{sa}(f_{1a})$ is not permitted. It is not easy to decide which class of these initial conditions corresponds to reality.

This problem may be tackled by a transformation which reduces the functional differential equations (3.26) to differential equations and enables us to introduce a set of possible initial conditions.

Bogolubov introduces the s-particle streaming operators

$$\mathscr{S}_s(\mathbf{r}_1, \ldots, \mathbf{v}_s\,;\,\tau) := \exp(i\tau\mathscr{L}_s) \tag{3.30}$$

where we have called the time variable τ to avoid confusion with the time t under consideration. This streaming operator describes the time development of all particle coordinates belonging to the s-group under the influence of internal interaction but free of interaction with the rest system.

The transformation of the functional equation into a differential equation with respect to τ follows from

$$\frac{\partial}{\partial\tau}\, \mathscr{S}_1(\tau) f_1(\mathbf{r}, \mathbf{v}; t) = i\mathscr{L}_1 \mathscr{S}_1(\tau) f_1(\mathbf{r}, \mathbf{v}; t) = -A_1^{(0)}(\mathscr{S}_1(\tau) f_1) \tag{3.31}$$

and hence

$$\frac{\partial}{\partial \tau} f_s(\mathscr{S}_1(\tau) f_1) = \frac{\delta f_s}{\delta f_1'} \circ \frac{\partial f_1'}{\partial \tau} = -\frac{\delta f_s}{\delta f_1} \circ A_1^{(0)}(f_1')$$
$$f_1' := \mathscr{S}_1(\tau) f_1 \tag{3.32}$$

Consequently the second equation (3.26) is equivalent to the differential equation

$$-\frac{\partial}{\partial \tau} {}^{(\nu+1)}f_s(\mathbf{r}_1, \ldots, \mathbf{v}_s \mid f_1') + i\mathscr{L}_s {}^{(\nu+1)}f_s$$
$$= \mathscr{C}_s {}^{(\nu+1)}f_s - \sum_{i=0}^{\nu} \frac{\delta\, {}^{(i)}f_s}{\delta f_1'} \circ A_1{}^{(\nu+1-i)}(f_1') \tag{3.33}$$

where the right-hand side of (3.33) is a known functional of f_1. In this variable Bogolubov chooses his boundary condition for $\tau = \infty$ to be

$$\mathscr{S}_s(-\tau) f_s(\ldots \mid \mathscr{S}_1(\tau) f_1) \xrightarrow[\tau\to\infty]{} \mathscr{S}_s(-\tau) \prod_{n=1}^{s} \mathscr{S}_1(\tau) f_1(\mathbf{r}_n, \mathbf{v}_n; t) \tag{3.34}$$

resulting in

$$\mathscr{S}_s(-\tau)\, {}^{(0)}f_s(\mathscr{S}_1(\tau) f_1) \xrightarrow[\tau\to\infty]{} \mathscr{S}_s(-\tau) \prod_{n=1}^{s} \mathscr{S}_1(\tau) f_1(\mathbf{r}_n, \mathbf{v}_n; t)$$
$$\mathscr{S}_s(-\tau)\, {}^{(\nu+1)}f_s(\mathscr{S}_1(\tau) f_1) \xrightarrow[\tau\to\infty]{} 0 \tag{3.35}$$

Since $\mathscr{L}_1$ does not contain the interaction potential, $\mathscr{S}_1$ becomes one in the case of a homogeneous system. In this case we have then

$$\mathscr{S}_s(-\tau) f_s(f_1) \xrightarrow[\tau\to\infty]{} \mathscr{S}_s(-\tau) \prod_{n=1}^{s} f_1(\mathbf{r}_n, \mathbf{v}_n; t) \tag{3.36}$$

Condition (3.35) is the generalization of Boltzmann's initial chaos assumption: If we follow the trajectories of the particles of the s-group free of interaction with the rest system back into the past, then (3.35) requires an uncorrelated state. Clearly this is the assumption which introduces irreversibility into the system of equations. The corresponding assumptions for the infinite future would result in a kinetic equation with "wrong time direction." This case is less attractive since one usually prefers causal considerations relative to teleological ones.

With respect to the evaluation of expressions (3.32) and (3.34) we refer to the literature for the reason given in the introduction of this section. For homogeneous systems with short-range interactions, a generalized

Boltzmann equation is found which accounts for contributions of higher correlations to the collision integral. For inhomogeneous systems, the collision term shows an interference with the streaming term.

3.2. Extension of Bogolubov's Theory to Plasmas

In Section 3.1 describing generally Bogolubov's approach on the basis of the assumption of functional dependence and weakening correlations we have seen how the BBGKY hierarchy can be solved in principle even for dense gases. The density expansion orders the results according to binary, ternary, and higher-order correlations.

Our interest in this investigation is focussed on fully ionized plasmas. We therefore study the application of the general approach to such plasmas. In doing so we encounter difficulties already when we consider the basic **synchronization assumption**. Whereas in a system with short range interactions the interaction time τ_{in} is clearly distinguished from the times τ_{c} and τ_{h} the distinction is less obvious in a plasma.

We may use for the interaction time

$$\tau_{\mathrm{in}} \approx \lambda_{\mathrm{D}}/v_{\mathrm{th}} \approx \omega_{\mathrm{p}}^{-1} \tag{3.37}$$

whereas in a homogeneous plasma all f_1 processes are governed by the relaxation time

$$\tau_{\mathrm{c}} = \Lambda/\omega_{\mathrm{p}} \ln \Lambda \tag{3.38}$$

Their ratio is therefore

$$\tau_{\mathrm{c}}/\tau_{\mathrm{in}} = \Lambda/\ln \Lambda \tag{3.39}$$

If, however, the system is inhomogeneous, then due to the collective effects the characteristic time for the f_1 variation may be given through the reciprocal plasma frequency so that we have

$$\tau_{\mathrm{h}} \approx \tau_{\mathrm{in}} \tag{3.40}$$

Equations (3.39) and (3.40) show that the application of Bogolubov's assumption of functional dependence is justified for a homogeneous plasma well below the critical density. However, for the case of an inhomogeneous plasma undoubtedly additional arguments beyond the order of magnitude consideration presented here have to be invoked if Bogolubov's synchronization hypothesis is to be used. In this connection we draw attention to the fact that in the derivation of kinetic equations

(p. 157) we also restricted our considerations to systems where the flux effect of the selfconsistent field is negligible in comparison to the correlation field effect which meant restriction to a homogeneous system in the above sense.

Bogolubov's **initial condition** of weakening correlations is not as stringent a limitation for the plasma application. Practically any theory considering many-body correlations in a Coulomb system yields inevitably the Debye screened two-particle correlation function which causes the correlation to disappear within a distance of one Debye length. This in some way would seem to justify the application of the Bogolubov initial condition for times $t > \omega_{\mathrm{p}}^{-1}$.

The **density expansion** cannot be transfered to the plasma situation since the density parameter Π_{d} is given by

$$\Pi_{\mathrm{d}} = nr_{\mathrm{c}}^{3} \approx n\lambda_{\mathrm{D}}^{3} \approx \Lambda \tag{3.41}$$

and in the subcritical region is always a quantity very much larger than one. All Π_{d} expansions consequently diverge.†

Looking for another type of suitable expansions we note that the coupling parameter Π_{co} in a plasma has the value

$$\Pi_{\mathrm{co}} = \phi_{\mathrm{c}}/mv_{\mathrm{c}}^{2} \approx e^{2}/\Theta\lambda_{\mathrm{D}} = r_{\mathrm{w}}/\lambda_{\mathrm{D}} \approx \Lambda^{-1} \tag{3.42}$$

which demonstrates that we have the case of weak interaction in the range below the critical density. Obviously in a plasma Π_{co} is a "good expansion parameter."

Bogolubov in his original paper (1946) already used the expansion in terms of Π_{co} to derive equations for the description of a homogeneous plasma. His results were the basis of Lenard's derivation of his kinetic equation (see Section 1.4). We prefer to present in the following the approach which has been used by Guernsey (1960) which incorporates the treatment of Bogolubov but is of greater formal generality in that it allows ordering a system simultaneously with respect to its strength of interaction and the multiplicity of collisions.

Guernsey's Double Expansion

Guernsey applies instead of the expansions in Π_{d} and Π_{co}, respectively, a double expansion in terms of the two parameters: Π_{co} measuring

† Actually the density expansion even for short range interactions as far as it pertains to four- and more- particle collisions indicates secularities and non-uniformities which can be cured only by sophisticated methods of singular perturbation theory (Dorfman and Cohen, 1965)

the strength of coupling, and $\Pi_{\mathrm{dc}} := \Pi_{\mathrm{d}}\Pi_{\mathrm{co}}$ measuring the effect of correlations. It can be shown (Wu and Rosenberg, 1962) that this expansion leads to the same result as an expansion in Π_{d} and Π_{co}.

Before proceeding with the double expansion, we remark that in a plasma the parameter $\Pi_{\mathrm{dc}} = \Pi_{\mathrm{d}}\Pi_{\mathrm{co}}$ according to (3.41) and (3.42) is obviously of order one. The basic divergence difficulty of the plasma density expansion can be avoided if the double expansion is followed by a summation over all powers in Π_{dc}, the double expansion being only an intermediate step introduced for the purpose of facilitating the calculations and achieving generality.

On the basis of Bogolubov's ansatz

$$f_s(\mathbf{r}_1, \ldots, \mathbf{v}_s; t) = f_s(\mathbf{r}_1, \ldots, \mathbf{v}_s \mid f_1) \tag{3.43}$$

we use in analogy to (3.18) and (3.19) the double expansion

$$f_s = \sum_{\mu,\nu=0}^{\infty} \Pi_{\mathrm{co}}^{\mu}\, \Pi_{\mathrm{dc}}^{\nu}\, f_s^{(\mu,\nu)}(f_1), \qquad s \geqslant 2 \tag{3.44}$$

and

$$\partial f_s/\partial t = \sum_{\mu,\nu=0}^{\infty} \Pi_{\mathrm{co}}^{\mu}\, \Pi_{\mathrm{dc}}^{\nu}\, A_s^{(\mu,\nu)}(f_1) \tag{3.45}$$

Again due to the functional dependence we have to expect a relation between the coefficients $A_s^{(\mu,\nu)}$ for $s \geqslant 2$ and $A_1^{(\mu,\nu)}$ which follows from

$$\sum_{\mu,\nu=0}^{\infty} \Pi_{\mathrm{co}}^{\mu}\, \Pi_{\mathrm{dc}}^{\nu}\, A_s^{(\mu,\nu)} = \frac{\partial f_s}{\partial t} = \sum_{i,j} \Pi_{\mathrm{co}}\, \Pi_{\mathrm{dc}} \frac{\delta f_s^{(i,j)}}{\delta f_1} \circ \frac{\partial f_1}{\partial t}$$

$$= \sum_{i,j,k,l} \Pi_{\mathrm{co}}^{i+k}\, \Pi_{\mathrm{dc}}^{j+l} \frac{\delta f_s^{(i,j)}}{\delta f_1} \circ A_1^{(k,l)} \tag{3.46}$$

$$A_s^{(\mu,\nu)} = \sum_{i=0}^{\mu} \sum_{j=0}^{\nu} \frac{\delta f_s^{(i,j)}}{\delta f_1} \circ A_1^{(\mu-i,\nu-j)}(f_1) \tag{3.47}$$

Since both parameters Π_{co} and Π_{dc} play a part now, we write the BBGKY hierarchy deviating from (3.22) in the symbolic form

$$\partial f_s/\partial t + \mathscr{A}_s f_s = \Pi_{\mathrm{co}} \mathscr{B}_s f_s + \Pi_{\mathrm{dc}} \mathscr{C}_s f_{s+1} \tag{3.48}$$

where the operator

$$\mathscr{A}_s = \sum_{i=1}^{s} \mathbf{v}_i \cdot \partial/\partial \mathbf{r}_i \tag{3.49}$$

describes the effect of the flux in the configuration space, and the operator

$$\mathscr{B}_s = \frac{v_c}{r_c} \sum_{i,j=1}^{s}{}' \frac{\partial \mathring{\phi}}{\partial \mathring{\mathbf{r}}_i} \cdot \frac{\partial}{\partial \mathring{\mathbf{v}}_i} \tag{3.50}$$

describes the effect of the interaction within the s-configuration. The operator $\mathscr{C}_s$ again describes the effect of the interaction of the s-configuration with the rest plasma [see (3.3)].

Introducing (3.44), (3.45), and (3.47) into (3.48) we find for $s \neq 1$ the dependence

$$\sum_{i=0}^{\mu} \sum_{j=0}^{\nu} \frac{\delta f_s^{(i,j)}}{\delta f_1} \circ A_1^{(\mu-i,\nu-j)} + \mathscr{A}_s f_s^{(\mu,\nu)} = \mathscr{B}_s f_s^{(\mu-1,\nu)} + \mathscr{C}_s f_{s+1}^{(\mu,\nu-1)} \tag{3.51}$$

and for the case $s = 1$ the relation

$$A_1^{(\mu,\nu)} + \delta_{0\nu} \delta_{0\mu} \mathscr{A}_1 f_1 = \mathscr{C}_1 f_2^{(\mu,\nu-1)} \tag{3.52}$$

These equations allow us a successive calculation of the coefficients. In the zero order we find from (3.52)

$$A_1^{(0,0)} + \mathscr{A}_1 f_1 = 0 \tag{3.53}$$

and from (3.51)

$$\frac{\delta f_s^{(0,0)}}{\delta f_1} \circ A_1^{(0,0)} + \mathscr{A}_s f_s^{(0,0)} = 0 \tag{3.54}$$

In the next order, we have to calculate the coefficients $A_s^{(0,1)}$, $A_s^{(1,0)}$ and the functions $f_s^{(0,1)}$, $f_s^{(1,0)}$. From (3.52) we find for the $A_1^{(0,1)}$ and $A_1^{(1,0)}$ the relations

$$A_1^{(1,0)} = 0, \qquad A_1^{(0,1)} = \mathscr{C}_1 f_2^{(0,0)} \tag{3.55}$$

For $f_s^{(0,1)}$ and $f_s^{(1,0)}$ we arrive from (3.51) at the equations

$$\frac{\delta f_s^{(1,0)}}{\delta f_1} \circ A_1^{(0,0)} + \mathscr{A}_s f_s^{(1,0)} = -\frac{\delta f_s^{(0,0)}}{\delta f_1} \circ A_1^{(1,0)} + \mathscr{B}_s f_s^{(0,0)} \tag{3.56}$$

$$\frac{\delta f_s^{(0,1)}}{\delta f_1} \circ A_1^{(0,0)} + \mathscr{A}_s f_s^{(0,1)} = -\frac{\delta f_s^{(0,0)}}{\delta f_1} \circ A_1^{(0,1)} + \mathscr{C}_s f_{s+1}^{(0,0)} \tag{3.57}$$

These equations (3.55)–(3.57) together with the initial conditions of weakening correlations are sufficient to determine the dependencies on f_1 needed in the first-order approximation.

The kinetic equation for the determination of f_1 reads in this order

$$\frac{\partial f_1}{\partial t} = A_1^{(0,0)} + A_1^{(1,0)} + A_1^{(0,1)} = -\mathscr{A}_1 f_1 + \mathscr{C}_1 f_2^{(0,0)} \tag{3.58}$$

where the $f_2^{(0,0)}$ can be expressed as a functional of f_1 through (3.53) and (3.54).

In higher orders ($\nu + \mu > 1$), the information is found in complete analogy. Of course, the system of equations becomes progressively complex.

As already stated (see p. 211) relevant results can be expected only by a summation over all values of ν. In the zero order ($\mu = 0$), this summation results in the Vlasov description. In the first order ($\mu = 1$ and all values of ν), Guernsey finds the equation derived by Lenard on the basis of Bogolubov's theory (see p. 210).

According to the aim declared in the beginning of this chapter we refer the reader for the details of the calculations of the coefficients and the various summation procedures to the work of Guernsey (1960) and Wu (1966).

It is particularly noteworthy that in Guernsey's theory as in the theory of Balescu and Lenard the kinetic equation is the result of a first-order expansion in the parameter Π_{co}. Therefore we may expect that this theory is valid for sufficiently large distances but must fail if we consider close encounters where the coupling parameter becomes of the order of magnitude one. Of course, Guernsey's derivation is also restricted to a spatially homogeneous plasma. An extension to inhomogeneous plasmas along the same lines has been given by Wu and Rosenberg (1962) to the first order in the coupling parameter and all orders of the correlation parameter.

Cluster Expansion Method Formalism

An independent method for the derivation of kinetic equations from the BBGKY hierarchy based on cluster expansions has been initiated by Rostoker and Rosenbluth (1960) and by Ichikawa (1960) and has subsequently been applied successfully by several authors (Fried and Wyld, 1961; Dupree, 1961; Guernsey, 1962).

To demonstrate the procedure, we start from the BBGKY hierarchy in the form (3.48) with the coupling and density parameter

$$\Pi_{\text{co}} = O(\Lambda^{-1}), \qquad \Pi_{\text{d}} = O(\Lambda), \qquad \Pi_{\text{dc}} = \Pi_{\text{d}}\Pi_{\text{co}} = O(1) \tag{3.59}$$

Rostoker and Rosenbluth apply to this hierarchy the cluster expansion (1.22) which reads in general form

$$f_s = \prod_{i=1}^{s} f_1(\mathbf{r}_i, \mathbf{v}_i; t) + \sum_{j,k}{}' \prod_{\substack{i=1 \\ i \neq j,k}}^{s} f_1 \, g_{jk} + \sum_{j,k,l}{}' \prod_{\substack{i=1 \\ i \neq j,k,l}}^{s} f_1 \, g_{jkl} + \cdots \tag{3.60}$$

This produces a new hierarchy for the one-particle distribution function f_1 and the correlation functions $g_{1\cdots s}$ which for the first two members is given by see [(1.20) and (1.21)]

$$\frac{\partial f_1}{\partial t} + \mathscr{A}_1 f_1 = \Pi_{\text{dc}} \mathscr{C}_1(1)[(f_1(\mathbf{r}_1, \mathbf{v}_1; t) f_1(\mathbf{r}_2, \mathbf{v}_2; t) + g_{12})] \tag{3.61}$$

and

$$\frac{\partial g_{12}}{\partial t} + \mathscr{A}_2 \, g_{12} + \Pi_{\text{co}} \mathscr{B}_2(f_1 f_1 + g_{12}) = \Pi_{\text{dc}} \mathscr{C}_2(f_1(\mathbf{r}_3, \mathbf{v}_3; t) g_{12} + g_{123})$$
$$+ \Pi_{\text{dc}} \mathscr{C}_1(1) f_1(\mathbf{r}_1, \mathbf{v}_1; t) g_{23} + \Pi_{\text{dc}} \mathscr{C}_1(2) f_1(\mathbf{r}_2, \mathbf{v}_2; t) g_{13} \tag{3.62}$$

where the argument (ν) of $\mathscr{C}_1(\nu)$ characterizes the variable of the differential part of this operator.

So far no additional information has been introduced. We have changed from the distributions f_s to the correlation functions $g_{1\cdots s}$ in analogy to p. 146 and 147.

The basic assumption of Rostoker and Rosenbluth—which in many respects leads to the same consequences as Bogolubov's assumption of functional dependence—can be written in the form

$$g_{1\cdots s} \Big/ \prod_{i=1}^{s} f_1(\mathbf{r}_i, \mathbf{v}_i; t) = O(\Lambda^{-s}) \tag{3.63}$$

In addition to this expansion the above named authors also make use of the expansion of the function f_1 and the correlation functions $g_{1\cdots s}$ in powers of the parameter (Λ^{-1}). With (3.63) this yields

$$\begin{aligned} f_1 &= {}^{(0)}f_1 + \Lambda^{-1}\,{}^{(1)}f_1 + \Lambda^{-2}\,{}^{(2)}f_1 + \cdots \\ g_{1\cdots s} &= \Lambda^{-s}\,{}^{(0)}g_{1\cdots s} + \Lambda^{-(s+1)}\,{}^{(1)}g_{1\cdots s} + \cdots \end{aligned} \tag{3.64}$$

Introducing the relations (3.64) into (3.61) and (3.62) and ordering with respect to the parameter Λ^{-1} yields in zero order the equation

$$\frac{\partial\, {}^{(0)}f_1}{\partial t} + \mathscr{A}_1\, {}^{(0)}f_1 = \mathscr{C}_1(1)[{}^{(0)}f_1(\mathbf{r}_1, \mathbf{v}_1; t)\, {}^{(0)}f_1(\mathbf{r}_2, \mathbf{v}_2; t)] \tag{3.65}$$

In first order, we find

$$\frac{\partial\, {}^{(1)}f_1}{\partial t} + \mathscr{A}_1\, {}^{(1)}f_1 = \mathscr{C}_1(1)[{}^{(0)}f_1(\mathbf{r}_1, \mathbf{v}_1; t)\, {}^{(1)}f_1(\mathbf{r}_2, \mathbf{v}_2; t) + {}^{(1)}f_1(\mathbf{r}_1, \mathbf{v}_1; t)\, {}^{(0)}f_1(\mathbf{r}_2, \mathbf{v}_2; t) + {}^{(0)}g_{12}] \tag{3.66}$$

and

$$\frac{\partial\, {}^{(0)}g_{12}}{\partial t} + \mathscr{A}_2\, {}^{(0)}g_{12} = \mathscr{B}_2[{}^{(0)}f_1(\mathbf{r}_1, \mathbf{v}_1; t)\, {}^{(0)}f_1(\mathbf{r}_2, \mathbf{v}_2; t)] + \mathscr{C}_2[{}^{(0)}f_1(\mathbf{r}_3, \mathbf{v}_3; t)\, {}^{(0)}g_{12}] + \mathscr{C}_1(1)[{}^{(0)}f_1(\mathbf{r}_1, \mathbf{v}_1; t)\, {}^{(0)}g_{23}] + \mathscr{C}_1(2)[{}^{(0)}f_1(\mathbf{r}_2, \mathbf{v}_2; t)\, {}^{(0)}g_{13}] \tag{3.67}$$

Trivially, the solutions of (3.66) and (3.67) allow us to calculate all distribution functions f_s up to the order Λ^{-1} with the help of the cluster expansion (3.60).

Although the procedure may in principle be continued to arbitrary orders in Λ^{-1} one easily recognizes that with increasing powers in Λ^{-1} the equations become untractable. Already for the second power the equations are so involved that so far no solutions have been achieved.

Cluster Expansion Method, Results

The zero-order equation (3.65) is of course the well-known Vlasov equation. The first-order equations produce for a homogeneous plasma the same kinetic equation as the theory of Bogolubov. At least to this extent, then, the equivalence of the Bogolubov assumption of functional dependence and the assumption (3.63) of decreasing correlations is justified.

Rostoker and Rosenbluth (1960) specifically apply the above formalism to the calculation of a test particle in a fully ionized system. Clearly this problem requires the simultaneous formulation of equations for two sets of reduced distribution functions referring to the test particle ω_s and the field particles f_s, respectively. To zeroth order, the equations for the test particle distribution function ${}^{(0)}\omega_1$ and for the field particles distribution function ${}^{(0)}f_1$ are identical. To first order, the time evolution of the distribution function of the test particle ${}^{(1)}\omega_1$ is described by an equation of the Fokker–Planck type

$$\frac{\partial\, {}^{(1)}\omega_1}{\partial t} + \mathbf{v}_1 \cdot \frac{\partial\, {}^{(1)}\omega_1}{\partial \mathbf{r}_1} = -\frac{\partial}{\partial \mathbf{v}_1} \cdot ({}^{(0)}\omega_1 \langle \Delta \mathbf{v} \rangle) - \frac{\partial}{\partial \mathbf{v}_1}\frac{\partial}{\partial \mathbf{v}_1} : ({}^{(0)}\omega_1 \langle \Delta \mathbf{v}\, \Delta \mathbf{v} \rangle) \tag{3.68}$$

where the Fokker–Planck coefficients are defined by

$$\langle \Delta \mathbf{v} \rangle = \frac{e_T}{m_T} \left(-n \int d\mathbf{r}_2 \int d\mathbf{v}_2 \frac{\partial \phi_{12}}{\partial \mathbf{r}_1} {}^{(1)}f_1(\mathbf{r}_2, \mathbf{v}_2 ; t) - \frac{\partial}{\partial \mathbf{v}_1} \cdot \underset{2}{\mathbf{\Delta}} \right) \tag{3.69}$$

and

$$\langle \Delta \mathbf{v})(\Delta \mathbf{v} \rangle = -2 \frac{e_T}{m_T} \underset{2}{\mathbf{\Delta}} \tag{3.70}$$

The index T refers to quantities of the test particle. $\underset{2}{\mathbf{\Delta}}$ is given by

$$\underset{2}{\mathbf{\Delta}}(\mathbf{r}_1, \mathbf{v}_1 ; t) = -n \int d\mathbf{r}_2 \int d\mathbf{v}_2 \frac{\partial \phi_{12}}{\partial \mathbf{r}_1})(\mathbf{P}_{12} \, {}^{(0)}f_1(\mathbf{r}_2, \mathbf{v}_2 ; t) \tag{3.71}$$

where the quantity $\mathbf{P}_{12}$ is defined by

$$\mathbf{P}_{12} \cdot \frac{\partial}{\partial \mathbf{v}_1} {}^{(0)}\omega_1(\mathbf{r}_1, \mathbf{v}_1 ; t) = \frac{{}^{(0)}g_{12}}{{}^{(0)}f_1(\mathbf{r}_2, \mathbf{v}_2 ; t)} \tag{3.72}$$

The function ${}^{(1)}f_1$ of the field particles entering the calculation of the coefficients is governed by the equation

$$\frac{\partial \, {}^{(1)}f_1}{\partial t} + \mathbf{v}_2 \cdot \frac{\partial \, {}^{(1)}f_1}{\partial \mathbf{r}_2} - \frac{e}{m} \frac{\partial}{\partial \mathbf{v}_2} \cdot \left\{ \frac{1}{V} \int d\mathbf{r}_1 \int d\mathbf{v}_1 \frac{\partial \phi_{12}}{\partial \mathbf{r}_1} {}^{(0)}\omega_1 \, {}^{(0)}f_1(\mathbf{r}_2, \mathbf{v}_2 ; t) \right.$$
$$\left. + n \int d\mathbf{r}_3 \int d\mathbf{v}_3 \frac{\partial \phi_{23}}{\partial \mathbf{r}_2} {}^{(1)}f_1(\mathbf{r}_3, \mathbf{v}_3 ; t) \, {}^{(0)}f_1(\mathbf{r}_2, \mathbf{v}_2 ; t) \right\} = 0 \tag{3.73}$$

The above results depend on the assumption that ${}^{(0)}f_1$ is Maxwellian and the pair correlation of the field particles corresponds to the equilibrium state.

We mention that the original treatment by Rostoker and Rosenbluth goes beyond what is sketched above, in that it provides an equation which shows time reversal invariance and only relaxes to the usual form.

The cluster expansion approach has also been applied successfully to inhomogeneous systems (e.g., Ichikawa, 1960; Guernsey, 1962). Both authors free themselves from the restriction of sufficient relaxation times and spatial homogeneity through the assumption of small deviations from equilibrium. The work on spatially inhomogeneous systems is particularly important—for instance in the case of plasma oscillations with collisions—since even small deviations from the homogeneous case can cause strong collective effects.

3.3. Multiple Time-Scales Formalism

A kinetic equation is an equation of the form

$$\partial f_1/\partial t = A(\mathbf{r}, \mathbf{v} \mid f_1) \tag{3.74}$$

for the first-order distribution function f_1, A designating a general functional dependence on f_1.

In Section 1 we derived the Boltzmann, Landau–Fokker–Planck, and Bogolubov–Lenard–Balescu kinetic equation, respectively, truncating the hierarchy simply on the basis of a rather crude discussion of the two typical parameters Π_{co}, Π_{d}.

In Section 3.1 on the other hand we displayed the more systematic approach of Bogolubov which provides—at least in principle—a procedure which allows to derive a kinetic equation including higher-order effects. However, Bogolubov's theory rests on the assumption of functional dependence $f_s(\dots \mid f_1)$ which means that this procedure does not justify but rather postulates the existence of a kinetic equation of type (3.74).

If one wants to arrive at a deeper understanding of the kinetic phase and to prove the existence of a kinetic equation then the postulate of functional dependence must be dropped. Of course, one can hardly hope that simply the straightforward application of expansions of the type

$$f_s = \sum \varepsilon^{\nu}\, {}^{(\nu)}f_s(\mathbf{r}_1, \dots, \mathbf{v}_s\,;\, t), \qquad \varepsilon = \Pi_{\mathrm{co}}\,, \Pi_{\mathrm{d}} \text{ respectively,} \tag{3.75}$$

yields the information which is desired. Indeed already Bogolubov (1946) demonstrated that such a naïve expansion in a smallness parameter performed on the equations of the hierarchy produces terms growing monotonously with time. That means the perturbation expansion leads to what astronomers call "secular terms." To remove the difficulty of such secular terms Poincaré (1892) devised improved perturbation techniques which expand both the dependent and the independent variables in powers of small parameters. In the case of nonlinear periodic systems, techniques for avoiding secular behavior have been developed by Van der Pol (see, e.g., Andronov *et al.*, 1966; and Krylov and Bogolubov, 1937).

In an extension of the work cited, Sandri (1963) and Frieman (1963) presented a new approach for obtaining irreversible equations to describe the approach to equilibrium in a system of many particles. Their investigations have enlightened considerably the understanding of the evolution of such systems.

The Problem

Since the conception of the problem is basic to the comprehension of the multiple time scales approach we permit ourselves to elaborate somewhat further on this point. Let us consider, as before, the hierarchy

$$\frac{\partial f_s}{\partial t} + \mathscr{A}_s f_s = \Pi_{\text{co}} \mathscr{B}_s f_s + \Pi_{\text{dc}} \mathscr{C}_s f_{s+1} \tag{3.76}$$

In the case of a subcritical Coulomb system ($\Pi_{\text{dc}} = 1$, $\Pi_{\text{co}} = \Lambda^{-1} = \varepsilon$) we may expand our distribution functions f_s in the form

$$f_s = \sum_{\nu=0}^{\infty} \varepsilon^{\nu} \, {}^{(\nu)}f_s \tag{3.77}$$

Introducing (3.77) into (3.76) and collecting terms of the same power in ε we arrive at the system

$$\frac{\partial \, {}^{(\nu)}f_s}{\partial t} + \mathscr{A}_s \, {}^{(\nu)}f_s = \mathscr{B}_s \, {}^{(\nu-1)}f_s + \mathscr{C}_s \, {}^{(\nu)}f_{s+1} \tag{3.78}$$

It is important to note that (3.78) is appropriate only if the application of the operators in these equations causes no change in the relative order of the term. Operators satisfying this condition we shall call "conform."

The operators $\mathscr{B}_s$ and $\mathscr{C}_s$ are of the order one if the time is measured in terms of the interaction time $\tau_{\text{in}} = r_{\text{c}}/v_{\text{c}}$ (see p. 202). In the Coulomb system we have $\tau_{\text{in}} \approx \omega_{\text{p}}^{-1}$ and the operators $\mathscr{B}_s$, $\mathscr{C}_s$, and $\partial/\partial t$ are conform only if f_s "varies on the scale" of the plasma oscillation time.

For time variations characterized by the collision time

$$\tau_{\text{c}} \approx \omega_{\text{p}}^{-1} \frac{\Lambda}{\ln \Lambda} = O\left(\frac{1}{\varepsilon} \tau_{\text{in}}\right) \tag{3.79}$$

the scheme (3.78) is therefore not appropriate. In this case we introduce a new time variable $t' = \varepsilon t$ so that the operator

$$\frac{\partial}{\partial t'} = \frac{1}{\varepsilon} \frac{\partial}{\partial t} \approx \frac{\tau_{\text{c}}}{\tau_{\text{in}}} \frac{\partial}{\partial t} = O\left(\frac{1}{\tau_{\text{in}}}\right) \tag{3.80}$$

is conform with $\mathscr{B}_s$ and $\mathscr{C}_s$. With this operator (3.76) reads

$$\varepsilon \frac{\partial f_s}{\partial t'} + \mathscr{A}_s f_s = \mathscr{B}_s f_s + \varepsilon \mathscr{C}_s f_{s+1} \tag{3.81}$$

and produces with the expansion (3.77) the system of differential equations

$$\frac{\partial\, {}^{(\nu-1)}f_s}{\partial t'} + \mathscr{A}_s\, {}^{(\nu)}f_s = \mathscr{B}_s\, {}^{(\nu)}f_s + \mathscr{C}_s\, {}^{(\nu-1)}f_{s+1} \tag{3.82}$$

which obviously differs from (3.78).

It seems clear from the above discussion that the form of the expansion in terms of a smallness parameter depends strongly not only on the type of system but also on the variation—in particular the time variation—of the phenomena under consideration.

False results are to be expected from the extension of an expansion valid for a characteristic time τ_1 to phenomena with a characteristic time τ_2 if there exists a relation between τ_1 and τ_2 depending on ε.

The Formalism

If one aims for an expansion only valid to describe variations in different time scales then it is necessary to account explicitly for the nonconformity of the time derivative operator. In this spirit the method of Sandri and Frieman adopts an expansion similar to procedures used in nonlinear mechanics. Instead of a single time variable they introduce multiple time scales which are related to the time t through

$$d\tau_\nu/dt = \varepsilon^\nu \tag{3.83}$$

where ε designates the smallness parameter (e.g., $\Pi_{\mathrm{co}}, \Pi_{\mathrm{d}}$).

In the above sense these time scales are supposed to distinguish explicitly those components of time derivatives which are different from each other by the order in ε. The freedom in the solution of (3.83) afforded by the choice of initial conditions permits us to treat the variables τ_ν as independent.

It is the logical consequence of the replacement of t by the set of variables τ_ν that we introduce "extended functions"

$$f_s(\mathbf{r}_1, \ldots, \mathbf{v}_s; t) = f_s(\mathbf{r}_1, \ldots, \mathbf{v}_s; \tau_0, \tau_1, \ldots) \tag{3.84}$$

and that the time derivative operator is written in the form

$$\partial/\partial t = \sum_\nu \varepsilon^\nu\, \partial/\partial\tau_\nu \tag{3.85}$$

where the operators $\partial/\partial\tau_\nu$ are now all to be considered conform operators. Again the distribution functions f_s are expanded in the form

$$f_s = \sum_{\nu} \varepsilon^{\nu} \, {}^{(\nu)}f_s(\ldots ; \tau_0 , \tau_1 ,\ldots) \tag{3.86}$$

The combination of (3.85) and (3.86) yields

$$\frac{\partial f_s}{\partial t} = \sum_{\nu} \varepsilon^{\nu} \sum_{\mu=0}^{\nu} \frac{\partial \, {}^{(\mu)}f_s}{\partial \tau_{\nu-\mu}} \tag{3.87}$$

We draw attention to the fact that the introduction of the extended functions has enlarged the domain of functions. Therefore a new requirement is needed to reduce the additional freedom. This condition is the requirement of "removal of secularity" which is a very essential point of the theory.

Introducing (3.87) into (3.76) and collecting the coefficients of ε^{ν} we arrive at the system

$$\sum_{\mu=0}^{\nu} \frac{\partial \, {}^{(\mu)}f_s}{\partial \tau_{\nu-\mu}} + \mathscr{A}_s \, {}^{(\nu)}f_s = \mathscr{B}_s \, {}^{(\nu-1)}f_s + \mathscr{C}_s \, {}^{(\nu)}f_{s+1} \tag{3.88}$$

for our plasma. Obviously these equations are not yet decoupled from the higher part of the hierarchy since f_{s+1} appears in the same order ν.

Since we are, of course, unable to deal with all orders s simultaneously, Sandri (1963) following Rostoker and Rosenbluth (1960) uses the assumption that the correlation functions $g_{1\cdots s}$ defined through the cluster expansion

$$f_s(1 ,\ldots, s) = \prod_{k=1}^{s} f_1(k) + \sum_{i,j=1}^{s}{}' g_{ij} \prod_{k=1}^{s}{}'' f_1(k) + \cdots + g_{1\cdots s} \tag{3.89}$$

are well ordered in the sense that

$$\frac{g_{1\cdots s}}{f_s} = O(\varepsilon^s) \tag{3.90}$$

holds. This assumption is by no means selfevident. In fact, for small particle distances one can expect (3.90) to fail.

With the cluster expansion (3.89) the hierarchy (3.88) can be transformed into a hierarchy for the function f_1 and the correlation functions $g_{1\cdots s}$. With the assumption (3.90) the truncation is achieved. The correlation function $g_{1\cdots s+1}$ introduced into the equation for $g_{1\cdots s}$ cannot cause a link to the higher part of the hierarchy since it is of the next smaller order in the expansion parameter.

With respect to the evaluation we again refer the reader to the original treatment by Sandri (1963) and summarize here the results:

Using the above expansion method and the requirement of exclusion of secular terms Sandri shows that the approach to the equilibrium state is described by a kinetic equation only if the correlation functions satisfy certain conditions with respect to their relative velocity variables. The condition—which is called the "absence of parallel motions"—reflects the physical background that if a system is to obey a kinetic equation there cannot be too many statistically independent particles with nearly equal velocities. It seems intuitively clear that such a condition should exist and that the relaxation into a kinetic phase will be hindered if the system contains a large number of particles which are at rest relative to each other.

The kinetic equation which evolves in first order if the initial coefficients satisfy the above condition is that of Bogolubov, except Sandri argues that his equation is completely convergent for Coulomb potentials, whereas the equation of Lenard and Balescu diverges for small distances.

The importance of the multiple time scales theory of Sandri and Frieman should not be seen so much in the production of improved kinetic equations but rather in their generally greater power and flexibility which yields deeper insight in the complex structures of the irreversible approach to equilibrium.

REFERENCES AND SUPPLEMENTARY READING

Hochstim, A. R., ed. (1969). "Kinetic Processes in Gases and Plasmas." Academic Press, New York.

Liboff, R. L. (1969). "Introduction to the Theory of Kinetic Equations." Wiley, New York.

Montgomery, D. C., and Tidman, D. A. (1964). "Plasma Kinetic Theory." McGraw-Hill, New York.

Wu, T. Y. (1966). "Kinetic Equations of Gases and Plasmas." Addison-Wesley, Reading, Massachusetts.

van Kampen, N. G., and Felderhof, B. U. (1967). "Theoretical Methods in Plasma Physics." Wiley, New York.

Section 1

Aono, O. (1968). *Phys. Fluids* **11**, 341.

Baldwin, D. (1962). *Phys. Fluids* **5**, 1523.

Balescu, R. (1960). *Phys. Fluids* **3**, 52.

Bogolubov, N. N. (1946). *J. Phys. (U.S.S.R.)* **10**, 256, 265.

Bogolubov, N. N. (1962). Problems of a dynamical theory in statistical physics. *In* "Studies in Statistical Mechanics" (J. De Boer and G. E. Uhlenbeck, eds.), Vol. I. North-Holland Publ., Amsterdam.

Boltzmann, L. (1872). *Wien. Ber.* **66**, 213.
Boltzmann, L. (1875). *Wien. Ber.* **72**, 427.
Boltzmann, L. (1896–1898). "Vorlesungen über Gastheorie," 2 Vol. Teubner, Leipzig.
Chandrasekhar, S. (1943). *Rev. Mod. Phys.* **15**, 1.
Dupree, T. H. (1961). *Phys. Fluids* **4**, 696.
Ecker, G., and Hölling, J. (1963). *Phys. Fluids* **6**, 70.
Fokker, A. D. (1914). *Ann. Phys.* **43**, 812.
Frieman, E. A. (1962). *Nucl. Fusion Suppl.* Pt. 2, 487.
Frieman, E. A., and Book, D. L. (1963). *Phys. Fluids* **6**, 1700.
Gasiorowicz, S., Neumann, M., and Riddel, R. J. (1956). *Phys. Rev.* **101**, 922.
Grad, H. (1958). *In* "Handbuch der Physik" (S. Flügge, ed.), Vol. XII, p. 205. Springer, Berlin.
Green, M. S. (1956). *J. Chem. Phys.* **25**, 836.
Green, M. S. (1958). *Physica (Utrecht)* **24**, 393.
Guernsey, R. L. (1962). *Phys. Fluids* **5**, 322.
Hubbard, J. (1961). *Proc. Roy Soc. Ser. A* **260,** 114.
Hubbard, J. (1961). *Proc. Roy. Soc. Ser. A* **261**, 371.
Kihara, T., and Aono, O. (1963). *J. Phys. Soc. Japan* **18**, 837, 1043.
Kirkwood, J. (1947). *J. Chem. Phys.* **15**, 72.
Kolmogoroff, A. (1931). *Math. Ann.* **104**, 415.
Landau, L. D. (1936). *J. Phys. (U.S.S.R.)* **10**, 154.
Landau, L. D. (1949). *Phys. Rev.* **77**, 567.
Lenard, A. (1960). *Ann. Phys. (New York)* **10**, 390.
Ludwig, G., Müller, W. J. C., and Schröter, J. (1964). *Physica (Utrecht)* **30**, 479.
Perkins, F. (1965). *Phys. Fluids* **8**, 1361.
Planck, M. (1917). *Sitzungsber. Preuss. Akad. Wiss. Phys. Math. Kl.,* 324.
Prigogine, I., and Balescu, R. (1959). *Physica (Utrecht)* **25**, 281, 302.
Prigogine, I., and Balescu, R. (1960). *Physica (Utrecht)* **26**, 145.
Prout, R. (1956). *Physica (Utrecht)* **22**, 509.
Prout, R., and Prigogine, I. (1956). *Physica (Utrecht)* **22**, 621.
Rosenbluth, M. N., McDonald, W. M., and Judd, D. L. (1957). *Phys. Rev.* **107,** 1.
Rostoker, N. (1960). *Nucl. Fusion* **1**, 101.
Rostoker, N., and Rosenbluth, M. N. (1960). *Phys. Fluids* **3**, 1.
Tchen, C. M. (1959). *Phys. Rev.* **114**, 394.
Temko, S. V. (1956). *Zh. Eksp. Teor. Fiz.* **31**, 1021.
Thompson, W. B., and Hubbard, J. (1960). *Rev. Mod. Phys.* **32**, 714.
Weinstock, J. (1963). *Phys. Rev.* **132**, 454.
Weinstock, J. (1964). *Phys. Rev.* **133**, 673.
Weinstock, J. (1967). *Phys. Fluids* **10**, 127.

SECTION 2

Bhatnagar, P. L., Gross, E. P., and Krook, M. (1954). *Phys. Rev.* **94**, 511.
Bopp, F., and Meixner, J. (1952). Compare A. Sommerfeld, "Thermodynamik und Statistik." Akademische Verlagsgesellschaft, Leipzig.
Cercignani, C. (1969). "Mathematical Methods in Kinetic Theory." Plenum, New York.
Chapman, S. (1916). *Phil. Trans. Roy. Soc. London Ser. A* **216**, 279.
Chapman, S. (1917). *Phil. Trans. Roy. Soc. London Ser. A* **217**, 115.
Chapman, S., and Cowling, T. G. (1952). "The Mathematical Theory of Non-Uniform Gases." Cambridge Univ. Press, London and New York.
Devanathan, C., and Bhatnagar, P. L. (1965). *Proc. Nat. Inst. Sci. India* 106.

Enskog, D. (1917). Thesis, Uppsala.
Enskog, D. (1921). *Arkiv Ast. Fys.* **16**, 1.
Grad, H. (1958). *In* "Handbuch der Physik" (S. Flügge, ed.), Vol. XII. Springer, Berlin.
Grad, H. (1963). *Phys. Fluids* **6**, 147.
Gross, E. P., and Jackson, E. A. (1959). *Phys. Fluids* **2**, 432.
Hilbert, D. (1912). *Math. Ann.* **72**, 562.
Hilbert, D. (1924). "Grundzüge einer allgemeinen Theorie der linearen Integralgleichungen." Wien.
Hirschfelder, J. O., Curtiss, C. F., and Bird, R. B. (1954). "Molecular Theory of Gases and Liquids." Wiley, New York.
Kirkwood, J. G., and Crawford, B. (1952). *J. Phys. Chem.* **56**, 1048.
Kogan, M. N. (1969). "Rarefied Gas Dynamics." Plenum, New York.
Liepmann, H. W., Narusimha, R., and Chahine, M. T. (1962). *Phys. Fluids* **5**, 1313.
Maecker, H., and Peters, T. (1956). *Z. Phys.* **144**, 586
Monchick, L., Munn, R. J., and Mason, E. A. (1966). *J. Chem. Phys.* **45**, 3051.
Schlüter, A. (1950). *Z. Naturforsch.* **5a**, 72.
Schlüter, A. (1951). *Z. Naturforsch.* **6a**, 73.
Suchy, K. (1964). *Ergeb. Exakt. Naturw.* **35**, 103.
Waldmann, L. (1958). *In* "Handbuch der Physik" (S. Flügge, ed.), Vol. XII. Springer, Berlin.

Section 3

Andronov, A. A., Vitt, A. A., and Khaikin, S. E. (1966). "Theory of Oscillators," Oxford Univ. Press, London and New York.
Balescu R. and Kuezell, A. (1964). *J. Math. Phys.* **5**, 1140.
Bogolubov, N. N. (1946). *J. Phys. (USSR)* **10**, 256, 265.
Bogolubov, N. N. (1966)"Studies in Statistical Mechanics" (J. DeBoer and G.E. Uhlenbeck, eds.), North-Holland Publ., Amsterdam.
Choh, S. T. and Uhlenbeck, G. E. (1958). "The kinetic theory of phenomena in dense gases", Thesis, University of Michigan.
Dorfman, J. R. and Cohen, E. G. D. (1965). *Phys. Lett.* **16**, 124.
Dupree, T. H. (1961). *Phys. Fluids* **4**, 696.
Fried, B. D. and Wyld, H. W. (1961). *Phys. Rev.* **121**, 1.
Frieman, E. A. and Book, D. L. (1963). *Phys. Fluids* **6**, 1700.
Frieman, E. A. (1963). *J. Math. Phys.* **4**, 410.
Gorman, D. and Montgomery, D. (1963). *Phys. Rev.* **131**, 7.
Guernsey, R. L. (1960). "The theory of fully ionized gases", Thesis, University of Michigan.
Guernsey, R. L. (1962). *Phys. Fluids* **5**, 322.
Ichikawa, Y. H. (1960). *Progr. Theoret. Phys.* **24**, 1085.
Kalman, G. and Ron, A. (1961). *Ann. Phys. (N.Y.)* **16**, 118.
Kaufman, A. N. (1960). "La Théorie des Gaz Neutres et Ionisés," (C. de Witt, J. F. Detoeuf, eds.), Paris.
Kritz, A. H., Ramanathan, G. V. and Sandri, G. (1970). *Phys. Rev. Lett.* **25**, 437.
Krylov, N. M. and Bogolubov, N. N. (1937). "Introduction to Nonlinear Mechanics," Kiev.
Montgomery, D. (1967). The foundations of classical kinetic theory, *in* "Lectures in Theoretical Physics," Vol. IX, C, Kinetic Theory (W. E. Brittin, ed.). University of Colorado Press, Denver, Colorado.

Oberman, L., Ron, A. and Dawson, J. (1962). *Phys. Fluids* **5**, 1514.

Poincaré, H. (1892). "Les Méthodes Nouvelles de la Méchanique Céleste," Vol. I, Chapter III, Paris.

Ron, A. and Kalman, G. (1960). *Ann. Phys.* **11**, 240.

Rostoker, N. and Rosenbluth, N. V. (1960). *Phys. Fluids* **3**, 1.

Sandri, G. (1961-1962). The foundations of nonequilibrium statistical mechanics, Notes from lectures given at Rutgers University, Spring term (1961.1962).

Sandri, G. (1963). *Phys. Rev. Lett.* **11**, 178.

Sandri, G. (1963). *Ann. Phys.* **24**, 380.

Sandri, G. (1966). ARAP Report No. 80, Princeton, New Jersey.

Schram, P. P. J. M. (1964). "Kinetic equations for plasmas", Thesis, Garching.

Thompson, W. B. and Hubbard, J. (1960). *Rev. Mod. Phys.* **32,** 714.

Uhlenbeck, G. E. and Ford, G. W. (1963). "Lectures in Statistical Mechanics." American Mathematical Society, Providence, Rhode Island.

Willis, C. R. (1962). *Phys. Fluids* **5**, 219.

Wu, T. Y. and Rosenberg, R. L. (1962). *Can. J. Phys.* **40**, 463.

Part 2 THE FULLY IONIZED SYSTEM IN THE GENERAL ELECTROMAGNETIC FIELD

So far our considerations were restricted to the Coulomb system with quasistatic interactions only. The following studies account for the electromagnetic field in general.

Chapter V Single-Particle Radiation

1. General Formulation of Electromagnetic Fields

The relations presented in the first section can be found in most textbooks on electrodynamics. We give the following concentrated summary essentially for the purpose of reference.

1.1. Basic Equations of Electrodynamics

The electrodynamic field is a vector field. The basic experiments of Coulomb, Biot-Savart, and Faraday showed that the electromagnetic phenomena *in vacuo* can be described by the electric field **E** and the magnetic induction **B**, which are governed by the relations

$$\begin{aligned} &(\text{a}) \quad \nabla \cdot \mathbf{E} = \varepsilon_0^{-1}\rho_{\text{tr}} \qquad &(\text{b}) \quad \nabla \cdot \mathbf{B} = 0 \\ &(\text{c}) \quad \nabla \times \mathbf{E} = -\partial \mathbf{B}/\partial t \qquad &(\text{d}) \quad \nabla \times \mathbf{B} = \mu_0 \mathbf{j}_{\text{tr}} \end{aligned} \tag{1.1}$$

ρ_{tr} designates the charge density and $\mathbf{j}_{\text{tr}}$ the current density, connected by

$$\frac{\partial}{\partial t}\rho_{\text{tr}} + \nabla \cdot \mathbf{j}_{\text{tr}} = 0 \tag{1.2}$$

Of course, ε_0 and μ_0 are determined through the choice of the system of units. In Chapters I–IV dealing with the Coulomb system, no generality was lost and formal simplicity was gained by the use of the electrostatic cgs system with $4\pi\varepsilon_0 = 1$.

In the case of the general fully ionized system, the constants ε_0 and μ_0 play a vital part, and to allow the use of any system of units we do not specify these constants.

For numerical calculations, we give the values of ε_0 and μ_0 in the MKSVA system of units

$$\varepsilon_0 = 8.859 \cdot 10^{-12} \text{ A sec/V m}, \qquad \mu_0 = 1.256 \cdot 10^{-6} \text{ V sec/Am} \tag{1.3}$$

It was Maxwell who noted that the set (1.1) could not be correct, since the necessary condition $\nabla \cdot (\nabla \times \mathbf{B}) = 0$ is not fulfilled. From (1.2) and (1.1d), he concluded that (1.1d) should be replaced by

$$\nabla \times \mathbf{B} = \mu_0(\mathbf{j}_{\text{tr}} + \varepsilon_0\, \partial\mathbf{E}/\partial t) \tag{1.4}$$

If we study the electromagnetic field not *in vacuo*, but in the presence of materials, then the laws given for **E** and **B** are still valid. However, we recognize that the interaction of the electromagnetic fields with the materials causes reactions which produce additional charges and currents which have to be taken into account. Considering the Poisson equation (1.1a), we have

$$\nabla \cdot \mathbf{E} = \varepsilon_0^{-1} \rho_{\text{tr}} = \varepsilon_0^{-1}(\rho_{\text{ex}} - \nabla \cdot \mathbf{P}) \tag{1.5}$$

where the true space charge density ρ_{tr} is the sum of the externally applied space charge ρ_{ex} and the negative divergence of the polarization of the material. The relation

$$\mathbf{P} = \epsilon_0 \chi_{\text{e}} \mathbf{E} \tag{1.6}$$

defines the electric susceptibility χ_{e} which, of course, cannot be taken from the laws of electrodynamics but must be calculated from the properties of the materials. With (1.6) and the definition

$$\mathbf{D} = \varepsilon_0 \mathbf{E} + \mathbf{P} = \varepsilon_0(1 + \chi_{\text{e}})\,\mathbf{E} =: \varepsilon_0 \varepsilon \mathbf{E} \tag{1.7}$$

we transform the Poisson equation to

$$\nabla \cdot \mathbf{D} = \rho_{\text{ex}} \tag{1.8}$$

Induced space charges and currents do not affect relations (1.1b) and (1.1c), and consequently these relations are not changed by the presence of materials.

Considering (1.1d) we remark that the true current density in a system with materials is composed of the externally applied current density $\mathbf{j}_{ex}$, the current density connected with the polarization of the material $\partial\mathbf{P}/\partial t$, and the current density related to the magnetization of the material $\mathbf{j}_m = \nabla \times \mathbf{M}$.

Therefore (1.4) is transformed into

$$\nabla \times \mathbf{B} = \mu_0(\mathbf{j}_{ex} + \partial\mathbf{P}/\partial t + \nabla \times \mathbf{M} + \epsilon_0\, \partial\mathbf{E}/\partial t) \tag{1.9}$$

We introduce the magnetic susceptibility χ_m by the relations

$$\mathbf{M} = \chi_m \mathbf{H} \tag{1.10}$$

and

$$\mathbf{B} = \mu_0(\mathbf{H} + \mathbf{M}) = \mu_0(1 + \chi_m)\, \mathbf{H} =: \mu_0\mu\mathbf{H} \tag{1.11}$$

Using (1.7), we find from (1.9) that

$$\nabla \times \mathbf{H} = \mathbf{j}_{ex} + \partial\mathbf{D}/\partial t \tag{1.12}$$

Summarizing, we represent Maxwell's equations describing the electromagnetic field in the presence of materials

$$\begin{array}{llll} \text{(a)} & \nabla \cdot \mathbf{D} = \rho & \text{(b)} & \nabla \cdot \mathbf{B} = 0 \\ \text{(c)} & \nabla \times \mathbf{E} = -\partial\mathbf{B}/\partial t & \text{(d)} & \nabla \times \mathbf{H} = \mathbf{j} + \partial\mathbf{D}/\partial t \\ \text{(e)} & \mathbf{j} = \sigma\mathbf{E} & \text{(f)} & \mathbf{D} = \varepsilon_0\varepsilon\mathbf{E} \\ \text{(g)} & \mathbf{B} = \mu_0\mu\mathbf{H} & & \end{array} \tag{1.13}$$

where we have added the relation connecting the current density with the field through the conductivity σ. We dropped the subscripts ex but want to keep in mind that the quantities ρ and $\mathbf{j}$ are the externally applied space charge and current densities.

The quantities ε, μ, and σ depend on the characteristics of the materials. Further they may depend on time and field. If the material constants ε, μ, and σ depend on the time or on the frequency of the process, respectively, and do not depend on the field quantity, the set of equations (1.13) still holds, if we study only one mode of the time

Fourier analysis, replacing the operator $\partial/\partial t$ by $i\omega$ and using the specific values ε_ω, μ_ω, and σ_ω† according to the relations

$$\begin{Bmatrix} \mathbf{D}(t) \\ \mathbf{B}(t) \\ \mathbf{j}(t) \end{Bmatrix} = \frac{1}{2\pi} \iint_{-\infty}^{+\infty} e^{i\omega(t-\tau)} \begin{Bmatrix} \varepsilon_0 \varepsilon_\omega \\ \mu_0 \mu_\omega \\ \sigma_\omega \end{Bmatrix} \begin{Bmatrix} \mathbf{E}(\tau) \\ \mathbf{H}(\tau) \\ \mathbf{E}(\tau) \end{Bmatrix} d\omega\, d\tau \tag{1.14}$$

Since we did not introduce convection currents, our equations are exact only for materials at rest.

1.2. The Electrodynamic Potentials

Equation (1.13b) shows that **B** may be presented through a vector potential **A**

$$\mathbf{B} = \nabla \times \mathbf{A} \tag{1.15}$$

From (1.13c) it follows that the electrical field can be written in the form

$$\mathbf{E} = -\partial \mathbf{A}/\partial t - \nabla \Phi \tag{1.16}$$

The remaining Maxwell equations then yield

$$\Delta \Phi + \partial \nabla \cdot \mathbf{A}/\partial t = -\rho/\varepsilon_0 \varepsilon \tag{1.17}$$

and

$$\Delta \mathbf{A} - \varepsilon_0 \varepsilon \mu_0 \mu\, \partial^2 \mathbf{A}/\partial t^2 - \nabla(\epsilon_0 \varepsilon \mu_0 \mu\, \partial \Phi/\partial t + \nabla \cdot \mathbf{A}) = -\mu_0 \mu \mathbf{j} \tag{1.18}$$

Whereas in the definition of the scalar potential Φ only an arbitrary constant is at our disposal, the definition of **A** leaves the choice of $\nabla \cdot \mathbf{A}$ open.

This freedom in the choice of the vector potential **A** allows us to change from a given set **A**, Φ, to another one **A**′, Φ'. We are free to add a gradient to **A**

$$\mathbf{A}' = \mathbf{A} - \nabla \psi \tag{1.19}$$

without changing **B**. According to (1.16), **E** remains unchanged if we use

$$\Phi' = \Phi + \partial \psi/\partial t \tag{1.20}$$

Equations (1.19) and (1.20) represent the so-called **gauge transforma-**

† It is clear from the definitions that the inverse transformation of ε_ω, μ_ω, σ_ω does not yield the time-dependent functions $\varepsilon(t)$, $\mu(t)$, $\sigma(t)$ as they might be taken formally from (1.7), (1.11), and (1.13e).

tions. They are a helpful means to ensure that a given physical relationship leads only to such consequences that can be expressed in the field formulation of interaction of particles and currents. In particular, if we choose $\nabla \cdot \mathbf{A} = 0$, we have the advantage that (1.17) takes the simple form of Poisson's equation. On the other hand, (1.18) would be quite complicated. This choice is called the **Coulomb gauge**. More commonly, the so-called **Lorentz gauge**

$$\nabla \cdot \mathbf{A} + \varepsilon_0 \varepsilon \mu_0 \mu \, \partial \Phi / \partial t = -\mu_0 \mu \sigma \Phi \tag{1.21}$$

is applied, since it has the advantage of symmetrizing the two equations governing the potentials. Introducing (1.21), we have

$$\Delta \mathbf{A} - \varepsilon_0 \varepsilon \mu_0 \mu \, \partial^2 \mathbf{A} / \partial t^2 - \mu_0 \mu \sigma \, \partial \mathbf{A} / \partial t = -\mu_0 \mu (\mathbf{j} - \sigma \mathbf{E}) =: -\mu_0 \mu \mathbf{j}_{\text{el}} \tag{1.22}$$

and

$$\Delta \Phi - \varepsilon_0 \varepsilon \mu_0 \mu \, \partial^2 \Phi / \partial t^2 - \mu_0 \mu \sigma \, \partial \Phi / \partial t = -\rho / \varepsilon_0 \varepsilon \tag{1.23}$$

It should be recalled that the current density $\mathbf{j}_{\text{el}}$ represents only that part of the current density that is produced by external electromotive forces. It does not contain contributions of the current density induced by the electric fields in conducting media and represented by the last term on the left-hand side of (1.22).

With the introduction of the D'Alembertian operator

$$\Box := \Delta - (1/c^2) \, \partial^2 / \partial t^2, \qquad c^{-2} = \epsilon_0 \varepsilon \mu_0 \mu \tag{1.24}$$

we have the relations

$$\Box \mathbf{A} - \mu_0 \mu \sigma \, \partial \mathbf{A} / \partial t = -\mu_0 \mu \mathbf{j}_{\text{el}} \tag{1.25}$$

$$\Box \Phi - \mu_0 \mu \sigma \, \partial \Phi / \partial t = -\rho / \varepsilon_0 \varepsilon \tag{1.26}$$

which we choose as the basis of our investigations.

Solutions for the Potentials

Considering the vacuum situation, we have to solve the equations

$$\Box \mathbf{A} = -\mu_0 \mathbf{j} \tag{1.27}$$

$$\Box \Phi = -\rho / \varepsilon_0 \tag{1.28}$$

Since both equations have the same structure, we may consider

$$\Box M(\mathbf{x}, t) = -a(\mathbf{x}, t) \tag{1.29}$$

There are various approaches to solve this equation. One is to consider the retarded potentials as given and prove by insertion into (1.29) that this equation is satisfied. We prefer a more deductive procedure by applying Fourier transforms:

$$\tilde{M}(\mathbf{x}, \omega) = (2\pi)^{-1/2} \int_{-\infty}^{+\infty} e^{-i\omega t} M(\mathbf{x}, t)\, dt \tag{1.30}$$

This yields, for (1.29),

$$\Delta\tilde{M}(\mathbf{x}, \omega) + (\omega^2/c^2)\, \tilde{M}(\mathbf{x}, \omega) = -\tilde{a}(\mathbf{x}, \omega) \tag{1.31}$$

The particular solution of the inhomogeneous equation (1.31) may be expressed with the help of Green's function derived from the equation

$$\Delta\tilde{G}(\mathbf{x}, \mathbf{x}', \omega) + (\omega^2/c^2)\, \tilde{G}(\mathbf{x}, \mathbf{x}', \omega) = -\delta(\mathbf{x} - \mathbf{x}') \tag{1.32}$$

through the integral

$$\tilde{M}(\mathbf{x}, \omega) = \int \tilde{G}(\mathbf{x}, \mathbf{x}', \omega)\, \tilde{a}(\mathbf{x}', \omega)\, d\mathbf{x}' \tag{1.33}$$

The Green's function satisfying (1.32) and vanishing at infinity is given by

$$\tilde{G}(\mathbf{x}, \mathbf{x}', \omega) = (1/4\pi r)\, e^{\pm i\omega r/c}, \qquad r = |\,\mathbf{r}\,| = |\,\mathbf{x} - \mathbf{x}'\,| \tag{1.34}$$

and the desired particular solution of (1.31) is therefore

$$\tilde{M}(\mathbf{x}, \omega) = (1/4\pi) \int \frac{\tilde{a}(\mathbf{x}', \omega)}{r}\, e^{\pm i\omega r/c}\, d\mathbf{x}' \tag{1.35}$$

Inverse Fourier transformation yields

$$M(\mathbf{x}, t) = \frac{1}{4\pi(2\pi)^{1/2}} \iint \frac{\tilde{a}(\mathbf{x}', \omega)}{r}\, e^{i\omega(t \pm r/c)}\, d\mathbf{x}'\, d\omega \tag{1.36}$$

or, respectively,

$$M(\mathbf{x}, t) = \frac{1}{4\pi} \int \frac{a(\mathbf{x}', t \pm r/c)}{r}\, d\mathbf{x}' \tag{1.37}$$

Our potentials are therefore determined by

$$\Phi(\mathbf{x}, t) = \frac{1}{4\pi\varepsilon_0} \int \frac{[\rho(\mathbf{x}', t)]}{r}\, d\mathbf{x}' \tag{1.38}$$

$$\mathbf{A}(\mathbf{x}, t) = \frac{\mu_0}{4\pi} \int \frac{[\mathbf{j}(\mathbf{x}', t)]}{r}\, d\mathbf{x}' \tag{1.39}$$

for arbitrary external current and space charge densities. (The brackets [ ] symbolize that the quantity embraced is to be taken at the time $(t \pm r/c)$.)

Formulas (1.38) and (1.39) resemble the time independent solutions, with the exception that in the current densities and space charges the time is replaced by $(t \pm r/c)$. This means that in calculating the potentials we have to sum over the contributions of currents and charges at different times in the future or in the past.

Now, whereas it seems very reasonable to sum over "retarded contributions" $(t - r/c)$, which take into account that the signal requires a finite time r/c to reach the point of observation, it seems unreasonable to consider contributions from future positions of the charges. These solutions with $(t + r/c)$—which are known as "advanced potentials"—violate the elementary experience of causality implying that no effect can precede its cause in time. For this reason they are generally ignored, although they are mathematically valid solutions and of interest in the study of the problems of renormalization. We are aware that in excluding these advanced contributions we go essentially beyond the content of the differential equations.

The Hertz Potential

Maxwell's equations imply the charge conservation

$$\nabla \cdot \mathbf{j} + \partial\rho/\partial t = 0 \tag{1.40}$$

Equation (1.40) is always fulfilled if we use for the representation of the densities the relations

$$\rho = -\nabla \cdot \mathbf{X}, \qquad \mathbf{j} = \partial\mathbf{X}/\partial t \tag{1.41}$$

Similarly, we may relate the potentials Φ and $\mathbf{A}$ to a vector field $\mathbf{Z}$ through

$$\Phi = -\nabla \cdot \mathbf{Z}, \qquad \mathbf{A} = (1/c^2)\,\partial\mathbf{Z}/\partial t \tag{1.42}$$

which then satisfies identically the Lorentz condition (1.21) *in vacuo*.

Equations (1.41), (1.42), and (1.29) yield for the interrelation of $\mathbf{Z}$ and $\mathbf{X}$:

$$\Box\mathbf{Z} = -(1/\varepsilon_0)\,\mathbf{X} \tag{1.43}$$

with the solution

$$\mathbf{Z}(\mathbf{x}, t) = \frac{1}{4\pi\varepsilon_0}\int \frac{[\mathbf{X}]}{r}\,d\mathbf{x}' \tag{1.44}$$

and the Fourier components

$$\tilde{\mathbf{Z}}_\omega(\mathbf{x}) = \frac{1}{4\pi\varepsilon_0} \int \frac{\tilde{\mathbf{X}}_\omega(\mathbf{x}')}{r} e^{-i\omega r/c} \, d\mathbf{x}' \tag{1.45}$$

Obviously the electromagnetic fields follow from the **Hertz potential Z** through

$$\mathbf{E} = \nabla \times (\nabla \times \mathbf{Z}) - (1/\varepsilon_0)\,\mathbf{X}, \qquad \mathbf{B} = (1/c^2)\,\nabla \times \partial\mathbf{Z}/\partial t \tag{1.46}$$

1.3. The General Electromagnetic Fields

Due to (1.15), the magnetic induction **B** follows from the vector potential (1.39) through

$$\tilde{\mathbf{B}}_\omega = \frac{\mu_0}{4\pi} \nabla \times \int \tilde{\mathbf{j}}_\omega(\mathbf{x}') \frac{e^{-i\omega r/c}}{r} \, d\mathbf{x}' \tag{1.47}$$

which after elementary calculations and inverse Fourier transformation results in

$$\mathbf{B}(\mathbf{x}, t) = \frac{\mu_0}{4\pi} \int \left(\frac{[\mathbf{j}] \times \mathbf{r}}{r^3} + \frac{1}{c} \frac{[\partial\mathbf{j}/\partial t] \times \mathbf{r}}{r^2} \right) d\mathbf{x}' \tag{1.48}$$

Similarly from (1.16) and the potentials (1.38) and (1.39), we have

$$\begin{aligned} 4\pi\varepsilon_0 \tilde{\mathbf{E}}_\omega &= -\int \tilde{\rho}_\omega \nabla \left(\frac{e^{-i\omega r/c}}{r} \right) d\mathbf{x}' - \frac{i\omega}{c^2} \int \tilde{\mathbf{j}}_\omega \frac{e^{-i\omega r/c}}{r} \, d\mathbf{x}' \\ &= \int \tilde{\rho}_\omega \frac{\mathbf{r}}{r^3} e^{-i\omega r/c} \, d\mathbf{x}' + \frac{i\omega}{c} \int \left(\tilde{\rho}_\omega \frac{\mathbf{r}}{r^2} - \frac{1}{cr} \tilde{\mathbf{j}}_\omega \right) e^{-i\omega r/c} \, d\mathbf{x}' \\ &= \tilde{\mathbf{E}}_{\omega\mathrm{C}} + \tilde{\mathbf{E}}_{\omega\mathrm{R}} \end{aligned} \tag{1.49}$$

The evaluation of the second term on the right-hand side of (1.49) is somewhat cumbersome. We therefore give only a sketch of the necessary calculations.

Applying the Fourier transform of the charge continuity equation

$$\nabla' \cdot \tilde{\mathbf{j}}_\omega + i\omega\tilde{\rho}_\omega = 0 \tag{1.50}$$

we may express this term $\tilde{\mathbf{E}}_{\omega\mathrm{R}}$ by $\tilde{\mathbf{j}}_\omega$ only:

$$\tilde{\mathbf{E}}_{\omega\mathrm{R}} = -\frac{1}{c} \int \left\{ \frac{\mathbf{r}}{r} (\nabla' \cdot \tilde{\mathbf{j}}_\omega(\mathbf{x}')) + \frac{i\omega}{c} \tilde{\mathbf{j}}_\omega(\mathbf{x}') \right\} \frac{e^{-i\omega r/c}}{r} \, d\mathbf{x}' \tag{1.51}$$

Application of standard procedures of the vector analysis produces

$$\tilde{\mathbf{E}}_{\omega \mathrm{R}} = \frac{1}{c} \int \{(\tilde{\mathbf{j}}_\omega \cdot \mathbf{r})\, \mathbf{r} - \mathbf{r} \times (\tilde{\mathbf{j}}_\omega \times \mathbf{r})\} \frac{e^{-i\omega r/c}}{r^4}\, d\mathbf{x}'$$

$$- \frac{i\omega}{c^2} \int \mathbf{r} \times (\tilde{\mathbf{j}}_\omega \times \mathbf{r}) \frac{e^{-i\omega r/c}}{r^3}\, d\mathbf{x}' \tag{1.52}$$

or, after inverse Fourier transformation,

$$\mathbf{E}(\mathbf{x}, t) = \frac{1}{4\pi\varepsilon_0} \left\{ \int \frac{[\rho]\mathbf{r}}{r^3}\, d\mathbf{x}' + \frac{1}{c} \int \frac{([\mathbf{j}] \cdot \mathbf{r})\, \mathbf{r} - \mathbf{r} \times ([\mathbf{j}] \times \mathbf{r})}{r^4}\, d\mathbf{x}' \right.$$

$$\left. + \frac{1}{c^2} \int \frac{([\partial \mathbf{j}/\partial t] \times \mathbf{r}) \times \mathbf{r}}{r^3}\, d\mathbf{x}' \right\} \tag{1.53}$$

As can be seen from (1.48) and (1.53), the fields are composed of contributions with a different dependence on the distance r. The first terms on the right-hand sides are the retarded induction and the retarded Coulomb field, respectively, varying with at least r^{-2}, whereas the last terms—the radiation terms—vary with r^{-1}.

Note that the general electromagnetic fields are not simply the retarded equivalents of the static or stationary solutions, in spite of the fact that this is true for the potentials Φ and $\mathbf{A}$. The reason is, of course, that due to the retardation the operator ∇ effectively also implies a time derivative of the retarded quantities. This, and only this, brings about the radiation fields.

1.4. The Liénard–Wiechert Potentials

We now specify our potentials (1.38) and (1.39) for the case of a charge distribution of narrow extension moving with a uniform velocity. This concept is used as a model for elementary charges. We do not assume such elementary charges to be "point charges" or "rigid structures" since such concepts cause mathematical divergencies and contradictions to relativistic requirements. A radius of our charge is conceived only in the sense that it indicates the breakdown of the validity of our macroscopic electrodynamics. This limit may or may not be identical with the classical electron radius

$$r_\mathrm{e} := e^2/4\pi\varepsilon_0 m_\mathrm{e} c^2$$

There are three basic facts on which our derivation of the potentials of a moving charge rests: (a) The extension of the charge distribution is infinitesimally small in comparison to the distance to the test point.

(b) All parts of the distribution move with the velocity $\mathbf{v}$. (c) The total charge is invariant: $\int \rho \, d\mathbf{x}' = e$.

The problem would be quite trivial if (1.38) and (1.39) contained ρ, $\mathbf{j}$, instead of $[\rho]$, $[\mathbf{j}]$, since we could then use $r_e \ll r$ and $\int \rho \, d\mathbf{x}' =$ constant $= e$. However, this is excluded because we have

$$\int [\rho(\mathbf{x}', t)] \, d\mathbf{x}' \neq e = \int \rho(\mathbf{x}', t) \, d\mathbf{x}' \tag{1.54}$$

Due to the motion, the same part of the charge distribution contributes signals from different places at different times.

It is not difficult to relate the signal source at $t' = t - r/c$ to the fraction de of the total charge e which contributes only at that time $(t - r/c)$. If the charge were at rest, then

$$(de)_{\mathbf{v}=0} = [\rho(\mathbf{x}', t)] \, d\sigma \, c \, dt' \tag{1.55}$$

would hold. $d\sigma$ denotes the surface element of a sphere surrounding the test point. The actual fraction de is smaller than $(de)_{\mathbf{v}=0}$ if the charge distribution moves in the direction of the "collecting sphere" toward the test point.

Therefore

$$\begin{aligned} de &= [\rho(\mathbf{x}', t)] \, d\sigma \, c \, dt' - [\mathbf{v} \cdot (\mathbf{r}/r) \, \rho(\mathbf{x}', t)] \, d\sigma \, dt' \\ &= [\rho] \left[1 - \frac{\mathbf{v} \cdot \mathbf{r}}{cr} \right] d\mathbf{x}' \end{aligned} \tag{1.56}$$

holds. Solving for the retarded charge density and introducing into (1.38) and (1.39), we find

$$\begin{Bmatrix} \Phi \\ \mathbf{A} \end{Bmatrix} = \frac{1}{4\pi} \int \begin{Bmatrix} \varepsilon_0^{-1} \\ \mu_0 \end{Bmatrix} \begin{Bmatrix} 1 \\ [\mathbf{v}] \end{Bmatrix} \frac{de}{[r - \mathbf{v} \cdot \mathbf{r}/c]} \tag{1.57}$$

Due to the assumption of an extension small in comparison to r, it follows that

$$\Phi = \frac{e}{4\pi\varepsilon_0} \left[\frac{1}{r - \mathbf{v} \cdot \mathbf{r}/c} \right], \qquad \mathbf{A} = \frac{e\mu_0}{4\pi} \left[\frac{\mathbf{v}}{r - \mathbf{v} \cdot \mathbf{r}/c} \right] \tag{1.58}$$

These are the **Liénard–Wiechert potentials.**

A more direct but formal derivation may be achieved by introducing a Dirac function in the form

$$\begin{Bmatrix} \Phi \\ \mathbf{A} \end{Bmatrix} = \frac{1}{4\pi} \begin{Bmatrix} \varepsilon_0^{-1} \\ \mu_0 \end{Bmatrix} \int \begin{bmatrix} \rho \\ \mathbf{j} \end{bmatrix} \frac{d\mathbf{x}'}{r} = \frac{1}{4\pi} \begin{Bmatrix} \epsilon_0^{-1} \\ \mu_0 \end{Bmatrix} \iint \begin{Bmatrix} \rho(\mathbf{x}', \tau) \\ \mathbf{j}(\mathbf{x}', \tau) \end{Bmatrix} \frac{\delta(\tau - t + r/c)}{r} \, d\mathbf{x}' \, d\tau \tag{1.59}$$

yielding

$$\begin{Bmatrix} \Phi \\ \mathbf{A} \end{Bmatrix} = \frac{e}{4\pi} \begin{Bmatrix} \varepsilon_0^{-1} \\ \mu_0 \end{Bmatrix} \int \begin{Bmatrix} 1 \\ \mathbf{v}(\tau) \end{Bmatrix} \frac{\delta(\tau - t + r/c)}{r(\tau)} \, d\tau \tag{1.60}$$

2. General Formulation of Radiation Fields

2.1. Multipole Expansion of the Field Calculated from the Hertz Vector

The ω-Fourier component of the Hertz vector [see (1.45)]

$$\tilde{\mathbf{Z}}_\omega = \frac{1}{4\pi\varepsilon_0} \int \tilde{\mathbf{X}}_\omega(\mathbf{x}') \frac{e^{-i\omega r(\mathbf{x},\mathbf{x}')/c}}{r(\mathbf{x}, \mathbf{x}')} \, d\mathbf{x}' \tag{2.1}$$

contains the Green's function $\tilde{G}(\mathbf{x}, \mathbf{x}', \omega)$ of (1.34).

In the following, we discuss the expression (2.1) on the basis of the expansion

$$(1/r)e^{-i\omega r/c} = -i(\omega/c) \sum_{l=0}^{\infty} (2l + 1)\, P_l(\cos\gamma)\, j_l(-(\omega/c)|\,\mathbf{x}'\,|)\, h_l(-(\omega/c)|\,\mathbf{x}\,|) \tag{2.2}$$

which holds for $|\,\mathbf{x}'\,| < |\,\mathbf{x}\,|$. γ is the angle formed by $\mathbf{x}$ and $\mathbf{x}'$, P_l the lth Legendre polynomial, and j_l, h_l, the spherical Bessel and Hankel functions

$$j_l(y) = (\pi/2y)^{1/2}\, J_{l+1/2}(y), \qquad h_l(y) = (\pi/2y)^{1/2} H_{l+1/2}(y) \tag{2.3}$$

where J and H are the Bessel and Hankel functions, respectively.

The expansion (2.2) offers real advantage only if asymptotic formulas can be engaged. This is possible if one of the following four cases holds:

$$\begin{array}{ll} \text{(a)} \quad |\,\mathbf{x}\,|, \quad |\,\mathbf{x}'\,| \gg \lambda\!\!\!^{-} & \text{(b)} \quad |\,\mathbf{x}\,| \gg \lambda\!\!\!^{-}, \quad |\,\mathbf{x}'\,| \ll \lambda\!\!\!^{-} \\ \text{(c)} \quad |\,\mathbf{x}\,| \ll \lambda\!\!\!^{-}, \quad |\,\mathbf{x}'\,| \gg \lambda\!\!\!^{-} & \text{(d)} \quad |\,\mathbf{x}\,|, \quad |\,\mathbf{x}'\,| \ll \lambda\!\!\!^{-} \end{array} \tag{2.4}$$

where $\lambda\!\!\!^{-} = c/\omega$ is the wavelength of the wave under consideration.

Case (a) is the radiation field limit of a high frequency charge distribution of large extension. It will not be considered since $j_l(-\,|\,\mathbf{x}'\,|/\lambda\!\!\!^{-})$ for $|\,\mathbf{x}'\,| \gg \lambda\!\!\!^{-}$ is a periodic function of the form

$$j_l\left(-\frac{|\,\mathbf{x}'\,|}{\lambda\!\!\!^{-}}\right) \rightarrow \frac{\lambda\!\!\!^{-}}{|\,\mathbf{x}'\,|} \sin\left(\frac{|\,\mathbf{x}'\,|}{\lambda\!\!\!^{-}} + \frac{l\pi}{2}\right) \tag{2.5}$$

and the introduction into (2.1) yields integrals over

$$\tilde{\mathbf{X}}_\omega(\mathbf{x}')\, P_l \sin(|\,\mathbf{x}'\,|/\lambda\!\!\!{}^{-} + \tfrac{1}{2}l\pi)$$

which exhibit no general trends.

Case (b) deals with the radiation field limit $|\,\mathbf{x}\,| \gg \lambda\!\!\!{}^{-}$ of a low-frequency localized charge distribution of small extension. This case is of great practical interest and allows to draw general conclusions from (2.1). It is the basis of the Hertz multipole expansion.

Cases (c) and (d) will not be studied here, since neither represent radiation fields due to the limitation $|\,\mathbf{x}\,| \ll \lambda\!\!\!{}^{-}$.

We investigate therefore case (b) with $|\,\mathbf{x}'\,| \ll \lambda\!\!\!{}^{-} \ll |\,\mathbf{x}\,|$, where we use the asymptotic formulas

$$\begin{aligned} j_l\left(-\frac{\omega}{c}\,|\,\mathbf{x}'\,|\right) &= \frac{2^l l!}{(2l+1)!}\left(-\frac{\omega}{c}\,|\,\mathbf{x}'\,|\right)^l \\ h_l\left(-\frac{\omega}{c}\,|\,\mathbf{x}\,|\right) &= (-i)^{l-1}\,\frac{e^{-i\omega|\mathbf{x}|/c}}{\omega\,|\,\mathbf{x}\,|/c} \end{aligned} \tag{2.6}$$

in (2.2), which yields

$$\tilde{\mathbf{Z}}_\omega = \sum_{l=0}^{\infty} \tilde{\mathbf{Z}}_\omega^{(l)} \tag{2.7}$$

with the multipole potential

$$\tilde{\mathbf{Z}}_\omega^{(l)} = \frac{i^l}{4\pi\varepsilon_0}\,\frac{e^{-i\omega|\mathbf{x}|/c}}{|\,\mathbf{x}\,|}\,\frac{2^l l!}{(2l)!}\int\left(\frac{\omega}{c}\,|\,\mathbf{x}'\,|\right)^l \tilde{\mathbf{X}}_\omega(\mathbf{x}')\,P_l(\cos\gamma)\,d\mathbf{x}' \tag{2.8}$$

In the preceding, we assumed tacitly that the origin of our coordinate system was chosen within our charge distribution, otherwise our discussion of the four cases would be meaningless. With this in mind, (2.8) shows that our multipole expansion gives the contributions in a decreasing order, since the magnitude of the pole $\tilde{\mathbf{Z}}_\omega^{(l)}$ is characterized by $\langle(|\,\mathbf{x}'\,|/\lambda\!\!\!{}^{-})^l\rangle$.

Since γ is the angle between $\mathbf{x}$ and $\mathbf{x}'$, (2.8) gives insight into the angular dependence only if the source distribution is linear in space and we choose the axis of our coordinate system in the direction of the line source. In general, however, we use the addition theorem of the spherical harmonics

$$P_l(\cos\gamma) = \sum_{m=0}^{l} a_m\,\frac{(l-m)!}{(l+m)!}\,P_l^{\,m}(\cos\vartheta)\,P_l^{\,m}(\cos\vartheta')\cos m(\varphi-\varphi') \tag{2.9}$$

$$a_0 = 1, \qquad a_m = 2 \qquad \text{for} \quad m \neq 0$$

where ϑ and ϑ' are the polar angles, φ and φ' the azimuthal angles of $\mathbf{x}$ and $\mathbf{x}'$, respectively.

This yields

$$\tilde{\mathbf{Z}}_\omega^{(l)} = \sum_{m=0}^{l} \tilde{\mathbf{Z}}_\omega^{(l,m)} \tag{2.10}$$

with

$$\tilde{\mathbf{Z}}_\omega^{(l,m)} = \frac{i^l}{4\pi\varepsilon_0} \frac{e^{-i\omega|\mathbf{x}|/c}}{|\mathbf{x}|} \frac{2^l l!}{(2l)!} a_m \frac{(l-m)!}{(l+m)!} P_l^m(\cos\vartheta)$$
$$\times \int (\omega\,|\,\mathbf{x}'\,|/c)^l\, \tilde{\mathbf{X}}_\omega(\mathbf{x}')\, P_l^m(\cos\vartheta') \cos m(\varphi - \varphi')\, d\mathbf{x}' \tag{2.11}$$

The Hertz potentials $\tilde{\mathbf{Z}}^{(l)}$ are related to the multipoles of the charge distribution. We prove this for the first orders starting with $\tilde{\mathbf{Z}}_\omega^{(0)}$, where we have

$$\tilde{\mathbf{Z}}_\omega^{(0)} = \frac{e^{-i\omega|\mathbf{x}|/c}}{4\pi\varepsilon_0\,|\,\mathbf{x}\,|} \int \tilde{\mathbf{X}}_\omega(\mathbf{x}')\, d\mathbf{x}' \tag{2.12}$$

Using

$$\nabla' \cdot (\tilde{\mathbf{X}}_\omega x_\alpha') = x_\alpha' \nabla' \cdot \tilde{\mathbf{X}}_\omega + \tilde{X}_{\omega\alpha} \tag{2.13}$$

the Gauss theorem, and (1.41), we find

$$\begin{aligned} \tilde{\mathbf{Z}}_\omega^{(0)} &= -\frac{e^{-i\omega|\mathbf{x}|/c}}{4\pi\varepsilon_0\,|\,\mathbf{x}\,|} \int \mathbf{x}'(\nabla' \cdot \tilde{\mathbf{X}}_\omega)\, d\mathbf{x}' = \frac{e^{-i\omega|\mathbf{x}|/c}}{4\pi\varepsilon_0\,|\,\mathbf{x}\,|} \langle \mathbf{x}'\tilde{\rho}_\omega(\mathbf{x}')\rangle \\ &= \frac{1}{4\pi\varepsilon_0} \frac{e^{-i\omega|\mathbf{x}|/c}}{|\,\mathbf{x}\,|} \tilde{\mathbf{D}}_\mathrm{e} \end{aligned} \tag{2.14}$$

where $\tilde{\mathbf{D}}_\mathrm{e}$ is the electric dipole moment of the charge distribution. We see that the lowest-order Hertz potential is determined by the electric dipole moment of the source. The first-order Hertz potential is given by

$$\tilde{\mathbf{Z}}_\omega^{(1)} = \frac{i}{4\pi\varepsilon_0} \frac{e^{-i\omega|\mathbf{x}|/c}}{|\,\mathbf{x}\,|} \int \frac{\omega}{c}\,|\,\mathbf{x}'\,|\,\tilde{\mathbf{X}}_\omega(\mathbf{x}')\, P_1(\cos\gamma)\, d\mathbf{x}' \tag{2.15}$$

or

$$\begin{aligned} \tilde{\mathbf{Z}}_\omega^{(1)} &= \frac{i}{4\pi\varepsilon_0} \frac{e^{-i\omega|\mathbf{x}|/c}}{|\,\mathbf{x}\,|^2} \frac{\omega}{c} \int (\mathbf{x}' \cdot \mathbf{x})\, \tilde{\mathbf{X}}_\omega(\mathbf{x}')\, d\mathbf{x}' \\ &= \frac{i}{4\pi\varepsilon_0} \frac{e^{-i\omega|\mathbf{x}|/c}}{|\,\mathbf{x}\,|^2} \frac{\omega}{c} \int \tilde{\mathbf{X}}_\omega(\mathbf{x}')\,)(\,\mathbf{x}' \cdot \mathbf{x}\, d\mathbf{x}' \end{aligned} \tag{2.16}$$

Let us decompose the tensor $\tilde{\mathbf{X}}_\omega)(\mathbf{x}'$ into its symmetric and antisymmetric parts

$$\tilde{\mathbf{X}}_\omega)(\mathbf{x}' = \frac{\tilde{\mathbf{X}}_\omega)(\mathbf{x}' + \mathbf{x}')(\tilde{\mathbf{X}}_\omega}{2} + \frac{\tilde{\mathbf{X}}_\omega)(\mathbf{x}' - \mathbf{x}')(\tilde{\mathbf{X}}_\omega}{2} \tag{2.17}$$

We find that

$$\int \tilde{\mathbf{X}}_\omega)(\mathbf{x}' \cdot \mathbf{x}\, d\mathbf{x}' = \tfrac{1}{2} \int \mathbf{x} \times (\tilde{\mathbf{X}}_\omega \times \mathbf{x}')\, d\mathbf{x}' + \tfrac{1}{2} \int \{\tilde{\mathbf{X}}_\omega(\mathbf{x}' \cdot \mathbf{x}) + \mathbf{x}'(\tilde{\mathbf{X}}_\omega \cdot \mathbf{x})\}\, d\mathbf{x}' \tag{2.18}$$

Remembering that $\tilde{\mathbf{j}}_\omega = i\omega\tilde{\mathbf{X}}_\omega$ holds, we get from the antisymmetric contribution

$$\begin{aligned} \tilde{\mathbf{Z}}^{(1)}_{\omega\,\mathrm{asym}} &= \frac{1}{4\pi\varepsilon_0} \frac{1}{c} \frac{e^{-i\omega|\mathbf{x}|/c}}{|\mathbf{x}|^2} \mathbf{x} \times \tfrac{1}{2} \int \tilde{\mathbf{j}}_\omega \times \mathbf{x}'\, d\mathbf{x}' \\ &= -\frac{1}{4\pi\varepsilon_0} \frac{1}{c} \frac{e^{-i\omega|\mathbf{x}|/c}}{|\mathbf{x}|^2} \mathbf{x} \times \tilde{\mathbf{D}}_{\mathrm{m}} \end{aligned} \tag{2.19}$$

where $\tilde{\mathbf{D}}_{\mathrm{m}}$ is the magnetic dipole moment

$$\tilde{\mathbf{D}}_{\mathrm{m}} = \tfrac{1}{2} \int \mathbf{x}' \times \tilde{\mathbf{j}}_\omega(\mathbf{x}')\, d\mathbf{x}' \tag{2.20}$$

of the current distribution. The antisymmetric part of the first-order Hertz potential represents therefore the field due to the magnetic dipole moment. Physically, the magnetic dipole is due to an oscillating circular current.

The symmetric contribution to the first-order Hertz vector

$$\tilde{\mathbf{Z}}^{(1)}_{\omega\,\mathrm{sym}} = \frac{i}{4\pi\varepsilon_0} \frac{e^{-i\omega|\mathbf{x}|/c}}{|\mathbf{x}|^2} \frac{\omega}{c} \tfrac{1}{2} \int \{\tilde{\mathbf{X}}_\omega)(\mathbf{x}' + \mathbf{x}')(\tilde{\mathbf{X}}_\omega\} \cdot \mathbf{x}\, d\mathbf{x}' \tag{2.21}$$

can be transformed using the relation

$$\nabla' \cdot (\tilde{\mathbf{X}}_\omega x_\alpha' x_\beta') = (\nabla' \cdot \tilde{\mathbf{X}}_\omega)\, x_\alpha' x_\beta' + \tilde{X}_{\omega\alpha} x_\beta' + \tilde{X}_{\omega\beta} x_\alpha' \tag{2.22}$$

the Gauss theorem, and (1.41) into

$$\tilde{\mathbf{Z}}^{(1)}_{\omega\,\mathrm{sym}} = \frac{i\omega}{8\pi\varepsilon_0 c} \frac{e^{-i\omega|\mathbf{x}|/c}}{|\mathbf{x}|^2} \mathbf{x} \cdot \underset{2}{\tilde{\mathbf{Q}}}_{\mathrm{e}} \tag{2.23}$$

where

$$\underset{2}{\tilde{\mathbf{Q}}}_{\mathrm{e}} = \int \mathbf{x}')(\mathbf{x}' \tilde{\rho}_\omega(\mathbf{x}')\, d\mathbf{x}' \tag{2.24}$$

is the electric quadrupole tensor of our charge distribution.

Summarizing, we state that the first-order Hertz vector represents the contributions from the magnetic dipole moment and the electric quadrupole moment, whereas the zero-order vector accounts for the effect of the electric dipole moment. Similar to the above consideration, one can show that, in general, the Hertz vector of order l represents the radiation of an electric $2^{(l+1)}$ multipole and a magnetic 2^l multipole.

The electrodynamic fields of the multipoles follow in straightforward fashion from relation (1.46). We present here only some of the relevant results:

(a) The magnetic (l, m)-multipole radiation of a particular source distribution is in general smaller than the electric one by about a factor v/c, where v characterizes the velocity of the source charges.

(b) In the radiation field region $\mathbf{B}$ and $\mathbf{E}$ are transversal to the radius vector each falling off with r^{-1}.

(c) The angular radiation distributions of an (l, m)-multipole are in general rather complicated and essentially given by the term

$$\tfrac{1}{2}(l-m)(l+m+1)|\,P_l^{m+1}\,|^2 + \tfrac{1}{2}(l+m)(l-m+1)|\,P_l^{m-1}\,|^2 + m^2\,|\,P_l^{m}\,|^2 \tag{2.25}$$

(d) The angular momentum radiated away per second by an (l, m)-multipole has only a z component (dL_z/dt). The ratio of this quantity and the energy radiated per second by the multipole $(d\varepsilon/dt)$ is given by

$$\frac{dL_z/dt}{d\varepsilon/dt} = \frac{m}{\omega} = \frac{m\hbar}{\hbar\omega} \tag{2.26}$$

This—although a classical result—has the quantummechanical interpretation that a photon $\hbar\omega$ emerging from a multipole (l, m) carries away the angular momentum $m\hbar$.

2.2. Electromagnetic Field of a Single Charge

The electromagnetic fields of a single charge are uniquely defined through the Liénard–Wiechert potentials

$$\Phi = \frac{e}{4\pi\varepsilon_0}\left[\frac{1}{r - \mathbf{v}\cdot\mathbf{r}/c}\right], \qquad \mathbf{A} = \frac{e}{4\pi\varepsilon_0}\frac{1}{c^2}\left[\frac{\mathbf{v}}{r - \mathbf{v}\cdot\mathbf{r}/c}\right] \tag{2.27}$$

and the relations

$$\mathbf{E} = -\nabla\Phi - \frac{\partial}{\partial t}\mathbf{A}, \qquad \mathbf{B} = \nabla \times \mathbf{A} \tag{2.28}$$

Here we recall that the symbol [ ] refers to the retarded time $t' = t - r/c$ and the operators ∇, $\partial/\partial t$, in (2.28) mean $\nabla|_t$, $\partial/\partial t|_{\mathbf{x}}$.

The difficulty in applying (2.28) to (2.27) arises from the fact that, in general, $\mathbf{v}$, $\mathbf{x}'$ are given as a function of t', so that we have $\Phi(x, t')$, $\mathbf{A}(\mathbf{x}, t')$, rather than $\Phi(\mathbf{x}, t)$, $\mathbf{A}(\mathbf{x}, t)$. To evaluate the relation (2.28) it seems therefore best to express the operators $\partial/\partial t|_{\mathbf{x}}$, $\nabla|_t$ by $\partial/\partial t'|_{\mathbf{x}}$, $\nabla|_{t'}$. This is done through the relations

$$r = c(t - t'), \qquad \partial r/\partial t'|_{\mathbf{x}} = -\mathbf{v}\cdot\mathbf{r}/r \tag{2.29}$$

We find

$$\left.\frac{\partial r}{\partial t}\right|_{\mathbf{x}} = c\left(1 - \left.\frac{\partial t'}{\partial t}\right|_{\mathbf{x}}\right) = \left.\frac{\partial r}{\partial t'}\frac{\partial t'}{\partial t}\right|_{\mathbf{x}} = -\left.\frac{\mathbf{v}\cdot\mathbf{r}}{r}\frac{\partial t'}{\partial t}\right|_{\mathbf{x}} \tag{2.30}$$

or

$$\left.\frac{\partial}{\partial t}\right|_{\mathbf{x}} = \left(1 - \frac{\mathbf{v}\cdot\mathbf{r}}{cr}\right)^{-1}\left.\frac{\partial}{\partial t'}\right|_{\mathbf{x}} \tag{2.31}$$

and

$$\nabla r|_t = -c\nabla t'|_t = \nabla r|_{t'} + \frac{\partial r}{\partial t'}\nabla t'|_t = \frac{\mathbf{r}}{r} - \left.\frac{\mathbf{r}\cdot\mathbf{v}}{r}\nabla t'\right|_t \tag{2.32}$$

or

$$\nabla t'|_t = -\frac{1}{c}\frac{\mathbf{r}}{r - \mathbf{r}\cdot\mathbf{v}/c} \tag{2.33}$$

which yields

$$\nabla|_t = \nabla|_{t'} - \frac{\mathbf{r}}{c(r - \mathbf{r}\cdot\mathbf{v}/c)}\frac{\partial}{\partial t'} \tag{2.34}$$

Applying (2.28), (2.31), and (2.34), the abbreviation

$$s = r - \mathbf{r}\cdot\mathbf{v}/c \tag{2.35}$$

and

$$\nabla s|_{t'} = \frac{\mathbf{r}}{r} - \frac{\mathbf{v}}{c} \tag{2.36}$$

we find for the electromagnetic fields

$$\frac{4\pi\varepsilon_0}{e}\mathbf{E} = \frac{1}{s^3}\left(\mathbf{r} - \frac{r}{c}\mathbf{v}\right)\left(1 - \frac{v^2}{c^2}\right) + \frac{1}{c^2 s^3}\left\{\mathbf{r}\times\left(\left(\mathbf{r} - \frac{r}{c}\mathbf{v}\right)\times\dot{\mathbf{v}}\right)\right\} \tag{2.37}$$

and

$$\frac{4\pi\varepsilon_0 c^2}{e}\mathbf{B} = \frac{\mathbf{v}\times\mathbf{r}}{s^3}\left(1 - \frac{v^2}{c^2}\right) + \frac{1}{cs^3}\frac{\mathbf{r}}{r}\times\left\{\mathbf{r}\times\left(\left(\mathbf{r} - \frac{r}{c}\mathbf{v}\right)\times\dot{\mathbf{v}}\right)\right\} \tag{2.38}$$

We recognize that

$$\mathbf{B} = \mathbf{r} \times \mathbf{E}/rc \tag{2.39}$$

and draw particular attention to the fact that—although we dropped the brackets [ ] for reasons of simplicity—all quantities used have to be retarded.

2.3. Angular and Frequency Distribution of the Energy Radiated from a Charge

Inspection of the r dependence of the terms in (2.37) and (2.38) shows that the second terms on the right-hand side are responsible for radiation. This information determines uniquely the energy transport in the radiation field via the Poynting vector.

However, at the moment we are not so much interested in learning how much electromagnetic energy passes through a surface element at a certain point of observation as in the knowledge of the energy loss of the particle due to radiation into a certain solid angle and within a certain frequency range per unit time t'.

Due to the conservation of energy in the radiation field, the electromagnetic energy passing through the element $d\Omega$ in a time interval dt must be identical with the energy loss of the electron due to emission into $d\Omega$ during the time interval dt' related by (2.31) to dt. If $\mathscr{E}$ is the electron energy, we therefore have

$$\begin{aligned} -d^2\mathscr{E} &= \hat{\mathbf{r}} \cdot \mathbf{S} r^2 \, d\Omega \, dt \\ &= \hat{\mathbf{r}} \cdot (\mathbf{E} \times \mathbf{H}) \, r^2 \, d\Omega \, dt \\ &= \varepsilon_0 c E^2 r^2 \, d\Omega \, dt \end{aligned} \tag{2.40}$$

where we observed that the Poynting vector $\mathbf{S}$ is parallel to $\mathbf{r}$.

Using (2.31), we find

$$w(\Omega) := -\frac{d^2\mathscr{E}}{dt' \, d\Omega} = \varepsilon_0 c E^2 r^2 \left(1 - \frac{\mathbf{v} \cdot \mathbf{r}}{cr}\right) \tag{2.41}$$

which yields

$$w(\Omega) = \frac{e^2}{16\pi^2 \varepsilon_0 c^3} \frac{\{\hat{\mathbf{r}} \times [(\hat{\mathbf{r}} - \mathbf{v}/c) \times \dot{\mathbf{v}}]\}^2}{(1 - \hat{\mathbf{r}} \cdot v/c)^5} \tag{2.42}$$

Integration of (2.42) over the solid angle results in

$$w^{\mathrm{T}} = \frac{e^2}{6\pi\varepsilon_0 c^3} \frac{\dot{v}^2 - (\mathbf{v} \times \dot{\mathbf{v}})^2/c^2}{(1 - v^2/c^2)^3} \tag{2.43}$$

for the total energy radiated by the electron per second.

The quantity w^{T} is identical with the total flux of radiation energy per second through a fixed closed surface containing our charge. Note, however, that the quantity $w(\Omega)$ is not identical with the energy flux $d^2\mathscr{E}/dt\,d\Omega$ measured by an observer at rest.

So far we studied the angular distribution. Now we consider in addition the spectral distribution of the radiation. To this end, we quote Parseval's theorem

$$\begin{aligned}\mathbf{W} &= \int_{-\infty}^{+\infty} \mathbf{E} \times \mathbf{H}\, dt = \int_{-\infty}^{+\infty} \tilde{\mathbf{E}}_\omega \times \tilde{\mathbf{H}}_{-\omega}\, d\omega \\ &= 2\int_0^{\infty} \tilde{\mathbf{E}}_\omega \times \tilde{\mathbf{H}}_\omega\, d\omega = 2\int_0^{\infty} \mathbf{S}_\omega\, d\omega\end{aligned} \tag{2.44}$$

where $\tilde{\mathbf{E}}_\omega$ and $\tilde{\mathbf{H}}_\omega$ are the Fourier transforms

$$\mathbf{E} = (2\pi)^{-1/2}\int \tilde{\mathbf{E}}_\omega e^{i\omega t}\, d\omega, \qquad \mathbf{H} = (2\pi)^{-1/2}\int \tilde{\mathbf{H}}_{\omega'} e^{i\omega' t}\, d\omega' \tag{2.45}$$

whereas $\mathbf{S}_\omega := \tilde{\mathbf{E}}_\omega \times \tilde{\mathbf{H}}_\omega$ is not the Fourier transform of $\mathbf{S} = \mathbf{E} \times \mathbf{H}$. $\mathbf{W}$ has the dimension of an energy per unit area, $\mathbf{S}$ that of energy per unit time and per unit area, and $\mathbf{S}_\omega$ that of energy per unit area and per unit frequency.

Since the Fourier analysis is unique, the radiated energy per unit frequency and unit area will show the intensity

$$2\,|\mathbf{S}_\omega| = 2\,|\tilde{\mathbf{E}}_\omega \times \tilde{\mathbf{H}}_\omega| = 2\varepsilon_0 c\tilde{\mathbf{E}}_\omega{}^2 \tag{2.46}$$

and the corresponding energy loss of the electron into a unit solid angle is given by

$$w(\Omega, \omega) = 2\varepsilon_0\, c\tilde{\mathbf{E}}_\omega{}^2\, r^2 = \frac{2e^2 r^2}{16\pi^2\varepsilon_0 c^3}\left|(2\pi)^{-1/2}\int_{-\infty}^{+\infty} e^{-i\omega t}\, \frac{\mathbf{r} \times [(\mathbf{r} - r\mathbf{v}/c) \times \dot{\mathbf{v}}]}{(r - \mathbf{v}\cdot\mathbf{r}/c)^3}\, dt\right|^2 \tag{2.47}$$

The motion of the particle is usually expressed as a function of t'. We therefore transform in (2.47) the time t into t'

$$w(\Omega, \omega) = \frac{e^2}{16\pi^3\varepsilon_0 c^3}\left|\int_{-\infty}^{+\infty} e^{-i\omega(t' - r/c)}\, \frac{\hat{\mathbf{r}} \times [(\hat{\mathbf{r}} - \mathbf{v}/c) \times \dot{\mathbf{v}}]}{(1 - \mathbf{v}\cdot\hat{\mathbf{r}}/c)^2}\, dt'\right|^2 \tag{2.48}$$

Our basic assumption limits the motion of a charge to a region small in comparison to the distance of the point of observation. Consequently we use

$$\mathbf{x} = \mathbf{x}' + \mathbf{r}, \qquad |\mathbf{x}'| \ll |\mathbf{r}|,\ r \approx x - \hat{\mathbf{r}}\cdot\mathbf{x}' \tag{2.49}$$

For the same reason, we consider $\hat{\mathbf{r}}$ as a constant in our approximation. Then the relation

$$\frac{1}{c}\frac{\hat{\mathbf{r}} \times [(\hat{\mathbf{r}} - \mathbf{v}/c) \times \dot{\mathbf{v}}]}{(1 - \mathbf{v} \cdot \hat{\mathbf{r}}/c)^2} = \frac{d}{dt'}\left(\frac{\hat{\mathbf{r}} - \mathbf{v}/c}{1 - \mathbf{v} \cdot \hat{\mathbf{r}}/c} - \hat{\mathbf{r}}\right) \tag{2.50}$$

holds.

Using partial integration in (2.48) with the conditions $\mathbf{v} = \mathbf{0}$ for $t \to \pm\infty$, we find from (2.50)

$$w(\Omega, \omega) = \frac{e^2\omega^2}{16\pi^3\varepsilon_0 c^3}\left|\int_{-\infty}^{+\infty} e^{-i\omega(t' - \hat{\mathbf{r}}\cdot\mathbf{x}'/c)}\hat{\mathbf{r}} \times (\hat{\mathbf{r}} \times \mathbf{v})\, dt'\right|^2 \tag{2.51}$$

Note that here (and in the following) the frequency must be taken positive in all formulas for spectral energy distributions since we have accounted for negative ω by the factor two in (2.44).

Equations (2.48) and (2.51) enable us to answer all questions concerning the spectral and angular distribution of the radiation from a moving charge provided that the point of observation is sufficiently distant.

The Dipole Approximation

An important special case of our formulas results in the low-velocity limit $v/c \ll 1$—the **dipole approximation.**

The angular distribution of the energy radiated from the electron per unit time is then given from (2.42) by

$$w(\Omega) = (e^2/16\pi^2\varepsilon_0 c^3)\, \dot{\mathbf{v}}^2 \sin^2 \vartheta, \qquad \frac{\dot{\mathbf{v}} \cdot \mathbf{r}}{\dot{v} r} = \cos \vartheta \tag{2.52}$$

and the corresponding total energy by

$$w^{\mathrm{T}} = (e^2/6\pi\varepsilon_0 c^3)\, \dot{v}^2 \tag{2.53}$$

The energy flux per solid angle and frequency unit at the point of observation is determined from (2.48) as

$$w(\Omega, \omega) = (e^2/16\pi^3\varepsilon_0 c^3)\left|\int_{-\infty}^{+\infty} e^{-i\omega t'}\hat{\mathbf{r}} \times (\hat{\mathbf{r}} \times \dot{\mathbf{v}})\, dt'\right|^2 \tag{2.54}$$

Finally, the total energy emitted in all directions per unit frequency is

$$w^{\mathrm{T}}(\omega) = (e^2/6\pi^2\varepsilon_0 c^3)\left|\int_{-\infty}^{+\infty} e^{-i\omega t'}\dot{\mathbf{v}}\, dt'\right|^2 \tag{2.55}$$

Equation (2.52) demonstrates why this approximation $v/c \ll 1$ is called the dipole approximation. Equations (2.53) and (2.55) are the well-known Larmor and Hertz formulas, respectively.

3. Bremsstrahlung

3.1. Classical Description

In our fully ionized system, we consider the effect of the acceleration caused by pair encounters of charged particles. The radiation emitted by these "free–free transitions" is called **bremsstrahlung**.

We base the following discussion on the dipole approximation with the radiation fields

$$\mathbf{E}_R = \frac{e}{4\pi\varepsilon_0 c^2} \frac{\mathbf{r} \times (\mathbf{r} \times \dot{\mathbf{v}})}{r^3}, \qquad \mathbf{B}_R = \frac{e}{4\pi\varepsilon_0 c^3} \frac{\dot{\mathbf{v}} \times \mathbf{r}}{r^2} \tag{3.1}$$

which hold in the range $v \ll c$ and therewith exclude relativistic effects.

Equations (3.1) show that the bremsstrahlung is determined by the acceleration of the center of charge. This is at rest or in uniform motion for partners with equal e/m, which consequently do not contribute to "pair-encounter dipole bremsstrahlung."

Further, for encounters of unlike particles the accelerations are inversely proportional to their masses, so that we may neglect the contribution of the ions in comparison to that of the electrons.

To calculate the radiation emitted during the passage of an electron near an ion of charge number Z, we have to know the kinematics of this motion. Fortunately this is a well-known problem. We know that the electron moves in a plane and with the symbols given in Figure V.1 the dependence of r on φ can be written in the form

$$\frac{1}{r} = \frac{1 - \epsilon \cos \varphi}{b_0(\epsilon^2 - 1)} \tag{3.2}$$

where the parameters b_0 and ϵ are related by

$$b_0 = Ze^2/4\pi\varepsilon_0 m v_s^2, \qquad \epsilon^2 - 1 = (b/b_0)^2 = (j/m v_s b_0)^2 \tag{3.3}$$

to the kinetic energy of the incident particle $\frac{1}{2}m v_s^2$ and the angular momentum j. v_s is the starting velocity of the incident electron.

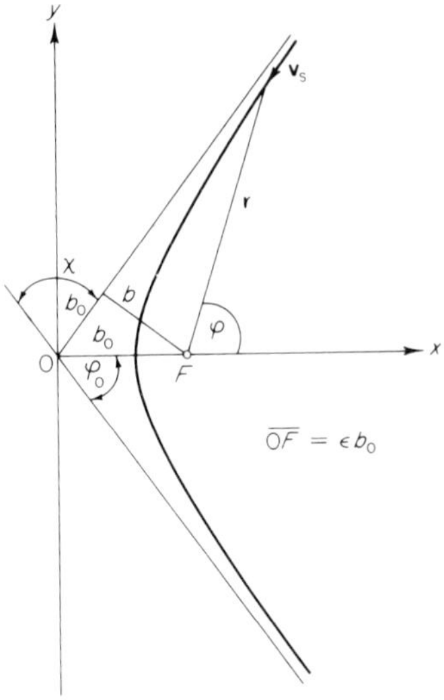

FIGURE V.1. Impact geometry for an attractive Coulomb force.

From Figure V.1 one readily deduces the relations

$$\begin{aligned} b &= b_0 \tan \varphi_0 \\ \epsilon^2 - 1 &= \tan^2 \varphi_0 = \cot^2(\chi/2) \\ \chi &:= \pi - 2\varphi_0 \end{aligned} \tag{3.4}$$

which show that b_0 is the collision parameter for 90° deflection of the incoming electron.

To calculate the spectral distribution of the energy emitted, we use (2.55). With the choice of the coordinate system indicated in Figure V.1, the quantities $\tilde{\ddot{x}}_\omega$ and $\tilde{\ddot{y}}_\omega$ required in (2.55) may be presented in the form

$$\begin{aligned} \tilde{\ddot{x}}_\omega &= (2\pi)^{-1/2} \int_{-\infty}^{+\infty} \ddot{x}(t) \cos \omega t \, dt \\ \tilde{\ddot{y}}_\omega &= -i(2\pi)^{-1/2} \int_{-\infty}^{+\infty} \ddot{y}(t) \sin \omega t \, dt \end{aligned} \tag{3.5}$$

where we made use of the symmetry properties of our system. Now in terms of r and φ, the components $\ddot{x}$ and $\ddot{y}$ are given by

$$\begin{pmatrix} \ddot{x} \\ \ddot{y} \end{pmatrix} = -\frac{Ze^2}{4\pi\varepsilon_0 m r^2} \begin{pmatrix} \cos \varphi \\ \sin \varphi \end{pmatrix} \tag{3.6}$$

Introducing this into (3.5), we have

$$\tilde{\tilde{x}}_\omega = -\frac{Ze^2}{4\pi\varepsilon_0 m v_s b}(2\pi)^{-1/2}\int_{\varphi_0}^{2\pi-\varphi_0} \cos\omega t \cos\varphi \, d\varphi$$

$$\tilde{\tilde{y}}_\omega = i\frac{Ze^2}{4\pi\varepsilon_0 m v_s b}(2\pi)^{-1/2}\int_{\varphi_0}^{2\pi-\varphi_0} \sin\omega t \sin\varphi \, d\varphi \tag{3.7}$$

where we substituted $d\varphi$ at the same time for dt through the law of conservation of angular momentum

$$r^2\, d\varphi/dt = v_s b \tag{3.8}$$

To evaluate the integrals in the relations (3.7), we would have to substitute φ for t which is an elementary but rather complicated procedure. It will prove valuable in the case of the straight-line approximation treated later in this section.

In the general treatment it seems better advised to apply a substitution from celestial mechanics which expresses the coordinates x and y as implicit functions of a suitably chosen parameter (Oster, 1961). The integration of (3.7) is then possible and yields for the spectral representation of the total emitted energy through (2.55)

$$w^{\mathrm{T}}(\omega, b, v_s) = \left(\frac{2e^2\omega^2}{3\pi^2\varepsilon_0 c^3}\right)\left(\frac{Ze^2}{4\pi\varepsilon_0 m v_s^2}\right)^2 \cdot \left\{\left[\frac{d}{du}K_{iQ}(u)\Big|_{Q\epsilon}\right]^2 + \frac{\epsilon^2-1}{\epsilon^2}[K_{iQ}(Q\epsilon)]^2\right\} e^{\pi Q} \tag{3.9}$$

where the parameter Q is given by

$$Q = \omega b_0/v_s \tag{3.10}$$

and $K_\nu(u)$ are the modified Bessel functions which are related to the Hankel functions through

$$K_\nu(u) = (\pi i/2)\exp(\pi\nu i/2)\, H_\nu^{(1)}(iu) \tag{3.11}$$

Equation (3.9) gives the spectral energy distribution in the dipole approximation under the assumption that radiation damping can be neglected.

With the energy spectrum for a single collision, we now formulate the **differential emissivity** of the electron due to ion collisions, which is defined by

$$\eta_\omega(v_\mathrm{S}) = (1/8\pi)\, n_+ v_\mathrm{S} \int_0^\infty \int_0^{2\pi} w^\mathrm{T}(\omega, b, v_\mathrm{S})\, b\, db\, d\varphi \tag{3.12}$$

where the factor $1/8\pi$ means the reduction to the unit solid angle and to one polarization direction. Introducing (3.9) into (3.12), we find for the differential emissivity

$$\eta_\omega(v_\mathrm{S}) = n_+ \left(\frac{e^2}{6\pi^2\varepsilon_0 c^3}\right)\left(\frac{Ze^2}{4\pi\varepsilon_0 m}\right)^2 \frac{1}{v_\mathrm{S}} [-QK_{iQ}(Q)\, K'_{iQ}(Q)]\, e^{\pi Q} \tag{3.13}$$

Of course, the general formalism presented so far should contain two widely used approximations: the **straight-line approximation** and the **low-frequency approximation**.

Straight-Line Approximation

By inspection of the integral

$$\int_{-\infty}^{+\infty} \dot{\mathbf{v}}(t)\, e^{-i\omega t}\, dt \tag{3.14}$$

we recognize readily the estimate

$$\begin{pmatrix} \tilde{x}_\omega \\ \tilde{y}_\omega \end{pmatrix} = -(2\pi)^{-1/2} \frac{Ze^2}{4\pi\varepsilon_0 m b^2} \int_{-\infty}^{+\infty} \begin{pmatrix} \cos\omega t \\ i\,(v_\mathrm{s} t)/b)\ \sin\omega t \end{pmatrix} \frac{dt}{(1+(v_\mathrm{S} t/b)^2)^{3/2}}$$

$$= -(2\pi)^{-1/2} \frac{Ze^2}{4\pi\varepsilon_0 m} \begin{pmatrix} 1/bv_\mathrm{S} \\ 0 \end{pmatrix} \quad \text{for} \quad \omega \ll v_\mathrm{S}/b \tag{3.15}$$

$$= \begin{pmatrix} 0 \\ 0 \end{pmatrix} \quad \text{for} \quad \omega \gg v_\mathrm{S}/b$$

The first part of this estimate follows from the multiplication of the maximum value of the integral with the halfwidth; the second has its justification in the interference for high frequencies.

We may therefore expect that photons of not too high energy are emitted mainly by grazing incidents. Speaking quantitatively, emission in the range of radiofrequencies originates preferentially from collisions with $b = O(10^{-2}$ cm) for temperatures of about a hundred-thousand degrees. Equation (3.3) together with (3.4) then yields

$$\epsilon^2 \approx \cot^2 \chi/2 \approx 10^{12} \tag{3.16}$$

which shows that the assumption of a straight line is good. In this approximation it is not difficult to evaluate the integral in (3.7). We here prefer to consider the equivalent transition $Q \to 0$ in (3.9).

In carrying out this limiting process, we have to observe that the quantity $Q\epsilon$ can have a finite value although Q goes to zero. This follows from (3.3) and (3.10). In the straight-line approximation

$$Q\epsilon \approx (\omega b_0/v_s)(b/b_0) = \omega b/v_s \tag{3.17}$$

holds. Using this result, the transition for our spectral distribution (3.9) gives

$$w^{\mathrm{T}}(\omega, b, v_s) = \left(\frac{2e^2\omega^2}{3\pi^2\varepsilon_0 c^3}\right)\left(\frac{Ze^2}{4\pi\varepsilon_0 m v_s^2}\right)^2 \left\{\left[\frac{dK_0(u)}{du}\bigg|_{\omega b/v_s}\right]^2 + \left[K_0\left(\frac{\omega b}{v_s}\right)\right]^2\right\} \tag{3.18}$$

This dependency is shown in Figure V.2.

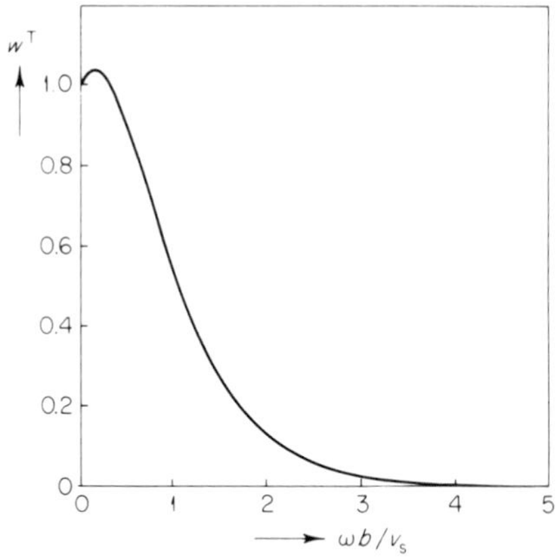

FIGURE V.2. Spectrum of radiation from an electron according to (3.18) in units of

$$\left(\frac{2e^2}{3\pi^2\varepsilon_0 c^3}\right)\left(\frac{Ze^2}{4\pi\varepsilon_0 m v_s^2}\right)^2 \frac{v_s^2}{b^2}.$$

The differential emissivity follows from (3.13) to

$$\eta_\omega(v_s) = n_+ \left(\frac{e^2}{6\pi^2\varepsilon_0 c^3}\right)\left(\frac{Ze^2}{4\pi\varepsilon_0 m}\right)^2 \frac{1}{v_s} \ln\left(\frac{2}{\gamma}\frac{4\pi\varepsilon_0 m v_s^3}{\omega Z e^2}\right) \tag{3.19}$$

with $\gamma = \exp C \approx 1.781$, where $C \approx 0.577$ is Euler's constant.

Low-Frequency Approximation

The low-frequency range is defined as that region of frequencies for which the time-dependent trigonometric functions in (3.7) can be treated constant. In this case, integration of these equations is trivial and yields

$$w^{\mathrm{T}}(\omega, b, v_{\mathrm{S}}) = \frac{Ze^2}{6\pi^2 \varepsilon_0 c^3} \left(\frac{2v_{\mathrm{S}}}{\epsilon}\right)^2 \tag{3.20}$$

Observe that this spectral distribution is independent of ω. That means we are approximating the emission process by a pulse of vanishing duration.

To describe quantitatively the approximation involved in the above definition of the low-frequency limit, we use in the general formula (3.9) an asymtotic representation of the modified Bessel functions for $\omega \rightarrow 0$. This yields the condition

$$(\underset{\sim}{Q}\epsilon)^{-1} \gg \ln(\underset{\sim}{Q}\epsilon) \tag{3.21}$$

for the low-frequency approximation which is fulfilled for

$$\underset{\sim}{Q}\epsilon \rightarrow 0 \tag{3.22}$$

Obviously this restriction is stronger than that for the straight-line approximation. In the latter case we had $\underset{\sim}{Q}\epsilon$ finite.

Since $\underset{\sim}{Q}\epsilon \rightarrow 0$ entails $\underset{\sim}{Q} \rightarrow 0$, we find for the differential emissivity in the low-frequency limit the same result (3.19) as in the straight-line approximation. Note that in contrast to the emissivity which one could derive from (3.20), Equation (3.19) depends on ω and is not divergent. This is the case because we did not introduce the assumption of on infinitesimal emission time.

3.2. Results of the Quantum-Mechanical Description

All the calculations presented so far dealt exclusively with classical radiation theory. This was implicit within the general policy set up in the definition of the model. With $\hbar = 0$, we were justified in neglecting the energy loss of the electron due to radiation in accordance with

$$\hbar\omega \ll \tfrac{1}{2} m v_{\mathrm{S}}^2 \tag{3.23}$$

and also allowed to use the classical path approximation for the electrons which is correct only if the condition

$$\lambda\!\!\!^{-} = \hbar / m v_{\mathrm{S}} \ll b \tag{3.24}$$

is fulfilled. These relations indicate that quantummechanical effects should be of importance as soon as radiation damping and emission from collisions with impact parameters

$$b \approx \hbar / m v_{\mathrm{s}} \tag{3.25}$$

contribute essentially. Since these are indeed important effects for the bremsstrahlung, we quote some of the quantummechanical results.

In this connection we remark that one can correlate each step in the quantummechanical discussion to the classical pendent for all radiation quantities which do not explicitly depend on the impact parameter.

It has become customary in radiation theory to account for the temperature, frequency, and other effects by the so-called **Gaunt factor**, which is defined in the following equation

$$\eta_\omega(v_{\mathrm{s}}) = n_+ \left(\frac{e^2}{6\pi^2 \varepsilon_0 c^3}\right)\left(\frac{Ze^2}{4\pi\varepsilon_0 m}\right)^2 \frac{1}{v_{\mathrm{s}}} \frac{\pi}{3^{1/2}} G(v_{\mathrm{s}}, \omega) \tag{3.26}$$

where in our classical treatment the factor $G(v_{\mathrm{s}}, \omega)$ would be $3^{1/2}/\pi$ times the logarithmic term in (3.19).

The quantummechanical treatment gives for the Gaunt factor the relation

$$\begin{aligned} G(v_{\mathrm{s}}, \omega) &= 3^{1/2}\pi \frac{x\, d\, |{}_2F_1(x)|^2/dx}{(e^{2\pi q_{\mathrm{s}}} - 1)(1 - e^{-2\pi q_{\mathrm{f}}})} \\ {}_2F_1(x) &= {}_2F_1(iq_{\mathrm{s}}, iq_{\mathrm{f}}, 1, -x) \end{aligned} \tag{3.27}$$

where the parameters q_{s} and q_{f} are defined by

$$q_{\mathrm{s}} = \frac{Ze^2}{4\pi\varepsilon_0 \hbar v_{\mathrm{s}}}, \qquad q_{\mathrm{f}} = \frac{Ze^2}{4\pi\varepsilon_0 \hbar v_{\mathrm{f}}} \tag{3.28}$$

x is given by

$$x = 4 q_{\mathrm{s}} q_{\mathrm{f}} / (q_{\mathrm{s}} - q_{\mathrm{f}})^2 \tag{3.29}$$

and ${}_2F_1$ denotes the generalized hypergeometric function in the usual notation as adopted by Watson (1962). v_{s} and v_{f} are the starting and final velocities of the electron which are connected by the Born condition

$$\tfrac{1}{2} m v_{\mathrm{s}}^2 - \tfrac{1}{2} m v_{\mathrm{f}}^2 = \hbar\omega \tag{3.30}$$

The general formula (3.27) has been given by Sommerfeld (1939). It is difficult to evaluate, although it has been widely used for the derivation of the bremsstrahlung of X-ray tubes and thermal fusion devices.

Again, we will consider the two limiting cases of low and high frequencies.

In the **low-frequency range** (straight-line approximation)

$$\omega Ze^2/4\pi\varepsilon_0 m v_s^3 = \omega b_0/v_s \ll 1 \tag{3.31}$$

the Gaunt factor is given by

$$G = (3^{1/2}/\pi)[\ln(2mv_s^2/\hbar\omega) - C - \mathrm{Re}\{\psi(1 + iq_s)\}] \tag{3.32}$$

where C is Euler's constant $C \approx 0.577$, and ψ the digamma function $\psi(x) = d \ln \Gamma(x)/dx$. For low electron velocities, the Gaunt factor simplifies to

$$G = (3^{1/2}/\pi) \ln((2/\gamma)(4\pi\varepsilon_0 m v_s^3/\omega Ze^2)) \tag{3.33}$$

with $\gamma = \exp C \approx 1.781$.

For high electron velocities we have

$$G = (3^{1/2}/\pi) \ln(2mv_s^2/\hbar\omega) \tag{3.34}$$

It is interesting to see that the low-velocity result (3.33) in (3.26) reproduces the classical result, whereas the high-velocity formula (3.34) in (3.26) exhibits typical quantum effects in the argument of the logarithm. At first sight this appears surprising.

The reason is that the limitation to the straight-path approximation restricts the low-velocity electrons to the range of impact parameters much larger than the corresponding de Broglie wavelength.

In the case of high electron velocities, the classical result is identical with the quantummechanical one if one chooses the de Broglie wavelength $\hbar/mv_s$ for the minimum impact parameter.

Elwert (1948) has shown numerically that the classical formula remains useful even for values of q_s as small as one.

In the **high-frequency range**, Elwert (1939) has given a formula which goes beyond the range of application of the Born approximation. The corresponding Gaunt factor is given by

$$G = \frac{3^{1/2}}{\pi} \frac{q_f}{q_s} \frac{1 - e^{-2\pi q_s}}{1 - e^{-2\pi q_f}} \ln \frac{q_f + q_s}{q_f - q_s} \tag{3.35}$$

Here, too, the quantities q_s and q_f are related by the Born condition (3.30) which introduces the frequency dependence into (3.35).

We introduce the **emission coefficient** which is related to the differential emissivity by

$$j_\omega = \int \eta_\omega(v) f(\mathbf{v})\, d\mathbf{v} \tag{3.36}$$

In analogy to the definition of the Gaunt factor used for the differential emissivity, it is customary to use an **average Gaunt factor** $\bar{G}(\Theta, \omega)$ for the emission coefficient defined by

$$j_\omega = n_- n_+ (Z^2 e^6 / 48\pi^4 \varepsilon_0{}^3 c^3 m^2)(m/2\pi\Theta)^{1/2}(\pi/3^{1/2})\, \bar{G}(\Theta, \omega) \tag{3.37}$$

The general form of this coefficient can be found by introducing (3.26) into (3.36) and comparing with (3.37). There is no analytical solution for this problem in general; expressions can be given for the two limiting cases treated above.

For the **low-frequency approximation** and high (but not relativistic) electron velocities, the Gaunt factor is determined by (Oster, 1961)

$$\bar{G}(\Theta, \omega) = (3^{1/2}/\pi) \ln[(4/\gamma)(\Theta/\hbar\omega)] \tag{3.38}$$

and for the low-frequency approximation and small electron velocities, we have

$$\bar{G}(\Theta, \omega) = (3^{1/2}/\pi) \ln[(2/\gamma)^{5/2}(\Theta/m)^{3/2}(4\pi\varepsilon_0 m/Ze^2\omega)] \tag{3.39}$$

It is interesting to note that the low-energy approximation holds down to the value $q_s \approx 1$.

In the **high-frequency approximation** one might want to use the Elwert formula. However, again no analytical solution can be found. In the Born approximation, developing the Elwert formula for large values of v_s and v_f, we have

$$G(v_s, \omega) = \frac{3^{1/2}}{\pi} \ln \frac{q_f + q_s}{q_f - q_s} \tag{3.40}$$

and find the result

$$\bar{G}(\Theta, \omega) = (3^{1/2}/\pi)\, e^{-\hbar\omega/2\Theta} K_0(\hbar\omega/2\Theta) \tag{3.41}$$

where K_0 is the modified Bessel function of order zero. For very high frequencies, the asymptotic presentation of K_0 yields

$$\bar{G}(\Theta, \omega) = (3/\pi)^{1/2}\, e^{-\hbar\omega/\Theta}(\hbar\omega/\Theta)^{-1/2} \tag{3.42}$$

Observe that in j_ω all frequency dependencies are included in the Gaunt factor $\bar{G}$.

The **total emission** per unit volume, time, and solid angle (for one polarization direction) is found from the emission coefficient by integration over the frequency. Again one introduces a Gaunt-factor $\bar{\bar{G}}$ through

$$j = n_- n_+ \left(\frac{2}{3}\pi\right)^{1/2} \frac{Z^2 e^6}{96\pi^4 \varepsilon_0{}^3 \hbar m c^2} \left(\frac{\Theta}{mc^2}\right)^{1/2} \bar{\bar{G}} \tag{3.43}$$

This quantity $\bar{\bar{G}}$ has been evaluated numerically by Greene (1959) from Sommerfeld's formula (3.27) and is presented in Figure V.3. In the

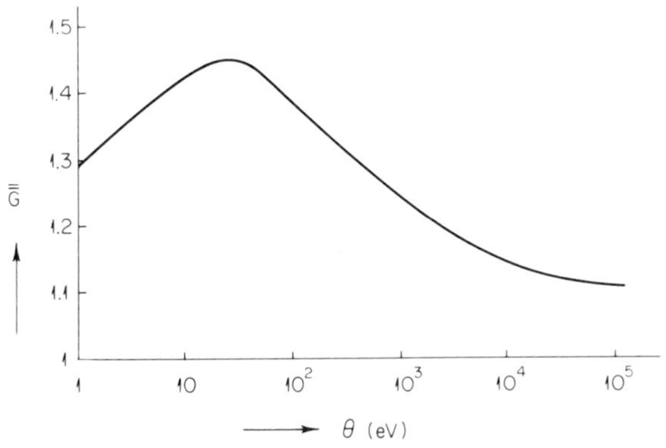

FIGURE V.3. $\bar{\bar{G}}$ for $Z = 1$ (Green, 1959).

Born approximation for high electron energies, it approaches the limiting value $\bar{\bar{G}} = 2 \cdot 3^{1/2}/\pi$.

4. Cyclotron Radiation

Cyclotron radiation is emitted by charged particles accelerated in magnetic fields. In this section, we take exception from our general rule not to study external magnetic fields, because the cyclotron radiation is of such principal interest and in its analytical structure closely related to the bremsstrahlung.

The dipole approximation of the cyclotron radiation in a constant magnetic field is rather trivial. The interesting features arise in the relativistic range. We therefore use the general formulation (2.51)

$$w(\omega, \Omega) = \frac{e^2\omega^2}{16\pi^3\varepsilon_0 c^3} \left| \int_{-\infty}^{+\infty} e^{-i\omega(t' - \hat{\mathbf{r}}\cdot\mathbf{x}'/c)}\, \hat{\mathbf{r}} \times (\hat{\mathbf{r}} \times \mathbf{v})\, dt' \right|^2 \tag{4.1}$$

To evaluate this relation we need the velocity $\mathbf{v}$ and the position $\mathbf{x}'$ as a function of time. The relativistic motion of a particle in any inertial system is described by

$$d\mathbf{p}/dt = -e(\mathbf{E} + \mathbf{v} \times \mathbf{B}) \tag{4.2}$$

with

$$\mathbf{p} = m_0\mathbf{v}(1 - \beta^2)^{-1/2} = \mathscr{E}\mathbf{v}/c^2 \tag{4.3}$$

where $\mathscr{E}$ is the total energy of our particle and $\boldsymbol{\beta} = \mathbf{v}/c$.

In the absence of an electric field, (4.3) requires that $|\mathbf{p}|$ be constant. We neglect radiation damping which is easily justified except for particles of extreme relativistic velocities.

With $p =$ constant, (4.2) describes a circular motion of the particle with the angular frequency

$$\omega_\mathrm{L} = \frac{eB}{m_0}(1 - \beta^2)^{1/2} =: \omega_{\mathrm{L}0}(1 - \beta^2)^{1/2} \tag{4.4}$$

where $\omega_{\mathrm{L}0}$ is the nonrelativistic Larmor frequency.

Allowing for constant velocity parallel to the magnetic field, the helix of our particle is described by

$$\begin{aligned} \mathbf{v} &= v_\perp \cos \omega_\mathrm{L} t'\mathbf{e}_1 + v_\perp \sin \omega_\mathrm{L} t'\mathbf{e}_2 + v_\parallel \mathbf{e}_3 \\ \mathbf{x}' &= (v_\perp/\omega_\mathrm{L}) \sin \omega_\mathrm{L} t'\mathbf{e}_1 - (v_\perp/\omega_\mathrm{L}) \cos \omega_\mathrm{L} t'\mathbf{e}_2 + v_\parallel t'\mathbf{e}_3 \end{aligned} \tag{4.5}$$

where we used the negative sign for the charge since we are interested in electrons. The symbols may be taken from Figure V.4. We choose the vector $\hat{\mathbf{r}}$ in the $\mathbf{e}_1$, $\mathbf{e}_3$ plane, so that we have

$$\hat{\mathbf{r}} = \sin \vartheta \mathbf{e}_1 + \cos \vartheta \mathbf{e}_3 \tag{4.6}$$

Introducing this into (4.1), we find for the exponential factor

$$\exp[-i\omega t' + i(\omega/c)\mathbf{x}' \cdot \hat{\mathbf{r}}] = \exp i\{(\omega/\omega_\mathrm{L})\beta_\perp \sin \vartheta \sin \omega_\mathrm{L} t' + (\beta_\parallel \cos \vartheta - 1)\omega t'\} \tag{4.7}$$

and for the double crossproduct

$$\hat{\mathbf{r}} \times (\hat{\mathbf{r}} \times \mathbf{v}) = \begin{pmatrix} v_\parallel \sin \vartheta \cos \vartheta - v_\perp \cos^2 \vartheta \cos \omega_\mathrm{L} t' \\ -v_\perp \sin \omega_\mathrm{L} t' \\ -v_\parallel \sin^2 \vartheta + v_\perp \sin \vartheta \cos \vartheta \cos \omega_\mathrm{L} t' \end{pmatrix} \tag{4.8}$$

Let us now group all the terms in (4.8) with respect to their time-

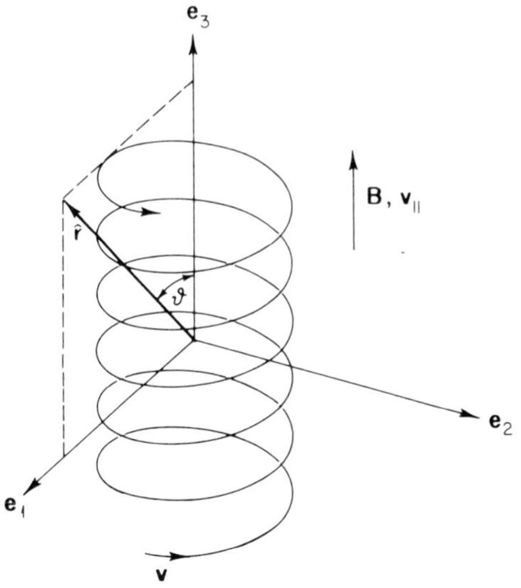

FIGURE V.4. Helix of an electron in a uniform magnetic field.

dependent factors. To evaluate (4.1), we have to deal with the integral

$$\mathbf{A} := \int_{-\infty}^{+\infty} \exp[-i\omega(t' - \hat{\mathbf{r}} \cdot \mathbf{x}'/c)]\, \hat{\mathbf{r}} \times (\hat{\mathbf{r}} \times \mathbf{v})\, dt'$$

$$= \int_{-\infty}^{+\infty} \exp[i\{(\omega/\omega_{\mathrm{L}})\, \beta_\perp \sin\vartheta \sin\omega_{\mathrm{L}} t' + (\beta_\parallel \cos\vartheta - 1)\, \omega t'\}]$$

$$\times (\mathbf{c}_1 + \mathbf{c}_2 \sin\omega_{\mathrm{L}} t' + \mathbf{c}_3 \cos\omega_{\mathrm{L}} t')\, dt' \tag{4.9}$$

where the quantities $\mathbf{c}_1$, $\mathbf{c}_2$, and $\mathbf{c}_3$ are vectors easily to be taken from (4.8). We apply the expansion

$$\exp(i\xi \sin\omega_{\mathrm{L}} t') = \sum_{-\infty}^{+\infty} \exp(im\omega_{\mathrm{L}} t')\, J_m(\xi), \qquad \xi := \frac{\omega}{\omega_{\mathrm{L}}}\, \beta_\perp \sin\vartheta \tag{4.10}$$

The evaluation of the $\mathbf{c}_1$ term is then trivial and results in a Dirac function. To calculate the $\mathbf{c}_2$ term, we first differentiate (4.10) with respect to ξ and then integrate yielding a Dirac function. With respect to the $\mathbf{c}_3$ term, we use integration by parts, considering together the $\mathbf{c}_3$ term and the first part of the exponential factor as a total differential. The result is

$$\mathbf{A} = 2\pi \sum_m \left\{ \mathbf{c}_1 J_m(\xi) - i\mathbf{c}_2 J_m'(\xi) + \mathbf{c}_3 \frac{\omega}{\omega_{\mathrm{L}}} \frac{1 - \beta_\parallel \cos\vartheta}{\xi} J_m(\xi) \right\}$$

$$\times\, \delta(m\omega_{\mathrm{L}} - \omega(1 - \beta_\parallel \cos\vartheta)) \tag{4.11}$$

We now decompose the vector quantities $\mathbf{c}_1$, $\mathbf{c}_2$, $\mathbf{c}_3$ into their contributions in the directions $\mathbf{e}_1$, $\mathbf{e}_2$, $\mathbf{e}_3$ and introduce (4.11) into (4.1).

Since we are interested in the emissivity of our system we want to calculate the emission per unit time. This was a simple problem in the case of the bremsstrahlung, since there each collision lasted only a very short time and could be treated as completed, and consequently the emissivity was given as the product of the emission during one collision process with the number of collisions per second.

Here, in the case of the cyclotron radiation, the emission process lasts from the time $-\infty$ to $+\infty$, since we neglect radiation damping. To study the transition from w to η, we consider an emission process of finite time T in the limit $T \to \infty$. The transition from the total energy emission to the emissivity is then characterized by

$$\left| \int_{-T/2}^{+T/2} e^{i\bar{\omega}t'}\, dt' \right|^2 \longrightarrow \left(\frac{1}{T} \int_{-T/2}^{+T/2} e^{i\bar{\omega}t'}\, dt' \right) \left(\int_{-T/2}^{+T/2} e^{-i\bar{\omega}t'}\, dt' \right) \tag{4.12}$$

In the limit $T \to \infty$, this yields

$$\int_{-\infty}^{+\infty} e^{-\bar{\omega}t'}\, dt' = 2\pi\, \delta(\bar{\omega}) \tag{4.13}$$

and the first term on the right-hand side approaches the value one for $T \to \infty$ and $\bar{\omega} = 0$

$$4\pi^2\, \delta^2(\bar{\omega}) \longrightarrow 2\pi\, \delta(\bar{\omega}) \tag{4.14}$$

Applying (4.14) to (4.1), (4.9), and (4.11), we find for the coefficient of emission

$$\eta_\omega(\omega, \mathbf{v}, \vartheta) = \frac{e^2\omega^2}{8\pi^2\varepsilon_0 c} \sum_{m=1}^{\infty} \begin{pmatrix} \dfrac{\cos\vartheta}{\sin\vartheta} (\beta_\| - \cos\vartheta)\, J_m(\xi) \\ i\beta_\perp\, dJ_m(\xi)/d\xi \\ (\cos\vartheta - \beta_\|)\, J_m(\xi) \end{pmatrix}^2 \times \delta(m\omega_\mathrm{L} - \omega(1 - \beta_\| \cos\vartheta)) \tag{4.15}$$

or, respectively,

$$\eta_\omega(\omega, \mathbf{v}, \vartheta) = \frac{e^2\omega^2}{8\pi^2\varepsilon_0 c} \sum_{m=1}^{\infty} \left\{ \left(\frac{\cos\vartheta - \beta_\|}{\sin\vartheta} \right)^2 J_m{}^2(\xi) + \beta_\perp{}^2 J_m'^2(\xi) \right\} \times \delta\, (m\omega_\mathrm{L} - \omega(1 - \beta_\| \cos\vartheta)) \tag{4.16}$$

where we omitted the terms for negative values of m since the corresponding negative ω are excluded (see p. 245).

Obviously the emissivity presents a line spectrum in which the line frequencies are given by

$$\omega_m = \frac{m\omega_L}{1 - \beta_\parallel \cos\vartheta} = \frac{m\omega_{L0}(1 - \beta^2)^{1/2}}{1 - \beta_\parallel \cos\vartheta} \tag{4.17}$$

as shown by the Dirac function in (4.16). The frequencies of the lines are multiples of the relativistic Larmor frequency accounting at the same time through the denominator for the Doppler effect caused by the motion parallel to the magnetic field.

The angular distributions of the various harmonics are quite complicated as can be seen from formula (4.16) and as demonstrated in Figure V.5 for a few terms of our expansion.

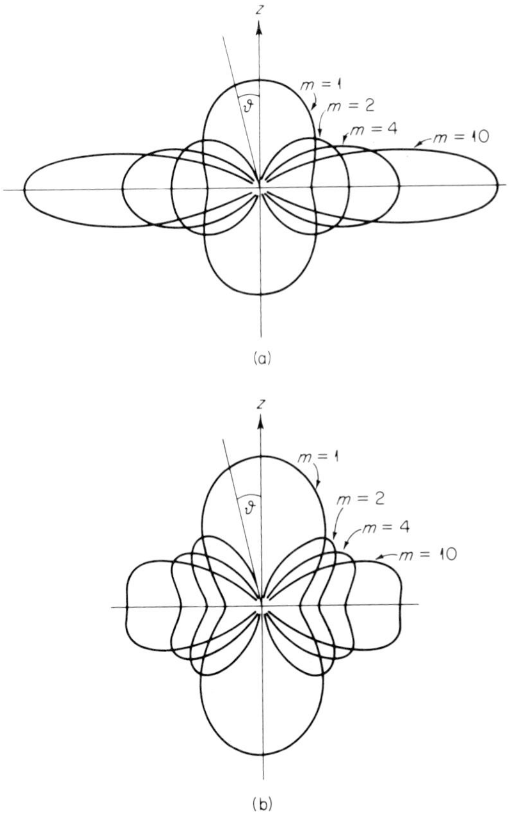

FIGURE V.5. Angular distributions of some harmonics of cyclotron radiation normalized to the same total intensity. (a) in the limit $\beta \to 0$, (b) $\beta = 0.9$.

The total emissivity of a given harmonic is determined from

$$\eta_m{}^{\mathrm{T}} = \frac{e^2\omega_{\mathrm{L0}}^2}{4\pi\varepsilon_0 c}\,\frac{1-\beta^2}{1-\beta_{\|}{}^2}\,\bar{\beta}\left\{J'_{2m}(\bar{\beta}) - \frac{1}{2}\,\frac{1-\beta^2}{\beta^2-\beta_{\|}{}^2}\int_0^{\bar{\beta}} J_{2m}(x)\,dx\right\} \tag{4.18}$$

with the abbreviation (see Schott, 1912)

$$\bar{\beta} = \frac{2m\beta_{\perp}}{(1-\beta_{\|}{}^2)^{1/2}} = 2m\left(\frac{\beta^2-\beta_{\|}{}^2}{1-\beta_{\|}{}^2}\right)^{1/2} \tag{4.19}$$

Finally, the total emission follows from the summation over all harmonics to

$$\eta^{\mathrm{T}} = \frac{e^2\omega_{\mathrm{L0}}^2}{6\pi\varepsilon_0 c}\,\frac{\beta_{\perp}{}^2}{1-\beta^2} = \frac{e^2\omega_{\mathrm{L0}}^2}{6\pi\varepsilon_0 c}\left\{(1-\beta_{\|}{}^2)\left(\frac{\mathscr{E}}{m_0c^2}\right)^2 - 1\right\} \tag{4.20}$$

Cyclotron Spectrum for a Thermal System

For a thermal system of electrons, the emission coefficient is defined by

$$j_m(\omega, \vartheta) = \int \eta_m(\omega, \boldsymbol{\beta}, \vartheta)\, f(\boldsymbol{\beta})\, d\boldsymbol{\beta} \tag{4.21}$$

where f is the distribution function. For the case $\beta_{\|} \ll \beta_{\perp}$, (4.16) simplifies to

$$\eta_m(\omega, \beta, \vartheta) = \frac{e^2\omega^2}{8\pi^2\varepsilon_0 c}\{\cot^2\vartheta\, J_m{}^2(m\beta) + \beta^2 J'^2_m(m\beta)\}\,\delta(\omega - m\omega_{\mathrm{L}})$$

$$=: \bar{\eta}_m(\beta, \vartheta)\,\delta(\omega - m\omega_{\mathrm{L}}) \tag{4.22}$$

and, with that, (4.21) is transformed into

$$j_m(\omega, \vartheta) = \int \delta(\omega - m\omega_{\mathrm{L}}(\beta_{\perp}))\,\bar{\eta}_m(\beta_{\perp}, \vartheta)\, f(\beta_{\perp})\, d\beta_{\perp} \tag{4.23}$$

In the range of electron velocities where terms of order higher than $\beta_{\perp}{}^2$ can be neglected, the Maxwellian distribution is given by

$$f(\beta_{\perp}) = n_{-}(m_0c^2/\Theta)\exp\{-\beta_{\perp}{}^2\, m_0c^2/2\Theta\} \tag{4.24}$$

where n_{-} is the electron number density.

Equations (4.23) and (4.24) were evaluated by Hirshfield, Baldwin,

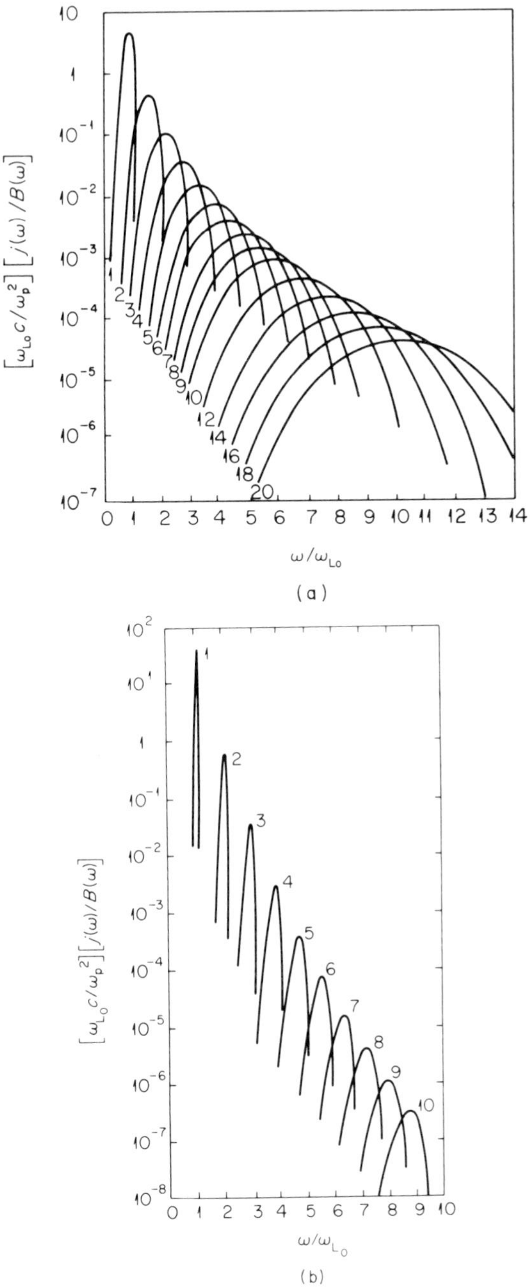

FIGURE V.6. Cyclotron radiation of a thermal system normalized by the blackbody flux. (a) $\Theta = 100$ keV; (b) $\Theta = 10$ keV.

and Brown (1961). We give their final results where the integration with respect to ϑ has been performed

$$j_m(\omega) = \frac{\omega_{L0}\omega_p{}^2 m_0}{2\pi^2 c}\frac{1}{x}\frac{m_0c^2}{\Theta}\exp\left\{-\frac{m_0c^2}{\Theta}\left(\frac{1}{x}-1\right)\right\}$$
$$\times\Big\{J_{2m}[2m(1-x^2)^{1/2}] - (1-x^2)^{-1/2}\,J_{2m+1}[2m(1-x^2)^{1/2}]$$
$$- x^{-2}(1-x^2)^{-1/2}\sum_{\nu=1}^{\infty} J_{2m+2\nu+1}[2m(1-x^2)^{1/2}]\Big\} \tag{4.25}$$

where ω_p is the electron plasma frequency and $x := \omega/m\omega_{L0}$.

Figures V.6 show the results of (4.25) for two temperatures normalized to the corresponding blackbody radiation

$$B(\omega) = \omega^2\Theta/2\pi^3c^2 \tag{4.26}$$

REFERENCES AND SUPPLEMENTARY READING

General References

Abramowitz, M., and Stegun, I. A., eds. (1965). "Handbook of Mathematical Functions." Dover, New York.

Becker, R., and Sauter, F. (1962). "Theorie der Elektrizität," Vol. I. Teubner, Stuttgart.

Bekefi, G. (1966). "Radiation Processes in Plasmas." Wiley, New York.

Jackson, J. D. (1962). "Classical Electrodynamics." Wiley, New York.

Landau, L. D., and Lifshitz, E. M. (1951). "The Classical Theory of Fields." Addison-Wesley, Reading, Massachusetts.

Landau, L. D., and Lifshitz, E. M. (1960). "Electrodynamics of Continuous Media." Addison-Wesley, Reading, Massachusetts.

Panofsky, W. K. H., and Phillips, M. (1955). "Classical Electricity and Magnetism." Addison-Wesley, Reading, Massachusetts.

Watson, G. N. (1962). "A Treatise on the Theory of Bessel Functions." Cambridge Univ. Press, London and New York.

Section 3

Berger, J. M. (1957). *Phys. Rev.* **105**, 35.

Brussaard, P. J., and van de Hulst, H. C. (1962). *Rev. Mod. Phys.* **34**, 507.

Elwert, G. (1939). *Ann. Phys. (Leipzig)* **34**, 178.

Elwert, G. (1948). *Z. Naturforsch. A* **3**, 477.

Ginzburg, V. L. (1961). "Propagation of Electromagnetic Waves in a Plasma." Addison-Wesley, Reading, Massachusetts.

Greene, J. (1959). *Astrophys. J.* **130**, 693.

Karzas, W. J., and Latter, R. (1961). *Astrophys. J. Suppl. Ser.* **6**, 167.

Kramers, H. A. (1923). *Phil. Mag.* **46**, 836.

Oster, L. (1961). *Rev. Mod. Phys.* **33**, 525.

Scheuer, P. A. G. (1960). *Mon. Not. Roy. Astron. Soc.* **120**, 231.

Sommerfeld, A. (1939). "Atombau und Spektrallinien," Vol. II. Vieweg, Braunschweig.

SECTION 4

Beard, D. B. (1959). *Phys. Fluids* **2**, 379.
Beard, D. B. (1960). *Phys. Fluids* **3**, 324.
Bekefi, G., Hirschfield, J. L. and Brown, S. C. (1961). *Phys. Rev.* **122**, 1037.
Drummond, W. E., and Rosenbluth, M. N. (1960). *Phys. Fluids* **3**, 45.
Hirshfield, J. L., and Brown, S. C. (1961). *Phys. Rev.* **122**, 719.
Hirshfield, J. L., Baldwin, D. E., and Brown, S. C. (1961). *Phys. Fluids* **4**, 198.
Oster, L. (1960). *Phys. Rev.* **119**, 1444.
Oster, L. (1961). *Phys. Rev.* **121**, 961.
Rosner, H. (1958). Rep. AFSWC-TR-58-47. Republic Aviation Corp., Farmingdale, Lond Island, New York.
Schott, G. A. (1912) "Electromagnetic Radiation." Cambridge Univ. Press, London and New York.
Trubnikov, B. A. (1961). *Phys. Fluids* **4**, 195.

Chapter VI Many-Particle Interactions with Electromagnetic Fields

1. Basic Equations

As we have seen, the BBGKY hierarchy can be used as the basis for the description of the Coulomb system. It is to be expected that a more general hierarchy can be derived for the fully ionized system in the general electromagnetic field in analogy to the procedure given in Chapter II. This hierarchy will be much more complicated due to the fact that the Coulomb interaction law is replaced by the set of Maxwell's equations. Deriving the hierarchy for the special distribution functions from the Klimontovich equations, a set of average field quantities defined via the specific distributions emerges. They require additional equations which follow from suitable averaging processes of Maxwell's equations.

In contrast to the procedure of Chapter II, we here use the velocity presentation, since in the presence of the Maxwell equations the Hamilton formalism does not display its full advantage. In the following applications we will use the velocity presentation anyhow to conform with the general practice in the literature.

We describe the density distribution of a single particle i in the phase space by

$$F_i({}^i\mathbf{r}\,,{}^i\mathbf{v}) = \delta({}^i\mathbf{r} - \mathbf{r}_i(t))\,\delta({}^i\mathbf{v} - \mathbf{v}_i(t)) \tag{1.1}$$

Differentiation yields

$$(\partial F_i/\partial t)_{{}^i\mathbf{r},{}^i\mathbf{v}} = -\mathbf{v}_i \cdot \partial F_i/\partial^i\mathbf{r} - \dot{\mathbf{v}}_i \cdot \partial F_i/\partial^i\mathbf{v} \tag{1.2}$$

where the acceleration is given by

$$\dot{\mathbf{v}}_i = (e_i/m_i)[\mathbf{E}(\mathbf{r}_i, t; \{\mathbf{r}_j \mid j \neq i\}) + \mathbf{v}_i \times \mathbf{B}(\mathbf{r}_i, t; \{\mathbf{r}_j, \mathbf{v}_j \mid j \neq i\})] \tag{1.3}$$

with electric and magnetic fields **E** and **B** depending not only on the set of coordinates $\{\mathbf{r}_j, \mathbf{v}_j \mid j \neq i\}$ of the other particles of the system, but also on the external experimental setup (ex). Introducing (1.3) into (1.2), we find

$$\frac{\partial F_i}{\partial t} + \mathbf{v}_i \cdot \frac{\partial F_i}{\partial\, {}^i\mathbf{r}} + \frac{e_i}{m_i}(\mathbf{E} + \mathbf{v}_i \times \mathbf{B}) \cdot \frac{\partial F_i}{\partial\, {}^i\mathbf{v}} = 0 \tag{1.4}$$

Now we multiply (1.4) by the ensemble distribution function $f_N(\mathbf{r}_1, \ldots, \mathbf{v}_N; t)$ and integrate over the whole phase space

$$\int d\mathbf{r}_1 \cdots d\mathbf{v}_N f_N(\mathbf{r}_1, \ldots, \mathbf{v}_N; t) \times \left\{\frac{\partial F_i}{\partial t} + \mathbf{v}_i \cdot \frac{\partial F_i}{\partial\, {}^i\mathbf{r}} + \frac{e_i}{m_i}(\mathbf{E} + \mathbf{v}_i \times \mathbf{B}) \cdot \frac{\partial F_i}{\partial\, {}^i\mathbf{v}}\right\} = 0 \tag{1.5}$$

In doing so we recall the relation (II.2.6)

$$f_N\left(\frac{\partial F_i}{\partial t}\right)_{{}^i\mathbf{r},{}^i\mathbf{v}} = \left(\frac{\partial f_N F_i}{\partial t}\right)_{{}^i\mathbf{r},{}^i\mathbf{v},\mathbf{r}_1,\ldots,\mathbf{v}_n} + \sum_j \left(\dot{\mathbf{r}}_j \cdot \frac{\partial}{\partial \mathbf{r}_j} + \dot{\mathbf{v}}_j \cdot \frac{\partial}{\partial v_j}\right)(f_N F_i) \tag{1.6}$$

where we used Liouville's theorem. Integration of (1.6) results in

$$\int d\mathbf{r}_1 \cdots d\mathbf{v}_N f_N \partial F_i/\partial t = \partial f_1({}^i\mathbf{r}; {}^i\mathbf{v}; t)/\partial t \tag{1.7}$$

since the last term from the right-hand side of (1.6) yields zero, which can be shown by partial integration and application of the equations of motion.

We introduce the reduced specific distribution functions

$$f_s({}^1\mathbf{r}, {}^1\mathbf{v}, \ldots, {}^i\mathbf{r}, {}^i\mathbf{v}, \ldots, {}^s\mathbf{r}, {}^s\mathbf{v}; t) = \int f_N F_1 \cdots F_s \, d\mathbf{r}_1 \cdots d\mathbf{v}_N \tag{1.8}$$

and the average fields $\langle \mathbf{M} \rangle_s$ of order s

$$\langle \mathbf{M} \rangle_s(\mathbf{r}, {}^1\mathbf{r}, {}^1\mathbf{v}, \ldots, {}^s\mathbf{r}, {}^s\mathbf{v}; t) = \frac{\int \mathbf{M}(\mathbf{r}, \ldots) f_N F_1 \cdots F_s \, d\mathbf{r}_1 \cdots d\mathbf{v}_N}{f_s({}^1\mathbf{r}, \ldots, {}^s\mathbf{v}; t)} \tag{1.9}$$

Using (1.7) and (1.9) in (1.5) yields

$$\frac{\partial f_1({}^i\mathbf{r}, {}^i\mathbf{v}; t)}{\partial t} + {}^i\mathbf{v} \cdot \frac{\partial f_1({}^i\mathbf{r}, {}^i\mathbf{v}; t)}{\partial\, {}^i\mathbf{r}}$$

$$+ (e_i/m_i) \int d\mathbf{r}_i \int d\mathbf{v}_i$$

$$\times [\langle\mathbf{E}\rangle_1(\mathbf{r} = \mathbf{r}_i, \mathbf{r}_i, \mathbf{v}_i; t) + \mathbf{v}_i \times \langle\mathbf{B}\rangle_1(\mathbf{r} = \mathbf{r}_i, \mathbf{r}_i, \mathbf{v}_i; t)]$$

$$\cdot f_1(\mathbf{r}_i, \mathbf{v}_i, t)\, \partial F_i/\partial\, {}^i\mathbf{v} = 0 \tag{1.10}$$

or, respectively,

$$\frac{\partial f_1({}^i\mathbf{r}, {}^i\mathbf{v}; t)}{\partial t} + {}^i\mathbf{v} \cdot \frac{\partial f_1({}^i\mathbf{r}, {}^i\mathbf{v}; t)}{\partial\, {}^i\mathbf{r}}$$

$$+ \frac{e_i}{m_i} \frac{\partial}{\partial\, {}^i\mathbf{v}} \cdot (\langle\mathbf{E}\rangle_1(\mathbf{r} = {}^i\mathbf{r}, {}^i\mathbf{r}, {}^i\mathbf{v}; t) + {}^i\mathbf{v} \times \langle\mathbf{B}\rangle_1(\mathbf{r} = {}^i\mathbf{r}, {}^i\mathbf{r}, {}^i\mathbf{v}; t))$$

$$\times f_1({}^i\mathbf{r}, {}^i\mathbf{v}; t) = 0 \tag{1.11}$$

Now we write down Maxwell's relations for the electromagnetic fields produced by all particles $j \neq i$ and external effects (ex) at point $\mathbf{r}$ and time t

$$\nabla \times \mathbf{B} = \mu_0 \left\{\epsilon_0\, \partial\mathbf{E}(\mathbf{r}, t)/\partial t + \mathbf{j}_{\text{ex}} + \sum_j{}' e_j \int d^j\mathbf{v}\ {}^j\mathbf{v} F_j(\mathbf{r}, {}^j\mathbf{v})\right\} \tag{1.12}$$

$$\nabla \times \mathbf{E} = -\partial\mathbf{B}/\partial t \tag{1.13}$$

$$\nabla \cdot \mathbf{E} = \varepsilon_0^{-1} \left\{\sum_j{}' e_j \int d^j\mathbf{v}\, F_j(\mathbf{r}, {}^j\mathbf{v}) + \rho_{\text{ex}}\right\} \tag{1.14}$$

$$\nabla \cdot \mathbf{B} = 0 \tag{1.15}$$

To find the averaged fields $\langle\mathbf{E}\rangle_1$, $\langle\mathbf{B}\rangle_1$, we multiply the relations (1.12)–(1.15) by the product $f_N F_i$ and integrate over Γ space. We have

$$\nabla \times \langle\mathbf{B}\rangle_1 (\mathbf{r}, {}^i\mathbf{r}, {}^i\mathbf{v}; t) f_1({}^i\mathbf{r}, {}^i\mathbf{v}; t)$$

$$= \mu_0 \left\{\mathbf{j}_{\text{ex}} f_1({}^i\mathbf{r}, {}^i\mathbf{v}; t) + \sum_j{}' e_j \int d\,{}^j\mathbf{v}\ {}^j\mathbf{v}\, f_2({}^i\mathbf{r}, {}^i\mathbf{v}, \mathbf{r}, {}^j\mathbf{v}; t)\right.$$

$$\left. + \varepsilon_0 \left\langle \frac{\partial\mathbf{E}}{\partial t} \right\rangle_1 (\mathbf{r}, {}^i\mathbf{r}, {}^i\mathbf{v}; t)\ f_1({}^i\mathbf{r}, {}^i\mathbf{v}; t)\right\} \tag{1.16}$$

$$\nabla \times \langle\mathbf{E}\rangle_1(\mathbf{r}, {}^i\mathbf{r}, {}^i\mathbf{v}; t)\ f_1({}^i\mathbf{r}, {}^i\mathbf{v}; t) = -\left\langle \frac{\partial\mathbf{B}}{\partial t} \right\rangle_1 (\mathbf{r}, {}^i\mathbf{r}, {}^i\mathbf{v}; t)\ f_1({}^i\mathbf{r}, {}^i\mathbf{v}; t) \tag{1.17}$$

$$\nabla \cdot \langle \mathbf{E} \rangle_1(\mathbf{r}, {}^i\mathbf{r}, {}^i\mathbf{v}; t) f_1({}^i\mathbf{r}, {}^i\mathbf{v}; t) = \varepsilon_0^{-1} \Big\{ \rho_{\text{ex}} f_1({}^i\mathbf{r}, {}^i\mathbf{v}; t) + \sum_j{}' e_j \int d{}^j\mathbf{v}\, f_2({}^i\mathbf{r}, {}^i\mathbf{v}, \mathbf{r}, {}^j\mathbf{v}; t) \Big\} \tag{1.18}$$

$$\nabla \cdot \langle \mathbf{B} \rangle_1(\mathbf{r}, {}^i\mathbf{r}, {}^i\mathbf{v}; t) f_1({}^i\mathbf{r}, {}^i\mathbf{v}; t) = 0 \tag{1.19}$$

Applying Liouville's theorem, we find for the time derivatives

$$\begin{aligned} \left\langle \frac{\partial}{\partial t} \mathbf{M} \right\rangle_1 &(\mathbf{r}, {}^i\mathbf{r}, {}^i\mathbf{v}; t)\; f_1({}^i\mathbf{r}, {}^i\mathbf{v}; t) \\ &= \int d\mathbf{r}_1 \cdots d\mathbf{v}_N\, F_i \left(\frac{\partial}{\partial t} f_N \mathbf{M}(\mathbf{r}, \ldots) \right)_{\mathbf{r}} \\ &= \left(\frac{\partial}{\partial t} f_1({}^i\mathbf{r}, {}^i\mathbf{v}; t) \langle \mathbf{M} \rangle_1(\mathbf{r}, {}^i\mathbf{r}, {}^i\mathbf{v}; t) \right)_{{}^i\mathbf{r}, {}^i\mathbf{v}, \mathbf{r}} \\ &\quad + {}^i\mathbf{v} \cdot \frac{\partial}{\partial\, {}^i\mathbf{r}} (f_1 \langle \mathbf{M} \rangle_1) + \int d\mathbf{r}_1 \cdots d\mathbf{v}_N\, F_i \dot{\mathbf{v}}_i \cdot \frac{\partial}{\partial \mathbf{v}_i} (f_N \mathbf{M}) \end{aligned} \tag{1.20}$$

With

$$\dot{\mathbf{v}}_i = (e_i/m_i) \Big\{ \mathbf{E}_{\text{ex}}(\mathbf{r}_i, \mathbf{v}_i; t) + \sum_j{}' \mathbf{E}_j{}^i(\mathbf{r}_i, \mathbf{v}_i, \mathbf{r}_j, \mathbf{v}_j) \Big\} \tag{1.21}$$

we arrive at

$$\begin{aligned} \left\langle \frac{\partial}{\partial t} \mathbf{M} \right\rangle_1 f_1 &= \left(\frac{\partial}{\partial t} + {}^i\mathbf{v} \cdot \frac{\partial}{\partial\, {}^i\mathbf{r}} + \frac{e_i}{m_i} \mathbf{E}_{\text{ex}} \cdot \frac{\partial}{\partial\, {}^i\mathbf{v}} \right) (f_1 \langle \mathbf{M} \rangle_1) \\ &\quad + (e_i/m_i) \sum_j{}' \int d{}^j\mathbf{r} \int d{}^j\mathbf{v}\, \mathbf{E}_j{}^i({}^i\mathbf{r}, {}^i\mathbf{v}, {}^j\mathbf{r}, {}^j\mathbf{v}) \\ &\quad \cdot \frac{\partial}{\partial\, {}^i\mathbf{v}} \left(f_2({}^i\mathbf{r}, {}^i\mathbf{v}, {}^j\mathbf{r}, {}^j\mathbf{v}; t) \langle \mathbf{M} \rangle_2(\mathbf{r}, {}^i\mathbf{r}, {}^i\mathbf{v}, {}^j\mathbf{r}, {}^j\mathbf{v}; t) \right) \end{aligned} \tag{1.22}$$

In (1.21), $\mathbf{E}_j{}^i$ represents the electric field and the Lorentz field exerted by the jth on the ith particle.

Applying (1.22) in (1.11) and (1.16)–(1.19), we find the following set of equations:

$$\begin{aligned} &\frac{\partial f_1({}^i\mathbf{r}, {}^i\mathbf{v}; t)}{\partial t} + {}^i\mathbf{v} \cdot \frac{\partial f_1}{\partial\, {}^i\mathbf{r}} + \frac{e_i}{m_i} \frac{\partial}{\partial\, {}^i\mathbf{v}} \\ &\quad \cdot \{[\langle \mathbf{E} \rangle_1(\mathbf{r} = {}^i\mathbf{r}, {}^i\mathbf{r}, {}^i\mathbf{v}; t) + {}^i\mathbf{v} \times \langle \mathbf{B} \rangle_1(\mathbf{r} = {}^i\mathbf{r}, {}^i\mathbf{r}, {}^i\mathbf{v}; t)]\, f_1({}^i\mathbf{r}, {}^i\mathbf{v}; t)\} = 0 \end{aligned} \tag{1.23}$$

$$\nabla \times \langle \mathbf{B} \rangle_1(\mathbf{r}, {}^i\mathbf{r}, {}^i\mathbf{v}; t) = \mu_0 \Big\{ \mathbf{j}_{\text{ex}} + \frac{1}{f_1} \sum_j{}' e_j \int d^j\mathbf{v}\, {}^j\mathbf{v}\, f_2({}^i\mathbf{r}, {}^i\mathbf{v}, \mathbf{r}, {}^j\mathbf{v}; t)$$

$$+ \frac{\varepsilon_0}{f_1} \left(\frac{\partial}{\partial t} + {}^i\mathbf{v} \cdot \frac{\partial}{\partial\, {}^i\mathbf{r}} + \frac{e_i}{m_i} \mathbf{E}_{\text{ex}} \cdot \frac{\partial}{\partial\, {}^i\mathbf{v}} \right) (f_1 \langle \mathbf{E} \rangle_1)$$

$$+ \frac{\varepsilon_0}{f_1} \frac{e_i}{m_i} \sum_j{}' \int d^j\mathbf{r} \int d^j\mathbf{v}\, \mathbf{E}_j{}^i \cdot \frac{\partial}{\partial\, {}^i\mathbf{v}} f_2({}^i\mathbf{r}, {}^i\mathbf{v}, {}^j\mathbf{r}, {}^j\mathbf{v}; t)$$

$$\times \langle \mathbf{E} \rangle_2(\mathbf{r}, {}^i\mathbf{r}, {}^i\mathbf{v}, {}^j\mathbf{r}, {}^j\mathbf{v}; t) \Big\} \qquad (1.24)$$

$$\nabla \times \langle \mathbf{E} \rangle_1(\mathbf{r}, {}^i\mathbf{r}, {}^i\mathbf{v}; t) = -\frac{1}{f_1} \left(\frac{\partial}{\partial t} + {}^i\mathbf{v} \cdot \frac{\partial}{\partial\, {}^i\mathbf{r}} + \frac{e_i}{m_i} \mathbf{E}_{\text{ex}} \cdot \frac{\partial}{\partial\, {}^i\mathbf{v}} \right) (f_1 \langle \mathbf{B} \rangle_1)$$

$$- \frac{e_i}{f_1 m_i} \sum_j{}' \int d^j\mathbf{r} \int d^j\mathbf{v}\, \mathbf{E}_j{}^i \cdot \frac{\partial}{\partial\, {}^i\mathbf{v}} [f_2({}^i\mathbf{r}, {}^i\mathbf{v}, {}^j\mathbf{r}, {}^j\mathbf{v}; t)$$

$$\times \langle \mathbf{B} \rangle_2(\mathbf{r}, {}^i\mathbf{r}, {}^i\mathbf{v}, {}^j\mathbf{r}, {}^j\mathbf{v}; t)] \qquad (1.25)$$

$$\nabla \cdot \langle \mathbf{E} \rangle_1(\mathbf{r}, {}^i\mathbf{r}, {}^i\mathbf{v}; t) = \varepsilon_0^{-1} \Big\{ \rho_{\text{ex}} + \frac{1}{f_1} \sum_j{}' e_j \int d\, {}^j\mathbf{v}\, f_2({}^i\mathbf{r}, {}^i\mathbf{v}, \mathbf{r}, {}^j\mathbf{v}; t) \Big\} \qquad (1.26)$$

$$\nabla \cdot \langle \mathbf{B} \rangle_1(\mathbf{r}, {}^i\mathbf{r}, {}^i\mathbf{v}; t) = 0 \qquad (1.27)$$

The system of equations (1.23)–(1.27) provides a set of partial differential equations to determine the functions $f_1({}^i\mathbf{r}, {}^i\mathbf{v}; t)$ and $\langle \mathbf{E} \rangle_1(\mathbf{r}, {}^i\mathbf{r}, {}^i\mathbf{v}; t)$, $\langle \mathbf{B} \rangle_1(\mathbf{r}, {}^i\mathbf{r}, {}^i\mathbf{v}; t)$.

Of course this set is not closed, since it contains the pair functions f_2, $\langle \mathbf{E} \rangle_2$, and $\langle \mathbf{B} \rangle_2$. To determine these pair functions, we must formulate a system of equations which will contain the third-order functions f_3, $\langle \mathbf{E} \rangle_3$, and $\langle \mathbf{B} \rangle_3$. Similar to the procedure used in the derivation of the BBGKY equations, we again arrive at a hierarchy. We restrain ourselves from writing these higher-order equations since they are very involved and will not be used in the following.

2. Solution with Collective Particle Correlations Only

2.1. Basic Equations and Linearization Procedure

In this approximation, the following relations hold and allow substantial simplifications of the $(f, \langle \mathbf{B} \rangle, \langle \mathbf{E} \rangle)$ hierarchy:

$$f_s({}^1\mathbf{r}, \ldots, {}^s\mathbf{v}; t) = \prod_{\nu=1}^{s} f_1({}^\nu\mathbf{r}, {}^\nu\mathbf{v}; t), \qquad s = 2, \ldots, N \qquad (2.1)$$

$$\langle \mathbf{E} \rangle_1 = \langle \mathbf{E} \rangle_1(\mathbf{r}; t), \qquad \langle \mathbf{B} \rangle_1 = \langle \mathbf{B} \rangle_1(\mathbf{r}; t) \tag{2.2}$$

$$\left\langle \left(\frac{\partial \mathbf{E}}{\partial t} \right)_{\mathbf{r}} \right\rangle_1 = \frac{\partial}{\partial t} \langle \mathbf{E} \rangle_1, \qquad \left\langle \left(\frac{\partial \mathbf{B}}{\partial t} \right)_{\mathbf{r}} \right\rangle_1 = \frac{\partial}{\partial t} \langle \mathbf{B} \rangle_1 \tag{2.3}$$

The last of these relations follows from (1.20) using (2.1) and (2.2) and the one-particle Liouville theorem. Applying these simplifications, we find from (1.11) and (1.16)–(1.19)

$$\frac{\partial f_1}{\partial t} + {}^i\mathbf{v} \cdot \frac{\partial f_1}{\partial\, {}^i\mathbf{r}} + \frac{e_i}{m_i} \{\langle \mathbf{E} \rangle_1 + {}^i\mathbf{v} \times \langle \mathbf{B} \rangle_1\} \cdot \frac{\partial f_1}{\partial\, {}^i\mathbf{v}} = 0 \tag{2.4}$$

$$\nabla \times \langle \mathbf{B} \rangle_1 = \mu_0 \left\{ \mathbf{j}_{\mathrm{ex}} + \sum_j{}' e_j \int d\,{}^j\mathbf{v}\; {}^j\mathbf{v}\, f_1(\mathbf{r}, {}^j\mathbf{v}; t) + \varepsilon_0 \frac{\partial \langle \mathbf{E} \rangle_1}{\partial t} \right\} \tag{2.5}$$

$$\nabla \times \langle \mathbf{E} \rangle_1 = -\frac{\partial \langle \mathbf{B} \rangle_1}{\partial t} \tag{2.6}$$

$$\nabla \cdot \langle \mathbf{E} \rangle_1 = \varepsilon_0^{-1} \left\{ \rho_{\mathrm{ex}} + \sum_j{}' e_j \int d\,{}^j\mathbf{v}\, f_1(\mathbf{r}, {}^j\mathbf{v}; t) \right\} \tag{2.7}$$

$$\nabla \cdot \langle \mathbf{B} \rangle_1 = 0 \tag{2.8}$$

We assume that all external sources are localized outside the volume under consideration.

Further, it is useful to introduce instead of the specific distribution functions f_1 the general distributions through [see Equation (II.2.19)]

$$f_\mu^{(1)}(\mathbf{r}, \mathbf{v}; t) = \sum_{j(\mu)} f_1({}^j\mathbf{r}, {}^j\mathbf{v}; t)\Big|_{\substack{{}^j\mathbf{r}=\mathbf{r} \\ {}^j\mathbf{v}=\mathbf{v}}} = N_\mu f_1 \tag{2.9}$$

where $j(\mu)$ designates all particle indices of the component μ and N_μ their total number.

Carrying out the summations in our set of differential equations then yields

$$\frac{\partial f_\mu^{(1)}}{\partial t} + \mathbf{v} \cdot \frac{\partial f_\mu^{(1)}}{\partial \mathbf{r}} + \left(\frac{e}{m} \right)_\mu \{\langle \mathbf{E} \rangle_1 + \mathbf{v} \times \langle \mathbf{B} \rangle_1\} \cdot \frac{\partial f_\mu^{(1)}}{\partial \mathbf{v}} = 0 \tag{2.10}$$

$$\nabla \times \langle \mathbf{B} \rangle_1 = \mu_0 \left\{ \sum_\nu e_\nu \int d\mathbf{v}'\, \mathbf{v}' f_\nu^{(1)} + \varepsilon_0\, \partial \langle \mathbf{E} \rangle_1 / \partial t \right\} \tag{2.11}$$

$$\nabla \times \langle \mathbf{E} \rangle_1 = -\partial \langle \mathbf{B} \rangle_1 / \partial t \tag{2.12}$$

$$\nabla \cdot \langle \mathbf{E} \rangle_1 = \varepsilon_0^{-1} \sum_\nu e_\nu \int d\mathbf{v}'\, f_\nu^{(1)} \tag{2.13}$$

$$\nabla \cdot \langle \mathbf{B} \rangle_1 = 0 \tag{2.14}$$

Equations (2.10)–(2.14) describe the particle distributions $f_\mu^{(1)}$ and the fields $\langle \mathbf{E} \rangle_1$, $\langle \mathbf{B} \rangle_1$ in the limit of vanishing individual particle correlations. For simplicity, we shall omit in the following the brackets at **E** and **B**, keeping in mind that **E** and **B** are average quantities.

Small Amplitude Theory

The small amplitude approximation is based on the assumption that the fields **E**, **B** may be considered as small causing only small deviations from the zero-order state of the system. We further assume that the zero-order terms of the development

$$f_\mu^{(1)} = {}^{(0)}f_\mu^{(1)} + {}^{(1)}f_\mu^{(1)} + \cdots \tag{2.15}$$

do not produce net charges and net currents:

$$\sum_\nu e_\nu \int d\mathbf{v}'\, {}^{(0)}f_\nu^{(1)} = 0, \qquad \sum_\nu e_\nu \int d\mathbf{v}'\, \mathbf{v}'\, {}^{(0)}f_\nu^{(1)} = 0 \tag{2.16}$$

If the zero-order state is a stationary one, it follows from (2.10) that the zero-order distribution depends on **v** only. We write†

$${}^{(0)}f_\mu^{(1)} =: f_{0\mu}(\mathbf{v}), \qquad {}^{(1)}f_\mu^{(1)} =: f_{1\mu}(\mathbf{r}, \mathbf{v}; t) \tag{2.17}$$

With these definitions, we rewrite our set of equations

$$\frac{\partial f_{1\mu}}{\partial t} + \mathbf{v} \cdot \frac{\partial f_{1\mu}}{\partial \mathbf{r}} + \left(\frac{e}{m}\right)_\mu \left\{\mathbf{E} + \mathbf{v} \times \mathbf{B}\right\} \cdot \frac{\partial f_{0\mu}}{\partial \mathbf{v}} = 0 \tag{2.18}$$

$$\nabla \times \mathbf{B} = \mu_0 \left\{\sum_\nu e_\nu \int d\mathbf{v}'\, \mathbf{v}' f_{1\nu} + \varepsilon_0\, \partial \mathbf{E}/\partial t\right\} \tag{2.19}$$

$$\nabla \times \mathbf{E} = -\partial \mathbf{B}/\partial t \tag{2.20}$$

$$\nabla \cdot \mathbf{E} = \varepsilon_0^{-1} \sum_\nu e_\nu \int d\mathbf{v}'\, f_{1\nu} \tag{2.21}$$

$$\nabla \cdot \mathbf{B} = 0 \tag{2.22}$$

† Note that here—in analogy to p. 96—we introduce for reasons of simplification a nomenclature for the general distribution functions which should not be confused with that for specific distribution functions.

2.2. The Conductivity Tensor and the Dielectric Tensor

Fourier transformation of (2.18) gives

$$i(\mathbf{k}\cdot\mathbf{v}-\omega)\,\tilde{\tilde{f}}_{1\mu}(\mathbf{k},\omega,\mathbf{v})+(e/m)_{\mu}\{\tilde{\tilde{\mathbf{E}}}(\mathbf{k},\omega)+\mathbf{v}\times\tilde{\tilde{\mathbf{B}}}(\mathbf{k},\omega)\}\cdot df_{0\mu}/d\mathbf{v}=0 \tag{2.23}$$

or

$$\tilde{\tilde{f}}_{1\mu}=\frac{i(e/m)_{\mu}}{\mathbf{k}\cdot(\mathbf{v}-\mathbf{u})}\frac{df_{0\mu}}{d\mathbf{v}}\cdot(\tilde{\tilde{\mathbf{E}}}+\mathbf{v}\times\tilde{\tilde{\mathbf{B}}}) \tag{2.24}$$

with the abbreviation

$$\mathbf{u}=(\omega/k)\,\hat{\mathbf{k}} \tag{2.25}$$

The current density is then determined by

$$\tilde{\tilde{\mathbf{j}}}=\sum_{\nu}e_{\nu}\int d\mathbf{v}'\,\mathbf{v}'\tilde{\tilde{f}}_{1\nu}=i\sum_{\nu}\frac{e_{\nu}^{2}}{m_{\nu}}\int d\mathbf{v}'\,\frac{\mathbf{v}'}{\mathbf{k}\cdot(\mathbf{v}'-\mathbf{u})}\left(\frac{df_{0\nu}}{d\mathbf{v}'}\cdot\underset{2}{\boldsymbol{\kappa}}\cdot\tilde{\tilde{\mathbf{E}}}\right) \tag{2.26}$$

with the definition for $\underset{2}{\boldsymbol{\kappa}}$

$$\underset{2}{\boldsymbol{\kappa}}\cdot\tilde{\tilde{\mathbf{E}}}=\tilde{\tilde{\mathbf{E}}}+\mathbf{v}\times\tilde{\tilde{\mathbf{B}}} \tag{2.27}$$

Equation (2.20) provides

$$i\mathbf{k}\times\tilde{\tilde{\mathbf{E}}}=i\omega\tilde{\tilde{\mathbf{B}}} \tag{2.28}$$

or

$$\tilde{\tilde{\mathbf{B}}}=\frac{1}{u}\hat{\mathbf{k}}\times\tilde{\tilde{\mathbf{E}}}=\frac{n}{c}\hat{\mathbf{k}}\times\tilde{\tilde{\mathbf{E}}} \tag{2.29}$$

where we defined the refractive index $n=c/u$.

Introducing (2.29) into (2.27), we find for the tensor $\underset{2}{\boldsymbol{\kappa}}$

$$\underset{2}{\boldsymbol{\kappa}}=\left(\underset{2}{\mathbf{I}}\left(1-\frac{v_{\parallel}}{u}\right)+\frac{\hat{\mathbf{k}})(\mathbf{v}}{u}\right) \tag{2.30}$$

Let us now define a new distribution function f_0 by

$$\omega_{\mathrm{p}0}^{2}f_0:=\sum_{\nu}(e_{\nu}^{2}/\varepsilon_0 m_{\nu})\,f_{0\nu} \tag{2.31}$$

which—in contrast to $f_{0\nu}$—is normalized to one, since we use

$$\omega_{\mathrm{p}0}^{2}=\sum_{\nu}(e_{\nu}^{2}/\varepsilon_0 m_{\nu})\int f_{0\nu}\,d\mathbf{v}' \tag{2.32}$$

According to (2.26), (2.30), and (2.31), the conductivity tensor is given by

$$\begin{aligned}\underset{2}{\boldsymbol{\sigma}} &= i\varepsilon_0\omega_{\mathrm{p}0}^2 \int d\mathbf{v}\, \frac{\mathbf{v})(\, df_0/d\mathbf{v}}{\mathbf{k}\cdot(\mathbf{v}-\mathbf{u})} \cdot \underset{2}{\boldsymbol{\kappa}} \\ &= i\varepsilon_0\omega_{\mathrm{p}0}^2 \int \frac{d\mathbf{v}}{\mathbf{k}\cdot(\mathbf{v}-\mathbf{u})}\, \mathbf{v}\Big)\Big(\frac{df_0}{d\mathbf{v}} \cdot \left(\underset{2}{\mathbf{I}}\left(1-\frac{v_{\parallel}}{u}\right) + \frac{\hat{\mathbf{k}})(\,\mathbf{v}}{u}\right)\end{aligned} \tag{2.33}$$

Insertion of $\tilde{\tilde{\mathbf{j}}} = \underset{2}{\boldsymbol{\sigma}} \cdot \tilde{\tilde{\mathbf{E}}}$ into (2.19) produces

$$\begin{aligned} i\mathbf{k} \times \tilde{\tilde{\mathbf{B}}} &= \mu_0\varepsilon_0\{-i\omega\tilde{\tilde{\mathbf{E}}} + \varepsilon_0^{-1}\underset{2}{\boldsymbol{\sigma}} \cdot \tilde{\tilde{\mathbf{E}}}\} \\ &= \frac{1}{c^2}\left\{\frac{1}{\varepsilon_0}\underset{2}{\boldsymbol{\sigma}} - i\omega\underset{2}{\mathbf{I}}\right\} \cdot \tilde{\tilde{\mathbf{E}}} \\ &=: -\frac{i\omega}{c^2}\underset{2}{\boldsymbol{\varepsilon}} \cdot \tilde{\tilde{\mathbf{E}}} \end{aligned} \tag{2.34}$$

and therewith the dielectric tensor

$$\underset{2}{\boldsymbol{\varepsilon}} = \underset{2}{\mathbf{I}} + (i/\varepsilon_0\omega)\,\underset{2}{\boldsymbol{\sigma}} \tag{2.35}$$

2.3. The Dispersion Relation

Equation (2.34) reads

$$\mathbf{k} \times \tilde{\tilde{\mathbf{B}}} = -(\omega/c^2)\,\underset{2}{\boldsymbol{\varepsilon}} \cdot \tilde{\tilde{\mathbf{E}}} \tag{2.36}$$

and (2.29) can be rewritten in the form

$$\mathbf{k} \times \tilde{\tilde{\mathbf{B}}} = (n/c)\,k[\hat{\mathbf{k}}\,)(\,\hat{\mathbf{k}} - \underset{2}{\mathbf{I}}] \cdot \tilde{\tilde{\mathbf{E}}} \tag{2.37}$$

With $\underset{2}{\mathbf{I}}_{\perp} = \underset{2}{\mathbf{I}} - \hat{\mathbf{k}}\,)(\,\hat{\mathbf{k}}$, we find

$$(nk\underset{2}{\mathbf{I}}_{\perp} - (\omega/c)\,\underset{2}{\boldsymbol{\varepsilon}}(\mathbf{k}, \omega)) \cdot \tilde{\tilde{\mathbf{E}}} = 0 \tag{2.38}$$

and with that the **dispersion relation**

$$\det\{n^2\underset{2}{\mathbf{I}}_{\perp} - \underset{2}{\boldsymbol{\varepsilon}}(\mathbf{k}, \omega)\} = \mathbf{0} \tag{2.39}$$

where the dielectric tensor according to (2.35) and (2.33) is given by

$$\underset{2}{\boldsymbol{\varepsilon}} = \underset{2}{\mathbf{I}} - \frac{\omega_{\mathrm{p}0}^2}{\omega^2} \int d\mathbf{v}\, \frac{u}{v_{\parallel} - u}\, \mathbf{v}\,)(\,\frac{df_0}{d\mathbf{v}} \cdot \left(\underset{2}{\mathbf{I}}\left(1 - \frac{v_{\parallel}}{u}\right) + \frac{\hat{\mathbf{k}})(\,\mathbf{v}}{u}\right) \tag{2.40}$$

There are two features of the dispersion relation which render the solution difficult: One is its obvious complexity. The other concerns the divergence of the denominator in the integrand of (2.40) for real values of u.

There is little we can do about the first difficulty except that we may eventually restrict our considerations to special distributions. With respect to the second problem, there are two areas of principal and practical interest where the divergence problem does not occur. The first is concerned with large real phase velocities $u \gtrsim c$ where the problem disappears, since we do not regard relativistic particles. This range is of interest because it shows the reaction of the plasma with respect to incoming electromagnetic waves. The second is concerned with phase velocities which have a finite imaginary value u_i. Such plasma waves are in general not useful for the transmission of signals. But they are interesting because of the occurrence of unstable growing modes leading to the rearrangement of particle velocities, as has been shown for the case of electrostatic waves by Bunemann (1959). In this way, complex wave solutions tend to have the same effect as molecular collisions in a normal gas.

$u_r \gtrsim c$. The real part of the phase velocity is assumed to be equal or larger than the velocity of light. This naturally implies $v \ll u$.

In this limit

$$\underset{2}{\boldsymbol{\kappa}} = \underset{2}{\mathbf{I}} \tag{2.41}$$

and

$$\underset{2}{\boldsymbol{\varepsilon}} = \underset{2}{\mathbf{I}} + \frac{\omega_{p0}^2}{\omega^2} \int d\mathbf{v}\, \mathbf{v}\,)(\,\frac{df_0}{d\mathbf{v}} \tag{2.42}$$

holds. Partial integration shows that in this case the dielectric tensor and conductivity tensor have the simple form

$$\underset{2}{\boldsymbol{\varepsilon}} = (1 - (\omega_{p0}^2/\omega^2))\, \underset{2}{\mathbf{I}}, \qquad \underset{2}{\boldsymbol{\sigma}} = (i\varepsilon_0/\omega)\, \omega_{p0}^2\, \underset{2}{\mathbf{I}} \tag{2.43}$$

and the dispersion relation reads

$$\det \left\{ \left(1 - \frac{\omega_{p0}^2}{\omega^2}\right) \underset{2}{\mathbf{I}} - n^2(\underset{2}{\mathbf{I}} - \hat{\mathbf{k}}\,)(\,\hat{\mathbf{k}}) \right\} = \det \left\{ \left(1 - \frac{\omega_{p0}^2}{\omega^2} - n^2\right) \underset{2}{\mathbf{I}} + n^2 \hat{\mathbf{k}}\,)(\,\hat{\mathbf{k}} \right\} = 0 \tag{2.44}$$

Choosing the z coordinate in the direction of $\mathbf{k}$, the following determinant represents the dispersion relation:

$$\begin{vmatrix} (1 - (\omega_{p0}^2/\omega^2) - n^2) & 0 & 0 \\ 0 & (1 - (\omega_{p0}^2/\omega^2) - n^2) & 0 \\ 0 & 0 & (1 - (\omega_{p0}^2/\omega^2)) \end{vmatrix} = 0 \tag{2.45}$$

The fact that the tensor in (2.45) is diagonal means that the longitudinal and transversal fields are decoupled. Since the corresponding tensor elements of the transversal components are identical but different from that of the longitudinal component, the transversal solutions have frequencies different from the longitudinal ones, and the transversal field may have any polarization direction.

The **longitudinal fields** have the frequency

$$\omega = \omega_{\mathrm{p0}} \tag{2.46}$$

prescribed by (2.45). The phase velocity therefore is given by

$$u = \omega_{\mathrm{p0}}/k \tag{2.47}$$

Our assumption $v \ll u$ implies that the above results are correct only in the range

$$k = \omega_{\mathrm{p0}}/u \ll (m/\Theta)^{1/2}\, \omega_{\mathrm{p0}} = k_{\mathrm{D}} \tag{2.48}$$

Note that our solution for electrostatic waves corresponds to the results we derived in Chapter III, Section 2.5. We do not expect Landau damping here, since it originates from particles with velocities close to the phase velocity.

The **transversal fields** have frequencies

$$\omega_{\mathrm{p0}}^2/\omega^2 = 1 - n^2 = 1 - c^2/u^2 \tag{2.49}$$

or

$$\omega^2/\omega_{\mathrm{p0}}^2 = u^2/(u^2 - c^2) \tag{2.50}$$

corresponding to phase velocities

$$u^2 = \frac{c^2}{1 - \omega_{\mathrm{p0}}^2/\omega^2} \tag{2.51}$$

We see therefore that for the transversal modes

$$\omega > \omega_{\mathrm{p0}}\,; \qquad u > c \tag{2.52}$$

holds. Of course the group velocity $u_{\mathrm{g}} = d\omega/dk$ derived from equation (2.49) in the form

$$\omega^2 = \omega_{\mathrm{p0}}^2 + k^2c^2 \tag{2.53}$$

is given by

$$u_{\mathrm{g}} = c^2/u = c(1 - \omega_{\mathrm{p0}}^2/\omega^2)^{1/2} \tag{2.54}$$

and is always smaller than the velocity of light.

The above results show that a plasma will react to transversal electromagnetic waves with a frequency above the plasma frequency like a substance with a refractive index smaller than one. If the frequency of the electromagnetic wave is very much larger than the plasma frequency ω_{p0}, then the refractive index is practically one. The plasma has no effect on the wave. If the frequency of the electromagnetic wave approaches the plasma frequency, the refractive index decreases, rapidly approaching zero.

There are no transversal solutions with a phase velocity $u > c$ and frequencies $\omega < \omega_{p0}$. Since we have no dissipation mechanism (σ is a purely imaginary quantity), incoming waves with these frequencies cannot be damped out. They must be totally reflected at the plasma surface. (This is clear from $u^2 < 0$, too.)

$u_i \neq 0$; $u_r = O(v_{th}) \ll c$. The waves to be studied in this section should be clearly distinguished from the familiar electromagnetic waves propagating with phase velocities equal to or larger than the speed of light which have been treated in the preceding paragraph. These latter ones can be considered simply as a form of radiowaves modified by the presence of the plasma. The waves which we study in this section have much smaller phase velocities of the order of the rms velocity of the particles under consideration. Further, their phase velocity is assumed to be complex.

Anticipating that the instability constant of such waves is in general of the order $\omega_{p0} v_{th}/c$, the development of these instabilities is much less violent than that of the electrostatic instabilities. In a range where both of these instabilities can occur, the electrostatic ones should therefore be dominant. However, the conditions which are needed for the existence of electromagnetic instabilities are less stringent than those for the existence of electrostatic ones. Therefore in many systems the electromagnetic instabilities are the only ones and consequently of importance.

Let us recall the dispersion relation as prescribed by (2.39) and (2.40). v_3 is the component of $\mathbf{v}$ in the direction of $\mathbf{k}$; v_1, v_2 are the components perpendicular to $\mathbf{k}$. The dispersion relation then reads

$$\det \left\{ \mathbf{I}_2 - \frac{c^2k^2}{\omega^2} \mathbf{I}_{2\perp} + \frac{\omega_{p0}^2}{\omega^2} \int (\mathbf{v}) \left(\frac{df_0}{d\mathbf{v}} - \frac{\mathbf{v})(\mathbf{v} \frac{\partial f_0}{\partial v_3} k}{v_3 k - \omega} \right) d\mathbf{v} \right\} =: \det \mathbf{A} = 0 \tag{2.55}$$

If we designate the coefficients of $\mathbf{A}$ by $A_{\mu\nu}$, then we may write

$$A_{\mu\nu} = \delta_{\mu\nu} \left(1 - \frac{c^2k^2}{\omega^2} - \frac{\omega_{p0}^2}{\omega^2} \right) - \frac{\omega_{p0}^2}{\omega^2} C_{\mu\nu} + \delta_{\mu 3} \delta_{\nu 3} \frac{c^2k^2}{\omega^2} \tag{2.56}$$

with

$$C_{\mu\nu} = k \int \frac{v_\mu v_\nu \, \partial f_0/\partial v_3}{v_3 k - \omega} \, d\mathbf{v} \tag{2.57}$$

Note that the tensors $\underset{2}{\mathbf{A}}$ and $\underset{2}{\mathbf{C}}$ are symmetric:

$$C_{\mu\nu} = C_{\nu\mu}, \qquad A_{\mu\nu} = A_{\nu\mu} \tag{2.58}$$

Now recalling that we are studying the range $u_r \simeq v_{\text{th}} \ll c$, we may neglect one in comparison to c^2k^2/ω^2, which physically means the neglect of the displacement current. This results in

$$A_{\mu\nu} = -(c^2/\omega^2)\{\delta_{\mu\nu}(k^2 + k_c^2) + k_c^2 C_{\mu\nu}\}, \qquad k_c := \omega_{p0}/c \tag{2.59}$$

for $(\mu, \nu) \neq (3,3)$ and

$$A_{33} = -(c^2/\omega^2)\{-(\omega^2/c^2) + k_c^2(1 + C_{33})\} \tag{2.60}$$

Electrostatic and electromagnetic waves will be decoupled provided that the relation

$$A_{13} = A_{23} = 0 \qquad \text{or} \qquad C_{13} = C_{23} = 0 \tag{2.61}$$

holds. This condition may also be rewritten in the form

$$C_{\substack{13\\23}} = k \int \frac{v_3 \, d(\langle v_{\substack{1\\2}} \rangle F_0(v_3))/dv_3}{v_3 k - \omega} \, dv_3 = 0 \tag{2.62}$$

where we have used

$$F_0(v_3) := \int f_0 \, dv_1 \, dv_2, \qquad \langle v_{\substack{1\\2}} \rangle = [1/F_0(v_3)] \int v_{\substack{1\\2}} f_0 \, dv_1 \, dv_2 \tag{2.63}$$

A sufficient condition for decoupling is then obviously given by

$$\langle v_{\substack{1\\2}} \rangle = 0 \qquad \text{for all } v_3 \tag{2.64}$$

In this case, the **electrostatic dispersion relation** reads

$$1 - (\omega_{p0}^2/\omega^2)(1 + C_{33}) = 0 \tag{2.65}$$

or with the identity

$$1 = \int f_0 \, d\mathbf{v} = -\int v_3 \frac{\partial f_0}{\partial v_3} \, d\mathbf{v} = -\int \frac{v_3(v_3 - u) \, \partial f_0/\partial v_3}{v_3 - u} \, d\mathbf{v} \tag{2.66}$$

we find

$$1 - \frac{\omega_{\text{po}}^2}{\omega^2} u \int \frac{v_3 \, \partial f_0/\partial v_3}{v_3 - u} \, d\mathbf{v} = 0 \tag{2.67}$$

By partial integration it can be shown that this result is identical with the dispersion relation for electrostatic waves [see Equation (III.2.18)].

Under the condition (2.64), the **dispersion relation for the electromagnetic waves** can be written in the form

$$(k^2/k_c^2 + 1 + C_{11})(k^2/k_c^2 + 1 + C_{22}) - C_{12}^2 = 0 \tag{2.68}$$

To solve this equation we could proceed by writing the integrals C_{11}, C_{22}, and C_{12} in the form

$$\int \frac{(d\chi_{\mu\nu}/dv_3)\, dv_3}{v_3 - u} = \int \frac{(v_3 - u_r)(d\chi_{\mu\nu}/dv_3)\, dv_3}{(v_3 - u_r)^2 + u_i^2} + iu_i \int \frac{(d\chi_{\mu\nu}/dv_3)\, dv_3}{(v_3 - u_r)^2 + u_i^2} \tag{2.69}$$

with $\chi_{\mu\nu} := \langle v_\mu v_\nu \rangle F_0(v_3)$, separating into real and imaginary parts, and finding from these equations the real and imaginary components of the frequencies or, respectively, phase velocities. This would result in a complicated numerical evaluation. We rather want to draw more general conclusions by an estimate of orders of magnitude.

Rewriting the coefficients $C_{\mu\nu}$

$$\begin{aligned} C_{\mu\nu} &= \int \frac{(d/dv_3)\langle v_\mu v_\nu \rangle F_0(v_3)\, dv_3}{v_3 - u} \\ &= \int \frac{\langle v_\mu v_\nu \rangle F_0(v_3)\, dv_3}{(v_3 - u)^2}, \qquad \mu = 1, 2, \quad \nu = 1, 2 \end{aligned} \tag{2.70}$$

and recalling that $|u|^2 \simeq \langle v^2 \rangle$, therefore

$$\langle v_\mu v_\nu \rangle \simeq |u|^2 \simeq |v_3 - u|^2, \qquad \int F_0(v_3)\, dv_3 = 1$$

$$C_{\mu\nu} = O(1), \qquad \mu = 1, 2, \quad \nu = 1, 2 \tag{2.71}$$

Equation (2.67) indicates

$$k_{\text{em}}^2 = O(k_c^2) \ll k_{\text{es}}^2 = O(\omega_{\text{po}}^2/\langle v^2 \rangle) \tag{2.72}$$

The wavevectors of the electromagnetic waves under consideration (k_{em}) are much smaller than those of the corresponding electrostatic phenomena (k_{es}). The information of (2.72) is useful for an estimate of the **coupling of electrostatic and electromagnetic waves**. To this end, we keep all terms in the dispersion relation

$$\begin{vmatrix} 1+(k_c^2/k^2)(1+C_{11}) & (k_c^2/k^2)\,C_{12} & (k_c^2/k^2)\,C_{13} \\ (k_c^2)k^2)\,C_{12} & 1+(k_c^2/k^2)(1+C_{22}) & (k_c^2/k^2)\,C_{23} \\ (k_c^2/k^2)\,C_{13} & (k_c^2/k^2)\,C_{23} & -(1/n^2)+(k_c^2/k^2)(1+C_{33}) \end{vmatrix} = 0 \tag{2.73}$$

Then we first consider waves with wavevectors close to $k_{es}^2 \simeq \omega_{po}^2/\langle v^2 \rangle$. Using the order of magnitude of the coefficients $C_{\mu\nu}$, we may rewrite the determinant (2.73) in the form

$$-(1/n^2) + (k_c^2/k^2)(1 + C_{33}) = O(k_c^4/k^4) \tag{2.74}$$

or

$$1 + C_{33}(k, \omega) - \omega^2/\omega_{po}^2 = O(k_c^2/k^2) \tag{2.75}$$

Using the expansion $k = k_{es} + \Delta k$, where k_{es} is the solution of (2.75) with the right-hand side equal to zero, we get

$$(\partial C_{33}/\partial k)\, \Delta k = O(k_c^2/k^2) \tag{2.76}$$

Assuming $\partial C_{33}/\partial k = O(C_{33}/k) = O(1/k)$, and remembering (2.72), we find in consistency with our ansatz

$$\Delta k/k_{es} = O(k_c^2/k_{es}^2) \ll 1 \tag{2.77}$$

This shows that the coupling of electrostatic and electromagnetic waves in a plasma hardly affects the dispersion relation and the frequencies of the electrostatic modes. On the other hand, for the electromagnetic phenomena the coupling terms cause an important change of the corresponding dispersion relation. Here we consider wave vectors $k^2 \simeq k_c^2$. Therefore in the dispersion relation (2.73), all coefficients are of the order of magnitude one, so that the coupling terms cannot be neglected.

2.4. Electromagnetic Instabilities

A detailed evaluation of the dispersion relation (2.73) considering stability aspects is cumbersome and cannot be presented within the framework of this investigation. We therefore quote the results of Kahn (1962) who studied the general case of distribution functions with central symmetry

$$f_0(\mathbf{v}) = f_0(-\mathbf{v}) \tag{2.78}$$

His calculations show that unstable transverse waves may occur in plasmas for a wide variety of velocity distribution functions.

An instability will certainly arise unless the conditions

$$\int_0^\infty v f_0(v, \vartheta, \varphi)\, dv = \text{constant} \tag{2.79}$$

and

$$\int_0^\infty v^2 f_0(v, \vartheta, \varphi)\, dv = \text{constant} \tag{2.80}$$

hold; that is, unless the number of particles moving into a given solid angle and their harmonic mean velocity are independent of direction.

Observe that these conditions neither mean that the velocity distribution has to be isotropic, nor do they require that there should be a pressure isotropy. To ensure isotropy, we would have to fulfill

$$\int_0^\infty v^n f_0(v, \vartheta, \varphi)\, dv = \text{constant} \tag{2.81}$$

for all values of n. Pressure isotropy in particular would require $n = 4$. Nevertheless, in practice a plasma would need to be specially prepared if its velocity distribution was to satisfy (2.79) and (2.80) and yet be anisotropic. This can hardly be expected of the distribution functions in a practical problem.

Kahn's investigation does not resolve all questions. It cannot be said that a plasma will definitely be stable if its distribution function satisfies the conditions (2.79) and (2.80). These conditions are only necessary. Further, it has not been possible to obtain results for velocity distributions without central symmetry. Kahn, however, believes that all such distributions are unstable.

In judging these results, let us recall that we have neglected correlations. This implies that the above claims are valid only when the times under consideration are small compared to the collision time, that is [see Appendix Eq. (3)]

$$\omega \gg \omega_{p0}(\ln \Lambda)/\Lambda \tag{2.82}$$

Now the phenomenon studied here deals with low frequencies, since we conclude from relation (2.72) $|\omega/u| = O(\omega_{p0}/c)$ and with our assumption $|u| = O(v_{th})$, we get with (2.82)

$$\omega/\omega_{p0} = O(v_{th}/c) \gg (\ln \Lambda)/\Lambda \tag{2.83}$$

It follows that the instabilities described above can be expected for low densities and high (but not relativistic!) temperatures only.

The physical mechanism behind the amplification of electromagnetic waves as described in the preceding has been studied by Fried (1959) and Furth (1963). Since one is interested only in the basic phenomenon and not in specific details, these authors base their discussions on a system of electrons in a simple experimental setup.

Let us start with an initial magnetic field disturbance of the type

$$B_z = Be^{ikx}, \qquad B_x = B_y = 0 \tag{2.84}$$

which represents wave modulation in the direction of the x coordinate. The magnetic field vector points in the direction of the z axis.

The electrons initially do not move in the z or x direction. In the y direction, all electrons have the same magnitude of velocity. The distribution function then reads

$$f_0(\mathbf{v}) = a\delta(v_x)\,\delta(v_z)\,\delta(v_y{}^2 - a^2), \qquad a > 0 \tag{2.85}$$

An electron with the initial velocity v_y will experience an acceleration by the magnetic field

$$\dot{v}_x = -(e/m)\,v_y B_z = -\omega_{\mathrm{L}} v_y \tag{2.86}$$

($\omega_{\mathrm{L}} = eB_z/m$ is the varying cyclotron frequency). As a consequence, there is a flux of the y component of momentum in direction of the x axis whose time derivative is given by

$$mv_y\dot{v}_x = -mv_y{}^2\omega_{\mathrm{L}} \tag{2.87}$$

Averaging with the distribution function $f = f_0 + f_1(\mathbf{r}, \mathbf{v}; t)$ and collecting the terms of first order, yields

$$\partial\langle v_y v_x\rangle/\partial t = -a^2\omega_{\mathrm{L}} \tag{2.88}$$

Since momentum has to be conserved, this flux (or pressure) causes a change in $\langle v_y\rangle$ given by the continuity equation

$$\partial\langle v_y\rangle/\partial t = -\partial\langle v_x v_y\rangle/\partial x \tag{2.89}$$

Maxwell's equation for the curl of **B** provides

$$ikB_z = -\mu_0 j_y = +\mu_0 en_-\langle v_y\rangle \tag{2.90}$$

if we neglect the displacement current. Using this relation to eliminate $\langle v_y\rangle$ from (2.88) and (2.89), we find

$$ik\,\partial^2 B_z/\partial t^2 = \mu_0 en_-\,\partial(a^2\omega_{\mathrm{L}})/\partial x = ik(\omega_{\mathrm{p}-}^2/c^2)\,a^2 B_z \tag{2.91}$$

This means that the initial perturbation B_z will grow exponentially at a rate $\omega_{p-}a/c$, which is in agreement with Kahn's conclusions.

The expression (2.91) is not identical with more general results obtained by Weibel (1959). Nevertheless this crude model with its special initial distribution can give a qualitative explanation of the phenomenon.

The fact that we have omitted $\partial\mathbf{E}/\partial t$ in (2.90) is equivalent to the neglect of one against c^2k^2/ω^2 in the dispersion relation. This shows that displacement currents do not contribute to the development of the instability. The mechanism is simply that the initial magnetic field causes such an acceleration of the electrons that the corresponding momentum flux affects the initial current distribution in such a way as to increase the fluctuating field. Physically speaking, we may consider the amplification of our initial disturbance as a pinching of the plasma into current sheets. If we drop the assumption of zero velocity v_x, then there will be an effect of superposition due to the x motion of the particles of our system. This effect counteracts the instability which we found without motion in the x direction.

3. Solutions Including Individual Electron-Ion Correlations

In the preceding section we studied the effect of a fully ionized system without individual particle correlations on the propagation of electromagnetic waves. Our interest now turns to the emission and absorption of such waves. Of course, individual correlations then cannot be neglected completely.

We have already discussed emission due to single-particle interactions. In this section, we investigate the emission and absorption as caused by the acceleration due to the simultaneous interaction with all plasma particles. Their field action is customarily split up into the collective field presented in Vlasov's equation and the individual correlation field represented in the rest of the BBGKY hierarchy. Both of these contributions may be termed "bremsstrahlung." The contribution of the collective field to the bremsstrahlung is, of course, a phenomenon unknown to the single-particle approach. On the other hand, the contribution to the bremsstrahlung from the individual correlation fields should contain the contribution calculated for the single-particle processes.

It is not possible to argue that the contribution to the radiation caused by the individual correlation fields is negligible in comparison to the contribution from the collective fields. In fact, since the collective field may have negligible amplitude, the opposite is probably true in many cases of practical importance.

In principle, we would therefore have to return to our general equations from Section 1 to formulate the next set of the hierarchy equations for f_2, $\langle E\rangle_2$, and $\langle B\rangle_2$ and make assumptions for higher-order correlations f_3, $\langle B\rangle_3$, and $\langle E\rangle_3$. Inspection of Section 1 is very discouraging in this respect.

Consequently, we choose a more ad hoc procedure. We determine the conductivity tensor of the system including the collective and correlational effects as well as possible. Then we engage Kirchhoff's law to find from the absorption coefficient—which is related to the conductivity tensor—the emission coefficient. Of course, this procedure is not rigorous, but it yields results which are sufficiently accurate for many applications.

Correlation effects of the particles can be taken into account in a comparatively simple way by use of the Fokker–Planck equation or similar methods, provided the "two-particle interaction time" τ_{in}, (Chapter IV, Section 1.3) can be considered much shorter than any other characteristic time of the problem, in particular, shorter than the oscillation time of the field. In our plasma, the "effective two-particle interaction time" is of the order of the reciprocal plasma frequency $\tau_{\text{in}} \simeq \omega_{\text{p}-}^{-1}$. We therefore expect that we can use the Fokker–Planck equation or similar methods provided that the condition $\omega \ll \omega_{\text{p}-}$ is fulfilled.

Unfortunately—as we have seen above—in the region $u \geqslant c$ which is of dominant practical interest, we have to deal with frequencies $\omega > \omega_{\text{p}-}$ and therefore cannot apply the Fokker–Planck equation. In this range we follow a procedure given by Dawson and Obermann (1962, 1963).

3.1. The Model of Dawson and Oberman

The application of this model is limited to the following range:

$$\omega > (2\pi/\tau_{\text{c}}) \approx \omega_{\text{p}-} \ln \Lambda/\Lambda, \qquad u \gg v_{\text{th}}$$

It is characterized by a number of assumptions:

(1) Our system is, as always, infinitely extended, spatially homogeneous, and consists of electrons and ions of opposite charge and equal number. There is a small uniform electric field $\mathbf{E}_0 e^{-i\omega t}$. The assumption

of spatial uniformity of the electric field implies the restriction to long wavelengths or, strictly speaking, to phenomena with a phase velocity large compared to the rms velocity.

(2) The ions do not move ($m_-/m_+ = 0$) and can be regarded as randomly distributed.

(3) Magnetic interactions are neglected since they are small of order v^2/c^2 in comparison to the electric interactions.

(4) Vlasov's equation is suitable to describe the electron dynamics. Therefore we require for the frequency

$$\omega \gg 2\pi/\tau_c \approx \omega_{p-}(\ln \Lambda)/\Lambda \tag{3.1}$$

where τ_c is the collision time.

In the dipole approximation, electron–electron correlations contribute neither to the conductivity nor to the absorption or emission. Within a coordinate system following the oscillatory motion of the free electrons in the external field, the zero-order distribution function of the electrons may be considered Maxwellian. That means the electron–electron correlations are neglected in general in our treatment except insofar as they are favorable to justify this assumption of the Maxwellian distribution.

We start with Vlasov's equation for the electrons ($m_- = m$, $e_- = -e$)

$$\frac{\partial f_-^{(1)}}{\partial t} + \mathbf{v} \cdot \frac{\partial f_-^{(1)}}{\partial \mathbf{r}} - \frac{e}{m} \left[\mathbf{E}_0 e^{-i\omega t} - \frac{\partial \Phi}{\partial \mathbf{r}}\right] \cdot \frac{\partial f_-^{(1)}}{\partial \mathbf{v}} = 0 \tag{3.2}$$

$$\Delta \Phi = \varepsilon_0^{-1} \left(e \int f_-^{(1)}\, d\mathbf{v} - e \sum_j \delta(\mathbf{r} - \mathbf{r}_j)\right) \tag{3.3}$$

Here $\mathbf{r}_j$ denotes the positions of the ions.

We now transform these equations to the coordinate frame which oscillates with the motion of a free electron in the external electric field. These transformation equations are

$$\mathbf{R} = \mathbf{r} - (e/m\omega^2)\,\mathbf{E}_0 e^{-i\omega t}, \qquad \mathbf{V} = \mathbf{v} + (ie/m\omega)\,\mathbf{E}_0 e^{-i\omega t} \tag{3.4}$$

Accordingly, the transformation of the derivatives is given by

$$\begin{aligned} &\left(\frac{\partial}{\partial \mathbf{r}}\right)_{\mathbf{v},t} = \left(\frac{\partial}{\partial \mathbf{R}}\right)_{\mathbf{V},t}, \qquad \left(\frac{\partial}{\partial \mathbf{v}}\right)_{\mathbf{r},t} = \left(\frac{\partial}{\partial \mathbf{V}}\right)_{\mathbf{R},t} \\ &\left(\frac{\partial}{\partial t}\right)_{\mathbf{r},\mathbf{v}} = \left(\frac{\partial}{\partial t}\right)_{\mathbf{R},\mathbf{V}} + \frac{ie}{m\omega}\mathbf{E}_0 e^{-i\omega t} \cdot \left(\frac{\partial}{\partial \mathbf{R}}\right)_{\mathbf{V},t} + \frac{e}{m}\mathbf{E}_0 e^{-i\omega t} \cdot \left(\frac{\partial}{\partial \mathbf{V}}\right)_{\mathbf{R},t} \end{aligned} \tag{3.5}$$

Therefore, Vlasov's equations transform into the following set:

$$\begin{aligned} &\frac{\partial f_-^{(1)}}{\partial t} + \mathbf{V} \cdot \frac{\partial f_-^{(1)}}{\partial \mathbf{R}} + \frac{e}{m}\frac{\partial \Phi}{\partial \mathbf{R}} \cdot \frac{\partial f_-^{(1)}}{\partial \mathbf{V}} = 0 \\ &\Delta\Phi = \varepsilon_0^{-1} e \left(\int f_-^{(1)}\, d\mathbf{V} - \sum_j \delta(\mathbf{R} - \boldsymbol{\xi} e^{-i\omega t} - \mathbf{r}_j)\right) \end{aligned} \tag{3.6}$$

with the abbreviation

$$\boldsymbol{\xi} = -(e/m\omega^2)\,\mathbf{E}_0 \tag{3.7}$$

Linearization of (3.6) yields [see (2.15) and (2.17)]

$$\begin{aligned} &\frac{\partial f_{1-}}{\partial t} + \mathbf{V} \cdot \frac{\partial f_{1-}}{\partial \mathbf{R}} + \frac{e}{m}\frac{\partial \Phi_1}{\partial \mathbf{R}} \cdot \frac{\partial f_{0-}}{\partial \mathbf{V}} = 0 \\ &\Delta\Phi_1 = (e/\varepsilon_0) \left(n_- + \int f_{1-}\, d\mathbf{V} - \sum_j \delta(\mathbf{R} - \boldsymbol{\xi} e^{-i\omega t} - \mathbf{r}_j)\right) \end{aligned} \tag{3.8}$$

where the zero-order distribution f_{0-} is assumed to be

$$f_{0-} = (n_-/(2\pi)^{3/2} V_{\text{th}}^3) \exp(-V^2/2V_{\text{th}}^2) \tag{3.9}$$

and where V_{th} designates the rms velocity of the electrons.

We now Fourier transform (3.8) in **R** space:

$$\begin{aligned} &\frac{\partial \tilde{f}_{1-}}{\partial t} + i\mathbf{k} \cdot \mathbf{V}\tilde{f}_{1-} + i\frac{e}{m}\tilde{\Phi}_1 \mathbf{k} \cdot \frac{df_{0-}}{d\mathbf{V}} = 0 \\ &k^2\tilde{\Phi}_1 = -(e/\varepsilon_0) \Big\{ (2\pi)^{3/2} n_-\, \delta(\mathbf{k}) \\ &\qquad + \int \tilde{f}_{1-}\, d\mathbf{V} - (2\pi)^{-3/2} \sum_j \exp(-i\mathbf{k} \cdot (\boldsymbol{\xi} e^{-i\omega t} + \mathbf{r}_j))\Big\} \end{aligned} \tag{3.10}$$

(When requiring neutrality of the unperturbed plasma, the first and the third term in the Fourier transformed Poisson equation cancel for $\mathbf{k} = 0$.)

We now transform (3.10) into the equivalent integral equation by formal application of the method of variation of the constant. With

$$L = i\mathbf{V} \cdot \mathbf{k}, \qquad R = -i(e/m)\,\mathbf{k} \cdot (df_{0-}/d\mathbf{V})\,\tilde{\Phi}_1[\tilde{f}_{1-}] \tag{3.11}$$

our differential equation has the general form

$$(\partial/\partial t + L)\,\tilde{f}_{1-} = R \tag{3.12}$$

The homogeneous equation has the solution $S_t = \exp(-Lt)$. We make the ansatz $\tilde{f}_{1-} = S_t g$ and find

$$\partial g/\partial t = S_t^{-1}R = S_{-t}R, \qquad \tilde{f}_{1-} = \int^t dt'\, S_{t-t'}\, R(t') \tag{3.13}$$

Requiring $\tilde{f}_{1-}\ (t = -\infty) = 0$ and substituting $\tau = t - t'$, we have

$$\tilde{f}_{1-} = \int_0^\infty d\tau\, S_\tau R[\tilde{f}_{1-}(t-\tau)] \tag{3.14}$$

Introducing this expression into the second equation of (3.10), we find

$$k^2\tilde{\Phi}_1 = -\frac{e}{\varepsilon_0}\Big\{-\frac{ie}{m}\int_0^\infty d\tau\, \tilde{\Phi}_1(t-\tau)\int d\mathbf{V} e^{-i\mathbf{k}\cdot\mathbf{V}\tau}\,\mathbf{k}\cdot\frac{df_{0-}}{d\mathbf{V}}$$
$$-(2\pi)^{-3/2}\sum_j \exp(-i\mathbf{k}\cdot(\boldsymbol{\xi}e^{-i\omega t}+\mathbf{r}_j))\Big\} \tag{3.15}$$

The integral over the velocity yields

$$\frac{kn_-}{(2\pi)^{1/2}V_{\text{th}}}\int_{-\infty}^{+\infty} dV\, e^{-i\tau kV}\frac{d}{dV}\exp\left(-V^2/2V_{\text{th}}^2\right)$$
$$= i\tau k^2\frac{n_-}{(2\pi)^{1/2}V_{\text{th}}}\int_{-\infty}^{+\infty} e^{-i\tau kV}\exp\left(-V^2/2V_{\text{th}}^2\right)dV$$
$$= i\tau k^2 n_-\exp(-\tfrac{1}{2}(V_{\text{th}}k\tau)^2) \tag{3.16}$$

and therewith

$$\tilde{\Phi}_1 = -\omega_{\text{p}-}^2\int_0^\infty d\tau\,\exp[-\tfrac{1}{2}(V_{\text{th}}k\tau)^2]\,\tau\tilde{\Phi}_1(t-\tau)$$
$$+(2\pi)^{-3/2}\,(e/\varepsilon_0 k^2)\sum_j\exp(-i\mathbf{k}\cdot(\boldsymbol{\xi}e^{-i\omega t}+\mathbf{r}_j)) \tag{3.17}$$

The conductivity is rigorously defined only in the limit $E_0 \to 0$ and therefore $\xi \to 0$. This allows an expansion of the last term on the right-hand side of (3.17)

$$\sum_j\exp(-i\mathbf{k}\cdot(\boldsymbol{\xi}e^{-i\omega t}+\mathbf{r}_j)) = (1-i\mathbf{k}\cdot\boldsymbol{\xi}e^{-i\omega t})\sum_j\exp(-i\mathbf{k}\cdot\mathbf{r}_j) \tag{3.18}$$

(This approximation needs a justification since very large values of k are relevant due to the singularities in the terms presenting the ions. A detailed investigation shows that the assumption $\xi k \ll 1$ gives correct results in the lowest order of ξ for all values of k.)

The decomposition of (3.18) into a constant and a time-varying term suggests splitting up $\tilde{\Phi}_1$ in the form $\tilde{\Phi}_1 = \tilde{\Phi}_c + \tilde{\Phi}_{k\omega}(t)$. Insertion into (3.17) yields for the constant term

$$\tilde{\Phi}_c = \frac{e}{(2\pi)^{3/2}\,\varepsilon_0\,(k^2 + \omega_{p-}^2/V_{th}^2)} \sum_j \exp\,(-\,i\mathbf{k}\cdot\mathbf{r}_j) \tag{3.19}$$

It is also obvious from the structure of (3.17) that an expression of the form $\tilde{\Phi}_{k\omega} = C\exp(-i\omega t)$ will fulfil the oscillatory part of the above equation. Introducing this ansatz, we find

$$\tilde{\Phi}_{k\omega} = \frac{-ie}{(2\pi)^{3/2}\,\varepsilon_0 k^2}\,\frac{\mathbf{k}\cdot\boldsymbol{\xi}}{A_{k\omega}}\,e^{-i\omega t}\sum_j \exp\,(-\,i\mathbf{k}\cdot\mathbf{r}_j) \tag{3.20}$$

with

$$A_{k\omega} = 1 + \omega_{p-}^2 \int_0^\infty \exp(i\omega\tau - (k\tau V_{th})^2/2)\,\tau\,d\tau \tag{3.21}$$

The evaluation of the potential should more correctly have started from the wave equation instead of Poisson's equation. This would have altered $\tilde{\Phi}_{k\omega}$ by a factor $k^2/(k^2 - \omega^2/c^2)$, whereas $\tilde{\Phi}_c$ would not have changed. Asymptotic formulas for $A_{k\omega}$ show, however, that the contributions to the total $\tilde{\Phi}_1$ come essentially from $k \approx \omega/V_{th} \gg \omega/c$ (see assumption (1)), so that the correction factor approaches one.

Having formulated the potential, we are now in a position to calculate the average force exerted by the electrons on the ions, which is identical and oppositely directed to the average force exerted by all ions on the electrons—a quantity needed in the formulation of the electron current density and the conductivity tensor.

The potential produced at the point $\mathbf{r}$ in the laboratory frame is identical with the potential at the point $\mathbf{R}$ in our moving coordinate system, and therefore given by the inversion formula of Fourier transformation through

$$\Phi_1(\mathbf{r}) = (2\pi)^{-3/2}\int d\mathbf{k}\,e^{i\mathbf{k}\cdot\mathbf{r}}\exp(i\mathbf{k}\cdot\boldsymbol{\xi}e^{-i\omega t})(\tilde{\Phi}_c + \tilde{\Phi}_{k\omega}) \tag{3.22}$$

Inserting (3.19) and (3.20), and omitting again terms of higher order in $\boldsymbol{\xi}$, we have

$$\Phi_1(\mathbf{r}) = \frac{e}{(2\pi)^3\,\varepsilon_0}\int\left\{\frac{1}{k^2 + \omega_{p-}^2/V_{th}^2} + i\mathbf{k}\cdot\boldsymbol{\xi}e^{-i\omega t}\left(\frac{1}{k^2 + \omega_{p-}^2/V_{th}^2} - \frac{1}{k^2 A_{k\omega}}\right)\right\}$$
$$\times \sum_j \exp\,[i\mathbf{k}\cdot(\mathbf{r} - \mathbf{r}_j)]\,d\mathbf{k} \tag{3.23}$$

The total average field exerted on the ions is then given by

$$\langle \mathbf{E} \rangle = -(1/n_+) \sum_i \nabla_i \Phi_1(\mathbf{r}_i) \tag{3.24}$$

where the summation covers all ions in a unit volume element. This average field is due to the electrons only, since the ion–ion interactions cancel through the summation.

Evaluating (3.24) with (3.23) and engaging the "random-phase approximation"

$$\sum_{i,j} \exp\,[i\mathbf{k} \cdot (\mathbf{r}_i - \mathbf{r}_j)] = n_+ \tag{3.25}$$

which follows from the random distribution of the ions, we arrive at the average electric field

$$\langle \mathbf{E} \rangle = \frac{e}{(2\pi)^3 \varepsilon_0} \int d\mathbf{k}\, \mathbf{k}(\mathbf{k} \cdot \boldsymbol{\xi})\, e^{-i\omega t} \left(\frac{1}{k^2 + \omega_\mathrm{p}{}^2/V_\mathrm{th}^2} - \frac{1}{k^2 A_{k\omega}} \right) \tag{3.26}$$

since the first term in the integrand of (3.23) yields no contribution to the integral (3.26).

The integration with respect to the angular coordinates in (3.26) is trivial and yields for $\mathbf{k}\,)(\,\mathbf{k}$ the identity tensor multiplied by $(4\pi/3)\,k^2$. So the result for the average force can be written in the form

$$\langle \mathbf{E} \rangle = \frac{e}{6\pi^2 \varepsilon_0}\, \boldsymbol{\xi}\, e^{-i\omega t} \int_0^{k_\mathrm{max}} dk\, k^4 \left(\frac{1}{k^2 + \omega_{\mathrm{p}-}^2/V_\mathrm{th}^2} - \frac{1}{k^2 A_{k\omega}} \right) \tag{3.27}$$

The usual cutoff $k_\mathrm{max} = 2\pi\varepsilon_0 m V_\mathrm{th}^2/e^2$ avoids the divergence for small impact parameters.

In the laboratory system, the current density—in this model originating from the electrons only—is given by

$$\mathbf{j} = -e \int \mathbf{v} f_{1-}\, d\mathbf{v} \tag{3.28}$$

Differentiation with respect to time yields, with Vlasov's equation,

$$\frac{\partial \mathbf{j}}{\partial t} = e\, \frac{\partial}{\partial \mathbf{r}} \cdot \int \mathbf{v}\,)(\,\mathbf{v} f_{1-}\, d\mathbf{v} - \frac{e^2}{m} \left(\mathbf{E}_0 e^{-i\omega t} - \frac{\partial}{\partial \mathbf{r}} \Phi_1 \right) \int \frac{df_{0-}}{\partial \mathbf{v}}\,)(\,\mathbf{v}\, d\mathbf{v} \tag{3.29}$$

Assumption (1) justifies the neglect of the first term on the right-hand side, so that partial integration of the remaining terms results in

$$\partial \mathbf{j}/\partial t = n_-(e^2/m)(\mathbf{E}_0 e^{-i\omega t} - \partial \Phi_1/\partial \mathbf{r}) \tag{3.30}$$

Averaging over all ion configurations as we have done when deriving (3.27) gives

$$\partial\langle\mathbf{j}\rangle/\partial t = (n\_e^2/m)(\mathbf{E}_0 e^{-i\omega t} + \langle\mathbf{E}\rangle) \tag{3.31}$$

Now let us recall that the coefficient of conductivity $\sigma(\omega)$ connects the field and the current density through

$$\tilde{\mathbf{j}}(\omega) = \sigma(\omega)\,\mathbf{E}_0(\omega) \tag{3.32}$$

Comparison with the Fourier transform of (3.31) yields for the complex conductivity

$$\sigma = i\varepsilon_0(\omega_{\mathrm{p}-}^2/\omega)\{1 - I(\omega)\} \tag{3.33}$$

with

$$I(\omega) = \frac{e^2}{6\pi^2\varepsilon_0 m\omega^2}\int_0^{k_{\max}} \left(\frac{1}{k^2 + \omega_{\mathrm{p}-}^2/V_{\mathrm{th}}^2} - \frac{1}{k^2 A_{k\omega}}\right) k^4\, dk \tag{3.34}$$

$I(\omega)$ is of the order $1/\omega\tau_{\mathrm{c}}$ and therefore small in the frequency region which we are considering.

The **resistivity coefficient** follows from the conductivity through

$$\rho := 1/\sigma = -(i\omega/\varepsilon_0\omega_{\mathrm{p}-}^2)\{1 + I(\omega)\} \tag{3.35}$$

In the limiting cases $\omega/\omega_{\mathrm{p}-} \ll 1$ or, respectively, $\omega/\omega_{\mathrm{p}-} \gg 1$, formula (3.35) can be evaluated analytically and yields for $\omega \ll \omega_{\mathrm{p}-}$

$$\rho = -\frac{i\omega}{\varepsilon_0\omega_{\mathrm{p}-}^2}\left\{\left(1 - \left(1 - \frac{\pi}{8}\right)\frac{e^2\omega_{\mathrm{p}-}}{24\pi\varepsilon_0 m V_{\mathrm{th}}^3}\right)\right.$$
$$\left. + i\left(\frac{2}{\pi}\right)^{1/2}\frac{e^2\omega_{\mathrm{p}-}^2}{24\pi\varepsilon_0 m V_{\mathrm{th}}^3\omega}\left(\ln\left(\frac{V_{\mathrm{th}}k_{\max}}{\omega_{\mathrm{p}-}}\right)^2 - 1\right)\right\} \tag{3.36}$$

or, respectively, for $\omega \gg \omega_{\mathrm{p}-}$,

$$\rho = -\frac{i\omega}{\varepsilon_0\omega_{\mathrm{p}-}^2}\left\{\left(1 + (2\pi)^{1/2}\frac{e^2\omega_{\mathrm{p}-}^2}{24\pi\varepsilon_0 m V_{\mathrm{th}}^3\omega}\right) + i\left(\frac{2}{\pi}\right)^{1/2}\frac{e^2\omega_{\mathrm{p}-}^2}{24\pi\varepsilon_0 m V_{\mathrm{th}}^3\omega}\right.$$
$$\left. \times\left(\ln\left(\frac{2V_{\mathrm{th}}^2 k_{\max}^2}{\omega^2}\right) - C\right)\right\} \tag{3.37}$$

where $C \approx 0.577$ is Euler's constant.

In addition, the above quoted authors computed the values of ρ in the whole range numerically. The results are shown for the real and imaginary part (**reactance**) in Figures VI.1 and VI.2. We see in Figure VI.1—

and it is obvious from (3.36)—that for frequencies below the plasma frequency the real part of the resistivity is a constant. Comparison with the results derived in the following section for low-frequency oscillations with $\omega \ll \omega_{p-}$ shows that our result here is identical with that of the high-frequency limit there. This is reasonable, since our present results should still hold in the range $1/\tau_c \ll \omega \leqslant \omega_{p-}$.

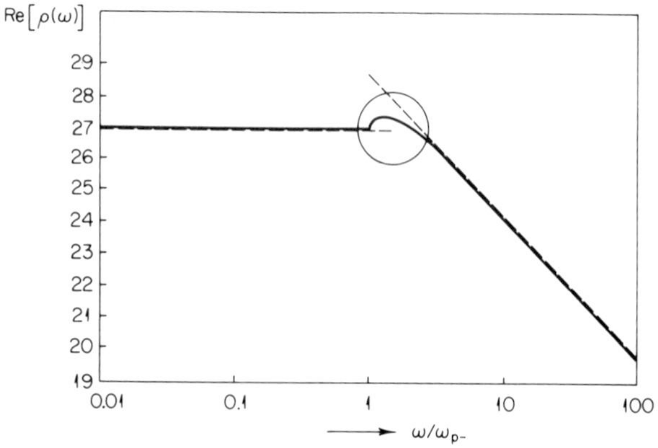

FIGURE VI.1. Resistivity $\text{Re}[\rho(\omega)]$ for $\omega > \nu_c$, in units $\rho_0 = [(32)^{1/2}e^2/3m_-(4\pi\varepsilon_0)^2] \times (m_-/\Theta)^{3/2} \simeq 5.8 \cdot 10^{-5}(\Theta[\text{eV}])^{3/2}\ \Omega m$ (Dawson and Oberman, 1962).

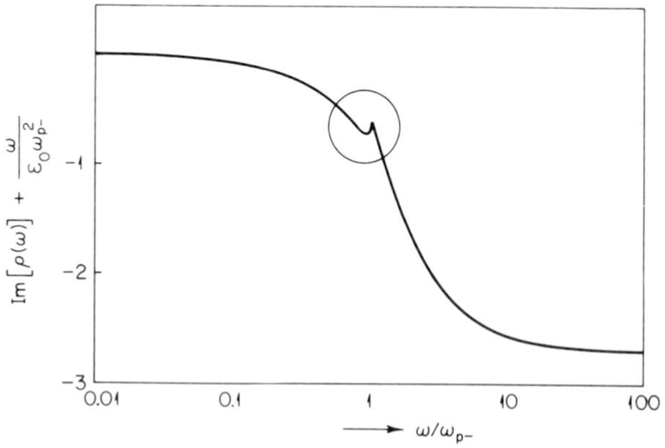

FIGURE VI.2. Reactance $\text{Im}[\rho(\omega)]$, for $\omega > \nu_c$ in the units ρ_0 given in Figure VI.1. The reactance $-\omega/\varepsilon_0\omega_{p-}^2$ of the free electrons is subtracted (Dawson and Oberman, 1962).

The resistivity shows a slight increase just above the plasma frequency. Dawson and Oberman demonstrated that this increase is due to the excitation of longitudinal plasma oscillations. The decrease in the resistivity for high frequencies can be interpreted as a contraction of the effective collision cross section: Collisions with large impact parameters are slow and do not contribute to high frequency fields. Therefore the maximum impact parameter is of the order V_{th}/ω.

Formula (3.37) and Figure VI.2 show that the variation of the reactance may be thought of as a change in the effective mass of the electrons due to the interaction with the ions. Observe that the reactance also shows a slight peak in the distribution near the plasma frequency, for the same reason as the enhancement of the resistivity.

Now that we have calculated the conductivity and resistivity of a plasma including correlation phenomena, we can also solve the dispersion relation (2.39)

$$\det(n^2 \underset{2}{\mathbf{I}}_{\perp} - \underset{2}{\mathbf{I}} - (i/\varepsilon_0\omega)\underset{2}{\boldsymbol{\sigma}}) = 0 \tag{3.38}$$

and find the **absorption coefficient** which is twice the negative imaginary part of the frequency calculated from the dispersion relation for a real wave vector k.

We introduce the conductivity σ from (3.33) into the dispersion relation.

For the **electrostatic longitudinal waves**, we find

$$\omega^2 = \omega_{\mathrm{p}-}^2(1 - I(\omega)) \tag{3.39}$$

and for the **transversal electromagnetic waves**

$$\omega^2 = k^2c^2 + \omega_{\mathrm{p}-}^2(1 - I(\omega)) \tag{3.40}$$

Taking advantage of the fact that $|I(\omega)| \ll 1$, which we already used above, we find for the electrostatic waves

$$\omega = \omega_{\mathrm{p}-}(1 - \tfrac{1}{2}I(\omega)) \tag{3.41}$$

and for the transversal waves

$$\omega = (k^2c^2 + \omega_{\mathrm{p}-}^2)^{1/2} - \tfrac{1}{2}(\omega_{\mathrm{p}-}^2/\omega)\, I(\omega) \tag{3.42}$$

The absorption coefficients are then

$$\alpha_{\parallel} = \omega_{\mathrm{p}-} \operatorname{Im}[I(\omega)] \tag{3.43}$$

$$\alpha_{\perp} = \omega_{\mathrm{p}-}^2 \operatorname{Im}[I(\omega)/\omega] \tag{3.44}$$

Figure VI.1 justifies the application of the asymptotic formula (3.37) down to frequencies $\omega \approx \omega_{p-}$. Insertion of the corresponding $I(\omega)$ into (3.43) leads to the absorption coefficient

$$\alpha_{\parallel} = \left(\frac{2}{27\pi}\right)^{1/2} \omega_{p-} \frac{\ln \Lambda}{\Lambda} \left(1 - \frac{C - \ln \frac{3}{2}}{2 \ln \Lambda}\right) \tag{3.45}$$

for longitudinal waves or, respectively,

$$\alpha_{\parallel} \simeq O(1/\tau_c) \tag{3.46}$$

The result is reasonable: The collision time is the relaxation time for directed disturbancies. Electron–electron encounters do not contribute because of the conservation of momentum. So the damping constant of the longitudinal waves is practically the reciprocal of the slowing-down time for electron–ion interaction.

It is not a surprise that we do not find Landau damping in the long wavelength range, where we could neglect the "diffusion term" in (3.29) which essentially is responsible for this effect.

For the absorption coefficient of the transversal waves, we have correspondingly

$$\alpha_{\perp} = \left(\frac{2}{27\pi}\right)^{1/2} \frac{\omega_{p-}^3}{\omega^2} \frac{\ln(\omega_{p-}\Lambda/\omega)}{\Lambda} \left(1 - \frac{C - \ln \frac{3}{2}}{2 \ln(\omega_{p-}\Lambda/\omega)}\right) \tag{3.47}$$

or, since we are in general interested in the range well below the critical density, that means $\ln \Lambda \gg 1$, the relation

$$\alpha_{\perp} \simeq O\left(\frac{\omega_{p-}^2}{\omega^2} \frac{1}{\tau_c}\right) \tag{3.48}$$

holds.

For the electromagnetic phenomena of primary interest characterized by $u > c$ and $\omega > \omega_{p-}$, we see that the damping of the transversal modes is smaller than that of the longitudinal modes. We further recognize that this is a frequency dependent effect, the physical background of which was given already when we discussed the resistivity.

With the knowledge of the absorption coefficient we will now turn to study the **emission coefficient**. In this respect, we are mainly interested in the transversal electromagnetic modes.

Kirchhoff's law relates the power emitted per unit volume to the product of the energy density and the absorption coefficient. It is based on the assumption of detailed balancing within each mode, or may be derived from the fluctuation–dissipation theorem.

Accordingly, we calculate the emission coefficient from the above value of $\alpha_{\perp}$ and the energy density of the transversal modes. To determine this density, we consider the number of transversal modes in the interval $(k, k + dk)$ per unit volume which is given by

$$dn_k = \pi^{-2} k^2 \, dk \tag{3.49}$$

(The $\mathbf{k}$ vectors possible in a cubic volume element $V = L^3$ form a lattice in $\mathbf{k}$-space with the lattice constant $2\pi/L$. There are two polarization directions for the transversal modes.)

The number density per frequency unit is then described by

$$dn_\omega/d\omega = \pi^{-2} k^2(\omega) \, dk/d\omega \tag{3.50}$$

where we may use the dispersion relation for transverse waves

$$k^2 = (\omega^2 - \omega_{\text{p}-}^2)/c^2 \tag{3.51}$$

Corrections in this dispersion equation due to absorption terms would be of the order of $\omega\tau_{\text{c}}$.

To find the energy density per unit volume and angular frequency, it is necessary to attribute to each mode a certain energy value. For a plasma close to equilibrium, we assume that this energy is Θ, which may be argued crudely as follows.

Let us consider a plasma in a large box which is in equilibrium. Let the boundary layer of the plasma be, on the one hand, diffuse enough to avoid substantial energy reflection of a ray entering the plasma, and, on the other hand, sharp enough so that we can apply Snellius' law. Then it is obvious that a spread of a perpendicular beam described by a solid angle $d\Omega_{\text{V}}$ in vacuum will change in a spread of solid angle $d\Omega_{\text{p}}$ in the plasma, where in accordance with the diffraction law

$$d\Omega_{\text{p}}/u^2 = d\Omega_{\text{V}}/c^2 \tag{3.52}$$

holds. Therefore

$$k_0^2 \, d\Omega_{\text{V}} = k^2 \, d\Omega_{\text{p}} \tag{3.53}$$

where k and $k_0 = \omega/c$ are the wave numbers in the plasma, respectively, *in vacuo*. Now the energy continuity may be written in the form

$$d\Omega_{\text{V}} \, \mathscr{E}_{\text{V}} \, c = d\Omega_{\text{p}} \, \mathscr{E}_{\text{p}} \, u_{\text{g}} \tag{3.54}$$

where $\mathscr{E}_{\text{V}}$, $\mathscr{E}_{\text{p}}$ gives the energy density per unit volume and angular frequency in the vacuum or in the plasma, respectively. u_{g} is the group velocity in the plasma. Division of (3.54) by (3.53) yields

$$\mathscr{E}_{\text{V}} c/k_0^2 = \mathscr{E}_{\text{p}} u_{\text{g}}/k^2 \tag{3.55}$$

and with the trivial relation

$$u_g \, dk = c \, dk_0 \tag{3.56}$$

we find

$$\mathscr{E}_p/k^2 \, dk = \mathscr{E}_V/k_0^2 \, dk_0 \tag{3.57}$$

The quantity $k^2 \, dk$ essentially gives the number n_k of modes in the range $(k, k + dk)$.

Equation (3.57) therefore states that the energy per mode is, within the assumption made, the same in the plasma and in the vacuum, namely Θ for each mode and polarization direction.

We then find from (3.51) and (3.57) for the energy density per frequency unit

$$\mathscr{E}_p(\omega) = \Theta(\omega^2/\pi^2c^3)(1 - (\omega_{p-}^2/\omega^2))^{1/2} \tag{3.58}$$

Kirchhoff's law then produces through the product of the energy density (3.58) and the absorption coefficient the **emission coefficient**

$$j(\omega) = \frac{\Theta}{\pi^2 c^3}\left(1 - \frac{\omega_{p-}^2}{\omega^2}\right)^{1/2} \omega_{p-}^2\omega \, \mathrm{Im}[I(\omega)]$$

$$\simeq \frac{1}{\tau_c}\frac{\omega_{p-}^2\Theta}{c^3}\left(1 - \frac{\omega_{p-}^2}{\omega^2}\right)^{1/2} \tag{3.59}$$

This coefficient gives the bremsstrahlung due to electron–ion collisions taking into account collective fields. Electron–electron collisions are neglected, which is consequent since we know that in the dipole approximation these correlations do not contribute to bremsstrahlung.

3.2. Description with the Fokker–Planck Equation

The application of this procedure is limited to the range

$$\omega < \omega_{p-}, \qquad u \gg v_{th}$$

It is well known that in a system with long-range forces, the cumulative effect of the numerous weak deflections resulting from distant encounters is more important than the influence of the few collisions with large deflections. The effect of the distant passages can be taken into account by the Landau collision term which—according to Section 1.3 of Chapter IV—is applicable in the range $\omega < \omega_{p-}$, $\ln \Lambda \gg 1$.

The first-order equation of the BBGKY hierarchy for the electrons including the correlation in the Landau form (IV.1.74) and (IV.1.77) generalized to several components reads

$$\frac{\partial f_-^{(1)}}{\partial t} + \mathbf{v} \cdot \frac{\partial f_-^{(1)}}{\partial \mathbf{r}} - \frac{e}{m_-} \mathbf{E} \cdot \frac{\partial f_-^{(1)}}{\partial \mathbf{v}} = \sum_\nu I(f_-^{(1)} \mid f_\nu^{(1)})$$

See pg. 161

$$I(f_-^{(1)} \mid f_\nu^{(1)}) = \Gamma_\nu \frac{\partial}{\partial \mathbf{v}} \cdot \int \frac{g_\nu^2 \mathbf{I} - \mathbf{g}_\nu \mathbf{g}_\nu}{g_\nu^3} \cdot \left(f_\nu^{(1)} \frac{\partial f_-^{(1)}}{\partial \mathbf{v}} - \frac{m_- f_-^{(1)}}{m_\nu} \frac{\partial f_\nu^{(1)}}{\partial \mathbf{v}_\nu} \right) d\mathbf{v}_\nu \tag{3.60}$$

$$\Gamma_\nu = -\frac{e^4(m_- + m_\nu)}{16\pi\varepsilon_0^2 m_-^2 m_\nu} \ln \left[\frac{4\pi\varepsilon_0 m_- m_\nu \langle g_\nu^2 \rangle \lambda_D}{e^2(m_- + m_\nu)} \right]$$

where we applied the terminology for the collision term of Chapter IV, Section 2.

In this equation, the sum over ν covers the electrons and the ions. The quantity $\mathbf{g}_\nu = (\mathbf{v} - \mathbf{v}_\nu)$ designates the relative velocity.

The physical model which we intend to study in the "low-frequency range" is the same we considered in the "high-frequency range" of the preceding section. Therefore assumptions (1)–(4) on p. 283 hold with the exception that, instead of the Vlasov equation prescribed under (4), we use the Fokker–Planck equation, and instead of the frequency range given, we have the frequency range shown at the beginning of this section. Moreover, we shall use the zero-order distribution Maxwellian in the laboratory system.

We linearize (3.60) [see (2.15) and (2.17)]

$$f_-^{(1)} = f_{0-} + f_{1-}, \qquad f_+^{(1)} = f_{0+} \tag{3.61}$$

This ansatz splits the collision term into

$$I(f_{0-} \mid f_{0-}) + I(f_{0-} \mid f_{0+}) + I(f_{0-} \mid f_{1-}) + I(f_{1-} \mid f_{0-}) + I(f_{1-} \mid f_{0+}) \tag{3.62}$$

The first two terms determine the zero-order distribution and vanish due to our assumption that f_0 is Maxwellian. The next two terms give the influence of electron–electron encounters on the disturbance and do not contribute in the dipole approximation. So we are concerned with the last term only.

The left-hand side of (3.60) is linearized in the usual way.

Assumption (2) simplifies the collision term with the ions to the form

$$I(f_{1-} \mid f_{0+}) = \Gamma_+ \frac{\partial}{\partial \mathbf{v}} \cdot \frac{v^2 \mathbf{I} - \mathbf{v}\mathbf{v}}{v^3} \cdot \frac{\partial f_{1-}}{\partial \mathbf{v}} \int f_{0+} \, d\mathbf{v}' \tag{3.63}$$

$$\Gamma_+ = -\frac{e^4}{16\pi\varepsilon_0^2 m_-^2} \ln \left(\frac{4\pi\varepsilon_0 m_- \langle v^2 \rangle \lambda_D}{e^2} \right)$$

The integral with respect to $\mathbf{v}'$ results in the ion density n_+. With the linearized left-hand side, we get

$$\frac{\partial f_{1-}}{\partial t} - \frac{e\mathbf{E}}{m_-} \cdot \frac{\partial f_{0-}}{\partial \mathbf{v}} = \Gamma_+ n_+ \frac{\partial}{\partial \mathbf{v}} \cdot \frac{v^2 \mathbf{I}_2 - \mathbf{v})(\mathbf{v}}{v^3} \cdot \frac{\partial f_{1-}}{\partial \mathbf{v}} \tag{3.64}$$

where we again have dropped the diffusion term $\mathbf{v} \cdot \partial f_{1-}/\partial \mathbf{r}$ due to the restriction to large phase velocities. f_{0-} is a Maxwellian distribution

$$f_{0-} = n_-(m_-/2\pi\Theta)^{3/2} \exp(-m_- v^2/2\Theta) \tag{3.65}$$

If we introduce the dimensionless variable

$$\mathbf{U} = \mathbf{v}(m_-/\Theta)^{1/2} \tag{3.66}$$

the kinetic equation transforms into

$$\frac{\partial f_{1-}}{\partial t} + \left(\frac{e^2}{m_- \Theta}\right)^{1/2} \mathbf{E} \cdot \mathbf{U} f_{0-} = \left(\frac{m_-}{\Theta}\right)^{3/2} \Gamma_+ n_+ \frac{\partial}{\partial \mathbf{U}} \cdot \frac{U^2 \mathbf{I}_2 - \mathbf{U})(\mathbf{U}}{U^3} \cdot \frac{\partial f_{1-}}{\partial \mathbf{U}} \tag{3.67}$$

The right-hand side presents the collision term as a product of a dimensionless divergence representing diffusion in velocity space with a collision frequency

$$2\nu_c := (m_-/\Theta)^{3/2} n_+ \Gamma_+ = \tfrac{3}{4}\omega_{p-}(\ln \Lambda)/\Lambda \tag{3.68}$$

We look for a solution of the form

$$\mathbf{E} = \mathbf{E}_0 e^{-i\omega t}, \qquad f_{1-} = \mathbf{E}_0 \cdot \mathbf{U} A(U)\, e^{-i\omega t} = \tilde{f}_{1-} e^{-i\omega t} \tag{3.69}$$

This ansatz yields

$$-i\omega \mathbf{U} \cdot \mathbf{E}_0 + \left(\frac{e^2}{m_- \Theta}\right)^{1/2} \mathbf{U} \cdot \mathbf{E}_0 \frac{f_{0-}}{A} = \frac{\nu_c}{2U^3} \frac{\partial}{\partial \mathbf{U}} \cdot (U^2 \mathbf{I}_2 - \mathbf{U})(\mathbf{U}) \cdot \frac{\partial}{\partial \mathbf{U}} \mathbf{U} \cdot \mathbf{E}_0 \tag{3.70}$$

Here we have used the specific quality of the tensor $U^2 \mathbf{I}_2 - \mathbf{U})(\mathbf{U}$ to project any vector into the plane perpendicular to $\mathbf{U}$.

As a consequence, we could shift the diffusion differential operator across the isotropic quantities (U^3, $A(U)$). With the identity

$$\partial/\partial \mathbf{U} \cdot (U^2 \mathbf{I}_2 - \mathbf{U})(\mathbf{U}) = 2\mathbf{U} - 3\mathbf{U} - \mathbf{U} = -2\mathbf{U} \tag{3.71}$$

we arrive at

$$A(U) = -i(e^2/m_- \Theta)^{1/2} f_{0-}(\omega + i\nu_c/U^3)^{-1} \tag{3.72}$$

Now we can calculate the current density by

$$\tilde{\mathbf{j}}_\omega = -e \int \mathbf{v}\tilde{f}_{1-}\, d\mathbf{v} = -e(\Theta/m_-)^2 \int \mathbf{U}A(U)\mathbf{E}_0 \cdot \mathbf{U}\, d\mathbf{U}$$
$$= -\tfrac{4}{3}\pi e(\Theta/m_-)^2 \int_0^\infty U^4 A(U)\, dU\, \mathbf{E}_0 \tag{3.73}$$

Inserting (3.72) and (3.65) and comparing with the definition of the conductivity σ gives

$$\sigma(\omega) = \frac{i}{3}\left(\frac{2}{\pi}\right)^{1/2} \varepsilon_0 \omega_{p-}^2 \int_0^\infty \frac{U^4 \exp(-U^2/2)}{\omega + i\nu_c/U^3}\, dU \tag{3.74}$$

In Figures VI.3 and VI.4, we plot the real and the imaginary part of the resistivity $\rho = 1/\sigma$ as calculated from (3.74).

With the knowledge of σ given in (3.74), we are in a position to determine the **absorption coefficient** from the solution of the dispersion relation

$$\det\,[n^2 \underset{2}{\mathbf{I}}_\perp - \underset{2}{\mathbf{I}}(1 + i\sigma/\varepsilon_0\omega)] = 0 \tag{3.75}$$

From this equation we find for the frequency of the **longitudinal field**

$$\omega_\parallel = -i\sigma/\varepsilon_0 \tag{3.76}$$

and with (3.74)

$$\omega_\parallel = \tfrac{1}{3}(2/\pi)^{1/2}\, \omega_{p-}^2 \int_0^\infty \frac{U^4 \exp(-U^2/2)}{\omega_\parallel + i\nu_c/U^3}\, dU \tag{3.77}$$

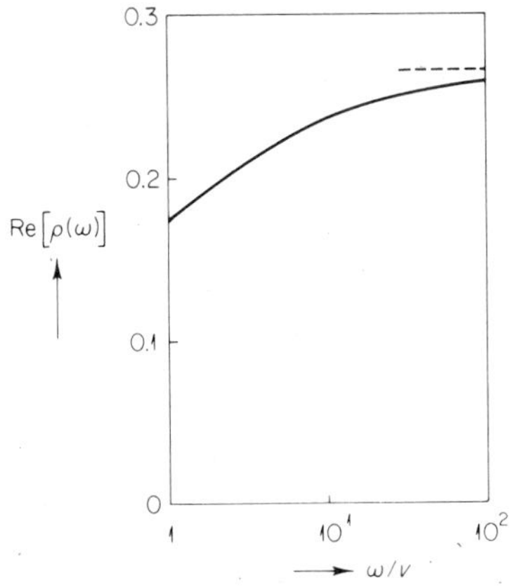

FIGURE VI.3. Resistivity $\mathrm{Re}[\rho(\omega)]$, for $\omega < \omega_{p-}$ in units of $\nu_c/\varepsilon_0\omega_{p-}$. The dotted line (high-frequency limit) corresponds to the low frequency limit of Figure VI.1.

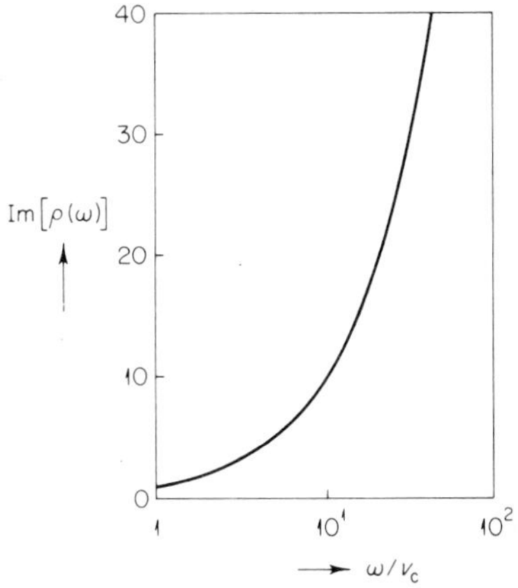

FIGURE VI.4. Reactance $\mathrm{Im}[\rho(\omega)]$, for $\omega < \omega_{p-}$ in the units of Figure VI.3.

Correspondingly, we find for the **transversal field**

$$\omega_\perp{}^2 = k^2c^2 - i\omega_\perp\sigma/\varepsilon_0 \tag{3.78}$$

and substituting (3.74)

$$\omega_\perp{}^2 = k^2c^2 + \tfrac{1}{3}\omega_{\text{p}-}^2(2/\pi)^{1/2}\int_0^\infty \frac{U^4\exp(-U^2/2)}{1+i\nu_{\text{c}}/\omega_\perp U^3}\,dU \tag{3.79}$$

As usual, the absorption coefficients are given through twice the negative imaginary part of ω.

An analytical evaluation is possible for the asymptotic cases $\omega \ll \nu_{\text{c}}$, $\omega \gg \nu_{\text{c}}$ $(\omega, \nu_{\text{c}} \ll \omega_{\text{p}-})$. In the discussion of these cases, we use the abbreviations

$$C_m := (2/\pi)^{1/2}\int_0^\infty U^m \exp(-U^2/2)\,dU$$

$$= \begin{cases} (m-1)(m-3)\cdots 1 & \text{for even } m \\ (2/\pi)^{1/2}(m-1)(m-3)\cdots 2 & \text{for odd } m \end{cases} \tag{3.80}$$

$\omega \ll \nu_{\text{c}}$. The conductivity σ is given by

$$\sigma = \tfrac{1}{3}(\varepsilon_0\omega_{\text{p}-}^2/\nu_{\text{c}})(C_7 + (i\omega/\nu_{\text{c}})\,C_{10}) \tag{3.81}$$

which for $\omega = 0$ yields essentially the usual short-range collision expression for the mobility of the electrons which is defined by $\mu_- := \sigma(\omega = 0)/en_-$

$$\mu_- = (C_7/3)(e\langle\lambda\rangle/m_-v_{\text{th}}) \tag{3.82}$$

where $\langle\lambda\rangle$ is an effective mean free path and v_{th} the rms velocity.

For the frequency of the longitudinal field, we have

$$\omega_\parallel = -i\omega_{\text{p}-}^2C_7\,\nu_{\text{c}}/(3\nu_{\text{c}}{}^2 + C_{10}\omega_{\text{p}-}^2) \tag{3.83}$$

Since $\nu_{\text{c}} \ll \omega_{\text{p}-}$, we may use approximately

$$\omega_\parallel \approx -i(C_7/C_{10})\,\nu_{\text{c}} = -i\alpha_\parallel/2 \tag{3.84}$$

which shows that $\omega_\parallel$ is a purely imaginary quantity, ν_{c} characterizing the relaxation time $2/\alpha_\parallel$ for disturbances.

For the transversal field, the frequency is determined by

$$\omega_\perp{}^2 = k^2c^2 - i(\omega_{\text{p}-}^2\omega_\perp/\nu_{\text{c}})(C_7 + i(\omega_\perp/\nu_{\text{c}})\,C_{10}) \tag{3.85}$$

This yields for the imaginary part

$$\text{Im}(\omega_\perp) = -\tfrac{1}{2}C_7(\omega_{\text{p}-}^2/\nu_{\text{c}}) = -\alpha_\perp/2 \tag{3.86}$$

Remembering that $\omega_{\mathrm{p}-}/\nu_{\mathrm{c}} = \ln \Lambda/\Lambda \ll 1$, we see that this solution is not consistent with our initial requirement $\omega \ll \nu_{\mathrm{c}}$. We conclude that the Landau equation is not apt to describe this phenomenon.

$\omega \gg \nu_{\mathrm{c}}$. Here the conductivity σ is given by

$$\sigma = \tfrac{1}{3} i \varepsilon_0 (\omega_{\mathrm{p}-}^2/\omega)(C_4 - i(\nu_{\mathrm{c}}/\omega)\, C_1) = \tfrac{1}{3}(2/\pi)^{1/2}\, \varepsilon_0 (\omega_{\mathrm{p}-}^2/\omega^2)\, \nu_{\mathrm{c}} + i \varepsilon_0 \omega_{\mathrm{p}-}^2/\omega \tag{3.87}$$

In the limit $\nu_{\mathrm{c}}/\omega \to 0$, this produces the conductivity of free electrons.

The frequency for the longitudinal field reads

$$\omega_{\parallel}^{\,2} = \omega_{\mathrm{p}-}^2 (1 - \tfrac{1}{3} i (2/\pi)^{1/2}\, \nu_{\mathrm{c}}/\omega_{\parallel}) \tag{3.88}$$

For $\nu_{\mathrm{c}} \to 0$, this dispersion formula reproduces the solution $\omega_{\parallel}^{\,2} = \omega_{\mathrm{p}-}^2$ for our system with $v_{\mathrm{th}} \ll \omega/k$.

The imaginary part of $\omega_{\parallel}$ determines the absorption coefficient for collision damping

$$\mathrm{Im}(\omega_{\parallel}) = -\tfrac{1}{6}(2/\pi)^{1/2}\, \nu_{\mathrm{c}} = -\alpha_{\parallel}/2 \tag{3.89}$$

The frequency for the transversal field is determined by

$$\omega_{\perp}^{\,2} = k^2 c^2 + \omega_{\mathrm{p}-}^2 - i \tfrac{1}{3}(2/\pi)^{1/2}\, (\omega_{\mathrm{p}-}^2 \nu_{\mathrm{c}}/\omega_{\perp}) \tag{3.90}$$

This result differs from the dispersion relation of a system without individual correlations only by the imaginary part. The absorption coefficient is given by the imaginary part of the frequency

$$\mathrm{Im}(\omega_{\perp}) = -\frac{1}{6}\left(\frac{2}{\pi}\right)^{1/2} \frac{\omega_{\mathrm{p}-}^2 \nu_{\mathrm{c}}}{k^2 c^2 + \omega_{\mathrm{p}-}^2} = -\frac{\alpha_{\perp}}{2} \tag{3.91}$$

However, since (3.90) indicates that $|\,\omega_{\perp}\,| \gg \omega_{\mathrm{p}-}$ holds, our presupposition for the application of the Landau equation is violated and the result therefore carries little weight.

4. Collective Description of Particle Dynamics—Individual Description of Emission Processes—Light Scattering

In Section 2, particle dynamics and their radiative effects were described collectively for all components. In this approximation the dielectric tensor describes the coherent response of the smeared-out continuous system to the electromagnetic waves.

In Section 3, electrons, their dynamics, and their emission again were described collectively. However, the ions were presented as individuals, and the electron correlations with these individuals were taken into account.

This section distinguishes itself from the two previous ones mainly by the fact that the electron emission is described individually (the emission of the ions being negligible in comparison). In this way we can account for light scattering. Particle dynamics, as far as it is taken into account, is treated collectively. As a consequence, bremsstrahlung, which has been treated in the previous section, is not covered again here.

4.1. Individual Electron Emission from the Plasma

Scattering is of interest particularly for light waves which penetrate the plasma relatively unaffected by the collective reaction and consequently have frequencies $\omega \gg \omega_{p0}$. We know that for such frequencies the coherent response of the plasma is negligible, and the dielectric tensor can be considered as unity. Consequently, the electrons may be considered as exposed to the external field given by

$$\mathbf{E} = \mathbf{E}_0 \exp[i(\mathbf{k}_0 \cdot \mathbf{x}' - \omega_0 t)] \tag{4.1}$$

Then the acceleration of a test electron is determined by

$$\ddot{\mathbf{x}}' = -\frac{e}{m_-} \mathbf{E}_0 \exp[i(\mathbf{k}_0 \cdot \mathbf{x}' - \omega_0 t)] \tag{4.2}$$

since we neglected acceleration by collective particle interactions due to $\varepsilon = 1$ and acceleration due to individual particle interactions as covered by the Bremsstrahlung in the previous section.

Equation (4.2) implies some more simplifications. It assumes that we can calculate nonrelativistically, i.e.,

$$\dot{\mathbf{x}}'^2 \ll c^2 \tag{4.3}$$

Moreover, it neglects the force exerted on the electron due to the motion in the magnetic field of the lightwave. This means that the following two more stringent conditions must hold:

$$v_{\text{th}} \ll c; \qquad \frac{eE_0}{m_- \omega_0 c} \ll 1 \tag{4.4}$$

where the second condition also says that the ratio of the maximum amplitude a of the electron oscillation in the electric field is to be much smaller than the wavelength:

$$k_0 a = a/\lambda\!\!\!^- \ll 1, \qquad a = eE_0/m_- \omega_0^2 \tag{4.5}$$

Finally (4.2) requires that the scattered intensity of the lightwave be much smaller than the incoming intensity, a condition which generally is very well satisfied, as one may see from the results.

With (4.2) we can calculate the emission of a test electron due to the acceleration in the external electromagnetic wave—that is, its scattering—via the Poynting vector. The Poynting vector $\mathbf{S}_\omega$ is defined by the Fourier components of the fields radiated from the accelerated electron. These fields are found from the Liénard–Wiechert potentials, which read in general form

$$\mathbf{A} = -\frac{e\mu_0}{4\pi}\frac{\dot{\mathbf{x}}'(t')}{r(t') - \dot{\mathbf{x}}'(t')\cdot\mathbf{r}(t')/c} \tag{4.6}$$

$$\Phi = -\frac{e}{4\pi\varepsilon_0}\frac{1}{r(t') - \dot{\mathbf{x}}'(t')\cdot\mathbf{r}(t')/c} \tag{4.7}$$

with the abbreviations

$$t' = t - r(t')/c, \qquad \mathbf{r} = \mathbf{x} - \mathbf{x}' \tag{4.8}$$

Here t and $\mathbf{x}$ refer to the time and point of observation, whereas $\mathbf{x}'$ gives the position of the source. The corresponding fields are

$$\mathbf{E} = -\frac{\partial\Phi}{\partial\mathbf{x}} - \frac{\partial\mathbf{A}}{\partial t}, \qquad \mathbf{B} = \frac{\partial}{\partial\mathbf{x}} \times \mathbf{A} \tag{4.9}$$

In particular, according to (V.2.39), we have

$$\mathbf{E} = (c/r)\,\mathbf{B} \times \mathbf{r} \tag{4.10}$$

for the radiative part of the electromagnetic field. This allows us to calculate the magnetic field only.

Since we are studying the wave region, we have $|\mathbf{x}| \gg |\mathbf{x}'|$ and may therefore use the approximation

$$t' \approx t - \frac{x}{c} + \frac{\mathbf{x}\cdot\mathbf{x}'(t')}{xc} \approx t - \frac{x}{c} + \frac{\mathbf{x}\cdot\mathbf{x}'(t - x/c)}{xc} \tag{4.11}$$

where we neglected all terms of higher than linear order in x'/x.

Neglecting also terms of order $\dot{x}'x'/cx$, we may write the vector potential in the simple form

$$\mathbf{A} \approx -\frac{e\mu_0}{4\pi}\frac{\dot{\mathbf{x}}'(t')}{r(t)} \approx -\frac{e\mu_0}{4\pi x}\dot{\mathbf{x}}'(t') \tag{4.12}$$

Of course, we cannot neglect this vector potential with the argument that relation (4.4) holds. Here the basic term is of the order v/c, whereas in (4.2) the magnetic field would only produce a correction of order v/c. From (4.12), we find for the radiative part of the magnetic field

$$\mathbf{B} = (e\mu_0/4\pi x)\,\ddot{\mathbf{x}}'(t') \times \partial t'/\partial \mathbf{x} \tag{4.13}$$

and using (4.11) we arrive at

$$\mathbf{B} = (e\mu_0/4\pi c x^2)\,\mathbf{x} \times \ddot{\mathbf{x}}'(t') \tag{4.14}$$

Now we introduce the acceleration from (4.2) and find

$$\mathbf{B} = \frac{\mu_0}{4\pi c}\frac{e^2}{m_-}\frac{\mathbf{E}_0 \times \mathbf{x}}{x^2}\exp\{i\mathbf{k}_0 \cdot \mathbf{x}'(t') - i\omega_0 t'\} \tag{4.15}$$

In the argument of $\mathbf{x}'$ (but not in the term $\omega_0 t'$) we may omit the third term of (4.11) for t', since according to (4.5) we can also neglect terms of order $\dot{x}'x'/c\lambda\!\!\!^{-}$.

Introducing the abbreviation

$$\mathbf{k} = \mathbf{k}_0 - (\omega_0/c)(\mathbf{x}/x) = \mathbf{k}_0 - |\,\mathbf{k}_0\,|\,\mathbf{x}/x \tag{4.16}$$

and using (4.11), we rewrite (4.15) in the form

$$\mathbf{B} = \frac{\mu_0}{4\pi c}\frac{e^2}{m_-}\frac{\mathbf{E}_0 \times \mathbf{x}}{x^2}\exp\left\{i\mathbf{k} \cdot \mathbf{x}'\left(t - \frac{x}{c}\right) - i\left(t - \frac{x}{c}\right)\omega_0\right\} \tag{4.17}$$

(It should be noted that $t - x/c$ is the argument of $\mathbf{x}'$.)

We see that within our approximation, the amplitude factor shows no first-order correction depending on the coordinate of the electron, but the phase factor does. This is decisive for the interference phenomena.

The wave vector $\mathbf{k}$ is obviously composed of two vectors of the same magnitude, the directions of which include the scattering angle ϑ. The magnitude of this vector is given by

$$|\,\mathbf{k}\,| = 2(\omega_0/c)\sin\vartheta/2 \tag{4.18}$$

Of course, in (4.17) we have to use the path of the test electron $\mathbf{x}'(t)$ as performed under the influence of the electromagnetic wave and the microfield.

For simple motions of the test particle, the content of (4.17) is transparent and easily interpretable. If for instance the particle is free and moving with constant velocity $\mathbf{v}$, then it is easy to see from the exponent of (4.17) that the magnetic field is that of a dipole radiation with a shifted frequency $\omega_0 - \mathbf{k}\cdot\mathbf{v}$ which is interpreted as resulting from two contributions. According to the decomposition (4.16), one contribution is associated with the motion of the test particle with respect to the electromagnetic wave, the other with the motion of the test particle source with respect to the point of observation. Both contributions are Doppler shifts.

If the path is more complicated due to the fluctuating electric micro-field, then we will not only expect frequency shifts but we will have a dispersion. In addition, of course, the various velocities of all particles in our plasma will cause a spread of the frequencies over the whole spectrum.

From the superposition principle of the electromagnetic waves, it follows that the magnetic field produced by all the electrons of our plasma can be described simply by a summation of (4.17) over all particles

$$\mathbf{B} = \frac{\mu_0}{4\pi c}\frac{e^2}{m_-}\frac{\mathbf{E}_0 \times \mathbf{x}}{x^2}\sum_j \exp\left\{i\mathbf{k}\cdot\mathbf{x}_j'\left(t-\frac{x}{c}\right) - i\left(t-\frac{x}{c}\right)\omega_0\right\} \tag{4.19}$$

Now we turn to the frequency discrimination.

In this connection, we refer to (V.2.46), where we have shown that the radiated energy per unit frequency and unit area is given by†

$$\mathbf{S}_\omega = \tilde{\mathbf{E}}_\omega \times \tilde{\mathbf{H}}_\omega{}^* = (1/\mu_0)\,\tilde{\mathbf{E}}_\omega \times \tilde{\mathbf{B}}_\omega{}^* \tag{4.20}$$

or with (4.10) by

$$\mathbf{S}_\omega = (c/\mu_0)\mid \tilde{B}_\omega \mid^2 \mathbf{r}/r \approx (c/\mu_0)\mid \tilde{B}_\omega \mid^2 \mathbf{x}/x \tag{4.21}$$

To reflect the result of a physical experiment we have to take the ensemble average of (4.21), yielding

$$\langle \mathbf{S}_\omega \rangle = (c/\mu_0)(\mathbf{x}/x)\langle \tilde{\mathbf{B}}_\omega \cdot \tilde{\mathbf{B}}_\omega{}^* \rangle \tag{4.22}$$

In view of the general use in the literature we relate the product of the Fourier spectral functions of **B** in (4.22) to the autocorrelation coefficient of **B**. This can be done quite generally if we restrict ourself to the case of a stationary system.

Let $\mathbf{B}(t)$ be a quantity depending on the variables of our statistical system. Then we define the autocorrelation coefficient (in time) as the ensemble average

$$\beta(\tau) := \langle \mathbf{B}(t+\tau)\cdot \mathbf{B}(t)\rangle \tag{4.23}$$

† Observe that we do not combine positive and negative frequencies and therefore have to omit the factor two.

This autocorrelation coefficient does not depend on time t, since we are considering a stationary system. Therefore we may write (4.23) in the form

$$\begin{aligned}\beta(\tau) &= (1/T)\int_{-T/2}^{T/2} dt\,\langle \mathbf{B}(t+\tau)\cdot\mathbf{B}(t)\rangle \\ &= (1/T)\int_{-T/2}^{T/2} dt\,\langle \mathbf{B}(t)\cdot\mathbf{B}(-(\tau-t))\rangle\end{aligned} \tag{4.24}$$

where we consider the limit $T\to\infty$. Now we apply the convolution theorem to find the Fourier transform $\tilde{\beta}(\omega)$ of $\beta(\tau)$. Remembering that when $\tilde{f}(\omega)$ is the Fourier transform of $f(\tau)$, then $\tilde{f}(-\omega)$ is that of $f(-\tau)$, and further that for real functions $f(\tau)$, the relation $\tilde{f}(-\omega) = \tilde{f}^*(\omega)$ holds, we arrive at

$$\tilde{\beta}(\omega) = [(2\pi)^{1/2}/T]\langle\tilde{\mathbf{B}}(\omega)\cdot\tilde{\mathbf{B}}^*(\omega)\rangle \tag{4.25}$$

Applying this result to (4.22), we find

$$\begin{aligned}\langle\bar{\mathbf{S}}_\omega\rangle &:= \langle\mathbf{S}_\omega\rangle/T \\ &= (c/2\pi\mu_0)(\mathbf{x}/x)\int_{-\infty}^{+\infty} d\tau\, e^{i\omega\tau}\langle\mathbf{B}_\mathrm{r}(t+\tau)\cdot\mathbf{B}_\mathrm{r}(t)\rangle\end{aligned} \tag{4.26}$$

where the index r designates the real part.

The quantity $\langle\bar{S}_\omega\rangle$ is the scattered radiation intensity per unit time, unit volume, and per unit frequency at ω. The formula (4.26) is an example of the Wiener–Khinchine formula.

If we define the differential scattering cross section by

$$\sigma_\omega(\omega_0, \Omega) = 2x^2\langle\bar{S}_\omega\rangle/c\varepsilon_0 E_0^{\,2} \tag{4.27}$$

then we find inserting (4.26) and the real part of (4.19)

$$\begin{aligned}\sigma_\omega &= (1/\pi)(e^2/4\pi\varepsilon_0 m\_c^2)^2\,(\mathbf{E}_0\times\mathbf{x}/E_0x)^2 \\ &\quad\times\int d\tau\, e^{i\omega\tau}\left\langle\sum_{j,l}\cos(\mathbf{k}\cdot\mathbf{x}_j'(t+\tau)-\omega_0(t+\tau))\cos(\mathbf{k}\cdot\mathbf{x}_l'(t)-\omega_0 t)\right\rangle\end{aligned} \tag{4.28}$$

(Due to the stationarity, the time shift x/c could be omitted.)

Applying the addition theorem of the trigonometric functions, we find the relation

$$\begin{aligned}&\cos(\mathbf{k}\cdot\mathbf{x}_j'(t+\tau)-\omega_0(t+\tau))\cos(\mathbf{k}\cdot\mathbf{x}_l'(t)-\omega_0 t) \\ &\quad= \tfrac{1}{2}\cos\{\mathbf{k}\cdot(\mathbf{x}_j'(t+\tau)+\mathbf{x}_l'(t))-\omega_0(2t+\tau)\} \\ &\qquad+\tfrac{1}{2}\cos\{\mathbf{k}\cdot(\mathbf{x}_j'(t+\tau)-\mathbf{x}_l'(t))-\omega_0\tau\}\end{aligned} \tag{4.29}$$

The ensemble average renders the first of the two terms on the right-hand side of (4.29) zero, since the probability of a state of the homogeneous system does not change if we add to all particle coordinates the same vector $\boldsymbol{\vartheta}$. We therewith find

$$\sigma_\omega = \frac{1}{2\pi}\left(\frac{e^2}{4\pi\varepsilon_0 m_- c^2}\right)^2 \left(\frac{\mathbf{E}_0 \times \mathbf{x}}{E_0 x}\right)^2$$
$$\times \int d\tau\, e^{i\omega\tau} \left\langle \sum_{j,l} \cos\{\mathbf{k}\cdot(\mathbf{x}_j'(t+\tau) - \mathbf{x}_l'(t)) - \omega_0\tau\}\right\rangle \tag{4.30}$$

This result could be used as a basis of the calculation of the scattered intensity.

Before we start with the evaluation we transform the result to a "density representation," which is common use in the literature and may be advantageous. To this end, we consider the microscopic particle density

$$n_-(\mathbf{x}', t) = \sum_j \delta(\mathbf{x}' - \mathbf{x}_j'(t)) \tag{4.31}$$

where the summation carries over all particles in a box of unit volume. Fourier expansion with respect to the space coordinate yields

$$\tilde{n}_-(\mathbf{k}, t) = (2\pi)^{-3/2} \sum_j \exp[-i\mathbf{k}\cdot\mathbf{x}_j'(t)] \tag{4.32}$$

and with that the relation

$$\langle \tilde{n}_-(\mathbf{k}, t+\tau)\, \tilde{n}_-^*(\mathbf{k}, t)\, e^{+i\omega_0\tau}\rangle$$
$$= (2\pi)^{-3} \sum_{j,l} \langle \cos\{\mathbf{k}\cdot(\mathbf{x}_j'(t+\tau) - \mathbf{x}_l'(t)) - \omega_0\tau\}\rangle$$
$$- i(2\pi)^{-3} \sum_{j,l} \langle \sin\{\mathbf{k}\cdot(\mathbf{x}_j'(t+\tau) - \mathbf{x}_l'(t)) - \omega_0\tau\}\rangle \tag{4.33}$$

With respect to τ, the first term on the right-hand side of (4.33) is an even function, the second term is an odd one. This can be seen by the relation

$$M(t; -\tau) = M(t+\tau; -\tau) = \pm M(t; \tau) \tag{4.34}$$

which follows from the argument of the trigonometric function (4.33), using the independence of the ensemble average with respect to time and interchanging the indices in the sum.

We rewrite (4.30) in the form

$$\sigma_\omega = \frac{(2\pi)^3}{4\pi}\left(\frac{e^2}{4\pi\varepsilon_0 m_- c^2}\right)^2\left(\frac{\mathbf{E}_0 \times \mathbf{x}}{E_0 x}\right)^2$$

$$\times \int d\tau \langle \tilde{n}_-(\mathbf{k}, t+\tau)\, \tilde{n}_-^*(\mathbf{k}, t)\rangle [e^{i(\omega-\omega_0)\tau} + e^{i(\omega+\omega_0)\tau}] \tag{4.35}$$

With the help of the relations (4.23) and (4.25), we may then also use the formulation

$$\sigma_\omega = \left(\frac{e^2}{4\pi\varepsilon_0 m_- c^2}\right)^2\left(\frac{\mathbf{E}_0 \times \mathbf{x}}{E_0 x}\right)^2 \frac{2\pi^2}{T}\{\langle|\,\tilde{\tilde{n}}_-(\mathbf{k}, \omega-\omega_0)|^2\rangle + \langle|\,\tilde{\tilde{n}}_-(\mathbf{k}, \omega_0+\omega)|^2\rangle\} \tag{4.36}$$

which is the desired density representation. Note that the autocorrelation coefficients $\langle|\,\tilde{\tilde{n}}_-(\mathbf{k}, \omega \pm \omega_0)|^2\rangle$ are depending on the wave vector $\mathbf{k} = \mathbf{k}_0 - \omega_0\mathbf{x}/cx$ (see 4.16).

Equation (4.36) shows that for a constant value of n_- there is no scattering, since we have

$$\tilde{\tilde{n}}_-(\mathbf{k}, \omega \pm \omega_0) = (2\pi)^2\, \delta(\mathbf{k})\, \delta(\omega \pm \omega_0) = (2\pi)^2\, \delta(\mathbf{k}_0 - \omega_0\mathbf{x}/cx)\, \delta(\omega \pm \omega_0) \tag{4.37}$$

This means that electromagnetic radiation is emitted only in the direction of the original wave vector and with the frequency of the original wave.

We see that scattering is only due to fluctuations. Density fluctuations with wave vector $\mathbf{k}_\text{f}$ and frequency ω_f contribute—as one can readily see from (4.36) and (4.16)—only to scattering processes which satisfy the condition

$$\mathbf{k}_0 = \mathbf{k}_\text{f} + \omega_0\mathbf{x}/cx, \qquad \omega_0 = |\,\omega \pm \omega_\text{f}\,| \tag{4.38}$$

As in the case of Raman scattering, these two conditions, after multiplication by Planck's constant, may be interpreted as the law for momentum and energy conservation, respectively, in the system of light quanta and plasmons. (Actually the momentum conservation would require ω instead of ω_0 on the right-hand side. This difference is caused by the neglect of terms of order $\dot{x}x'/c\lambda$ in our treatment.)

It is easy to see that for positive frequencies the second term on the right-hand side in the brackets of (4.36) may be neglected, since the fluctuation spectrum does hardly contain contributions for frequencies $\omega_\text{f} > \omega_0$, where according to our assumption $\omega_0 \gg \omega_{p0}$ holds. In general, the density fluctuations have frequencies below or near the plasma frequency. Correspondingly, the first term does not contribute to negative frequencies.

Therefore, to determine the differential scattering cross section, we focus our interest on the calculation of the term

$$(1/T)\langle| \tilde{n}_-(\mathbf{k}_f, \omega_f)|^2\rangle \tag{4.39}$$

Of course we could just as well start from the autocorrelation coefficient presented in (4.35).

4.2. Scattering for a Correlation-Free System

The problem is easily tractable if we neglect correlations. We then assume a straight path and write the Fourier transform of the particle number density according to (4.32)

$$\tilde{n}_-(\mathbf{k}_f, \omega_f) = (2\pi)^{-3/2} \sum_j \exp(-i\mathbf{k}_f \cdot \mathbf{x}_j')(2\pi)^{-1/2} \int dt \exp[i(\omega_f - \mathbf{k}_f \cdot \dot{\mathbf{x}}_j')t] \tag{4.40}$$

or, respectively,

$$\tilde{n}_-(\mathbf{k}_f, \omega_f) = (2\pi)^{-1} \sum_j \exp(-i\mathbf{k}_f \cdot \mathbf{x}_j')\, \delta(\omega_f - \mathbf{k}_f \cdot \dot{\mathbf{x}}_j') \tag{4.41}$$

With this, we find

$$\langle| \tilde{n}_-(\mathbf{k}_f, \omega_f)|^2\rangle = (2\pi)^{-1} \langle n_-\rangle\langle(\delta(\omega_f - \mathbf{k}_f \cdot \mathbf{v}))^2\rangle_{\mathbf{v}} \tag{4.42}$$

taking into account that for a homogeneous system the factor in front of the Dirac function of (4.41) cancels all terms in the product except those with identical indices. Therefore we have the total number density of the electrons $\langle n_-\rangle$ in (4.42).

Observe that the ensemble average on the right-hand side of (4.42) is to be taken only with respect to the velocity space. If we denote the velocity distribution by $f(\mathbf{v})$—normalized to one—then we may evaluate the integral

$$(1/T) \int (\delta(\omega_f - \mathbf{k}_f \cdot \mathbf{v}))^2 f(\mathbf{v})\, d\mathbf{v} = (1/k_f)\, F(\omega_f/k_f) \lim_{\substack{T\to\infty \\ \bar{\omega}\to 0}} \delta(\bar{\omega})/T \tag{4.43}$$

where the function F is defined by

$$F(u) = \int f(\mathbf{v})\, \delta(u - \mathbf{k}_f \cdot \mathbf{v}/k_f)\, d\mathbf{v} \tag{4.44}$$

The value of the limit is, according to (4.24), given by

$$(2\pi)^{-1} \lim_{\substack{T\to\infty \\ \bar{\omega}\to 0}} (1/T) \int_{-T/2}^{+T/2} e^{i\bar{\omega}t}\, dt = 1/2\pi \tag{4.45}$$

We then have

$$(1/T)\langle|\tilde{\tilde{n}}_-(k_f, \omega_f)|^2\rangle = [\langle n_-\rangle/(2\pi)^2][1/k_f]\,F(\omega_f/k_f) \tag{4.46}$$

and after introducing this into (4.36), we find for the differential scattering cross section

$$\sigma_\omega(\omega_0, \Omega) = \frac{1}{2}\left(\frac{e^2}{4\pi\varepsilon_0 m_- c^2}\right)^2 \left(\frac{\mathbf{E}_0 \times \mathbf{x}}{E_0 x}\right)^2 \frac{\langle n_-\rangle}{|\mathbf{k}_0 - \omega_0\mathbf{x}/cx|}$$
$$\times\left\{F\left(\frac{\omega - \omega_0}{|\mathbf{k}_0 - \omega_0\mathbf{x}/cx|}\right) + F\left(\frac{\omega + \omega_0}{|\mathbf{k}_0 - \omega_0\mathbf{x}/cx|}\right)\right\} \tag{4.47}$$

Integration with respect to the frequency ω and using the normalization of F yields the differential scattering cross section of all electrons in the unit volume summed over all frequencies through

$$\sigma(\omega_0, \Omega) = \langle n_-\rangle \left(\frac{e^2}{4\pi\varepsilon_0 m_- c^2}\right)^2 \left(\frac{\mathbf{E}_0 \times \mathbf{x}}{E_0 x}\right)^2 = \langle n_-\rangle \left(\frac{e^2}{4\pi\varepsilon_0 m_- c^2}\right)^2 \sin^2\delta \tag{4.48}$$

where δ is the angle formed by $\mathbf{E}_0$ and $\mathbf{x}$ (not the deflection angle formed by $\mathbf{k}_0$ and $\mathbf{x}$).

Finally, the total scattering cross section of a single electron is found by division by $\langle n_-\rangle$ and integration with respect to the solid angle of the point of observation:

$$\sigma^{\mathrm{T}} = (8\pi/3)(e^2/4\pi\varepsilon_0 m_- c^2)^2 = (8\pi/3)\, r_e^2 \tag{4.49}$$

This is the familiar Thompson scattering cross section, where r_e denotes the classical electron radius.

On the basis of this result, one might be tempted to conclude that the neglect of the particle correlation was not so bad an approximation after all.

4.3. Scattering in a System with Collective Correlations

Unfortunately experimental evidence from the scattering of electromagnetic waves from the ionosphere proves our result of the preceding section to be quite wrong. The Doppler shift of this radiation from the ionosphere appeared not to be that of the electrons but rather that of the ions, although our basic assumption that an electron scatters much more than an ion is absolutely correct.

The discrepancy must be caused by the correlation between the electrons and ions. Qualitatively it can be understood as follows:

Let us consider an electron moving through the electron–ion system. It causes a displacement of the neighboring electrons, whereas the slowly reacting positive ions are not affected. Consequently the electron is accompanied by a positive space charge due to the electron hole. The total system of the electron charge and the positive screening charge caused by the electron hole is neutral. Therefore no scattering will be observed.

On the other hand, a slowly moving ion displaces as well the electrons as the ions in its neighborhood. The shielding is about half due to the ions and half due to the electrons. As a consequence, the scattering of this system amounts to about one-half effective electron charge which moves with the velocity of the ion and shows the specific electron charge.

Of course for fast ions, the shielding by the other ions may be less then half, and therefore fast ions can show more scattering then slow ones. This latter effect explains why the Doppler shift observed from the plasma is not simply given by the Gaussian distribution which one should expect in accord with a Maxwellian distribution of the velocities. Since the plasma is able to support strong collective fluctuations near the electron and ion plasma frequencies, we will also expect scattering peaks near these frequencies.

It should be noted that the above interpretation is valid only in the range

$$\lambda \gg \lambda_D \tag{4.50}$$

since otherwise the polarization cloud as a whole does not react in phase with the test particle, and the simple neutralization argument is not valid.

A rigorous description of the phenomena including correlations is untractable here. Even an approximate consideration using a heuristic model (testparticle approach) is still rather cumbersome (Rosenbluth and Rostoker, 1962; Bernstein, Trehan and Weenink, 1964) and we restrict ourselves to quoting the results.

The test particle concept is applicable to a plasma which is essentially homogeneous and stationary, so that the basic assumption

$$L \gg \lambda_D, \qquad 1/\tau \ll \omega_{p0} \tag{4.51}$$

(L is the characteristic length, and τ the characteristic time) is fulfilled. For the calculation of the autocorrelation coefficient, the influence of the electromagnetic wave produces only second-order effects. Therefore the above condition (4.51) does not infer the requirement $\lambda \gg \lambda_D$.

The test particle model claims that the effective electron density is that due to the point electrons taken to be uncorrelated and moving along straight-line trajectories with constant velocity, plus the contribution of the statistically steady polarization clouds they carry along, plus the contribution of the polarization clouds associated with the ions. The ions are regarded as moving along straight-line trajectories with constant velocity. This model has been justified by various authors (e.g., Rostoker and Rosenbluth, 1960).

The theory of the testparticle model quoted yields an expression for the time–space Fourier component of the density $\tilde{\tilde{n}}_-(\mathbf{k}, \omega)$ which determines the normalized differential scattering cross section

$$\bar{\sigma}_\omega(\omega_0) := \left(\frac{e^2}{4\pi\varepsilon_0 m_- c^2}\,\frac{\mathbf{E}_0 \times \mathbf{x}}{E_0 x}\right)^{-2} \frac{\sigma_\omega(\omega_0\,,\,\Omega)}{4\pi^3} = \lim \frac{1}{2\pi T} \langle |\,\tilde{\tilde{n}}_-(\mathbf{k}, \omega - \omega_0)|^2\rangle \tag{4.52}$$

[In accordance with the statement on p. 306 we could omit the second term on the right-hand side of (4.36.)]

The result of the test particle calculation presents itself in the form

$$\bar{\sigma}_\omega(\omega_0) = -(\langle n_-\rangle/\pi)\, k^2\lambda_{D-}^2 \,\operatorname{Im} \Xi(\omega, k) \tag{4.53}$$

with

$$\Xi(\omega, k) = \frac{1}{\omega}\, \frac{\left(1 - \dfrac{\omega_{\mathrm{p}+}^2}{k^2}\displaystyle\int \frac{[dF_+(u)/du]\, du}{u - (\omega/k) - i0}\right)\left(\dfrac{\omega_{\mathrm{p}-}^2}{k^2}\displaystyle\int \frac{[dF_-(u)/du]\, du}{u - (\omega/k) - i0}\right)}{1 - \dfrac{\omega_{\mathrm{p}+}^2}{k^2}\displaystyle\int \frac{[dF_+(u)/du]\, du}{u - (\omega/k) - i0} - \dfrac{\omega_{\mathrm{p}-}^2}{k^2}\displaystyle\int \frac{[dF_-(u)/du]\, du}{u - (\omega/k) - i0}} \tag{4.54}$$

F_+ and F_- are the one-dimensional ion or, respectively, electron Maxwellian distributions normalized to one. The path of integration is to be extended below the singular point of the integrand.

The qualitative behavior of the scattering intensity can be understood from a discussion of the characteristic properties of the function Ξ given in (4.54). We refer the reader in this respect to the literature (Bernstein, Trehan, Weenink, 1964).

It is convenient to fix the parameters $\omega_{\mathrm{p}-}$ and m_-/m_+, and to vary the scattering angle ϑ which corresponds to what might be conveniently done in an experiment. Variation of this scattering angle is reflected in a variation of the quantity.

$$k = 2(\omega_0/c) \sin \vartheta/2 \tag{4.55}$$

Let us first contemplate the range $k^2\lambda_{D-}^2 \gg 1$. Under these circumstances, the spectrum is essentially that due to the Doppler shift of uncorrelated electrons. (Observe that we violate the condition (4.50).)

A typical example is shown schematically by curve (a) in Figure VI.5 for the case $kv_{th-} = 3\omega_{p-}$. Since kv_{th-} is about the halfwidth (ω_h) of the Doppler distribution, we expect $\omega_h \approx 3\omega_{p-}$.

As we decrease the scattering angle ϑ, we also decrease the value k. In Figure VI.5 curve b refers to the value $kv_{th-} = \omega_{p-}$. This curve is still dominated by the electrons. It shows a narrower halfwidth and a greater height. The center of the line begins to flatten as it increases in height and narrows. The area under the curve giving the total scattering intensity is the same as for $kv_{th-} = 3\omega_{p-}$

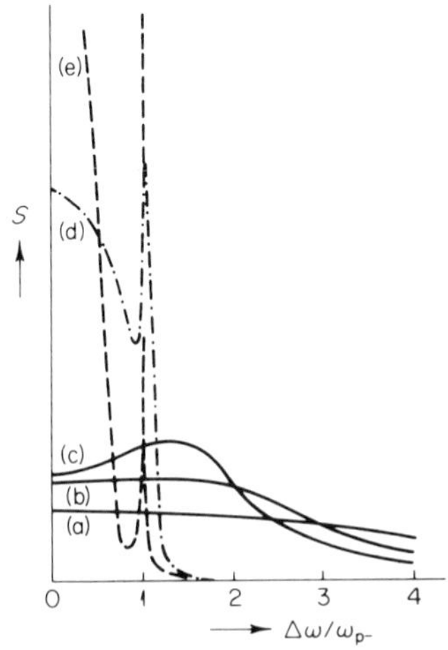

FIGURE VI.5. Schematic variation with $k\lambda_{D-}$ of the scattered intensity $S(k, \Delta\omega)$ versus frequency shift. [See text for explanation of curves (a)–(e)].

Curves c and d in Figure VI.5 for the values $kv_{th-}/\omega_{p-} = 0.3, 0.2$, respectively, show the characteristic behavior with still further decrease in k. First a maximum develops to the right of the electron plasma frequency. This peak near the plasma frequency narrows and rises approaching the plasma frequency from the right. At the same time, a second peak develops in the origin. The two peaks effectively detach. The one at the origin becomes higher and narrower with a halfwidth corresponding to the ion thermal spread of the order kv_{th+}. The area under the central peak is approximately identical with that for the curve $kv_{th-}/\omega_{p-} = 3$. The detached peak close to the electron plasma frequency narrows and rises rapidly with decreasing kv_{th-}.

If we choose still smaller values in k, the central peak narrows as should be expected, since it is dominated by ion Doppler broadening. It also develops a maximum away from the center as demonstrated by curve a in Figure VI.6. This offside maximum which curve a exhibits is connected with the presence of ion oscillations in the plasma.

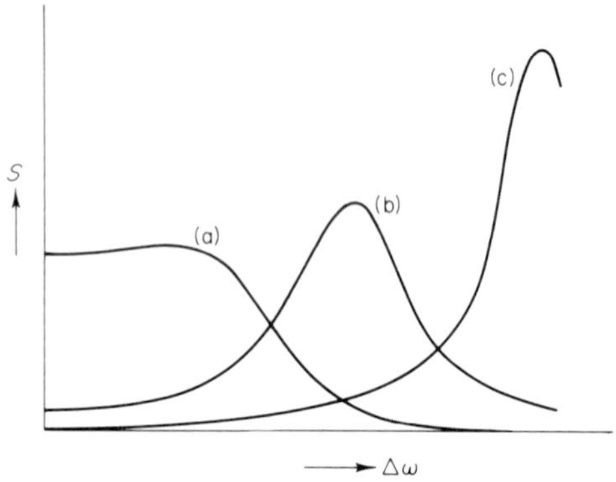

FIGURE VI.6. The scattered intensity $S(k, \Delta\omega)$ for different electron and ion temperatures: (a) $\Theta_- = \Theta_+$; (b) $\Theta_- = 5\Theta_+$; (c) $\Theta_- = 10\Theta_+$ (schematically).

The above interpretation can be corroborated by the study of plasmas which have an electron temperature different from that of the ion temperature. These results are shown in curves b and c of Figure VI.6 for temperature ratios of values five and ten. As we see, with increasing electron temperature the maximum narrows and moves to the wings. Again collective ion fluctuations are responsible. According to our calculations in Chapter III "large amplitude fluctuations" should be expected due to ion acoustic waves with the wave velocity $\omega_f/k = (\Theta_-/m_+)^{1/2}$.

The importance of such collective responses of the plasma is also recognizable in the result shown in Figure VI.7. Here the scattering profiles have been plotted schematically for situations where the electron and ion velocity distributions show a small separation in their center. That means the electrons drift with respect to the ions. The most striking phenomenon is now that the curves are asymmetric. One of the peaks increases rapidly with increasing drift velocity. We recall that according to our previous calculations (see Chapter III) electrons drifting through the ions tend to produce collective instabilities. In the curves of Figure VI.7, we have not reached this unstable point yet, because for the point of instability the peak on the right-hand sides of the curves would grow infinitely high.

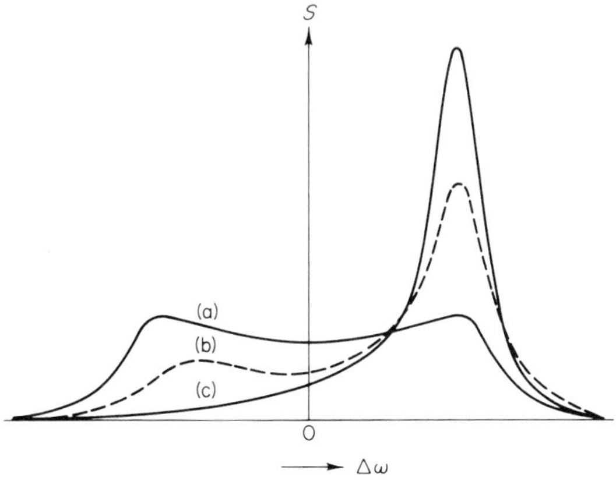

FIGURE VI.7. The scattered intensity for electrons with a drift motion u relative to the ions. (a) $u = 0$; (b) $u = \frac{1}{2}v_{\text{th}-}$; (c) $u = v_{\text{th}-}$.

Summary

We summarize the results for light scattering in the test particle approximation of a plasma with correlations:

1. For large values $k\lambda_{D-}$ the scattering profile can be described by the Doppler shift of uncorrelated electrons. This is not surprising since the situation violates the condition $\lambda \gg \lambda_{D-}$.

2. With decreasing $k\lambda_{D-}$ the line profile develops two maxima, one to the right of the electron plasma frequency, the other in the center. As $k\lambda_{D-}$ further decreases, the two peaks detach. The peak near the center shows a halfwidth corresponding to the Doppler shift of the thermal motion of the ions. The formation of these peaks is due to two effects. One is that the excitation of collective phenomena near the plasma frequency causes strong scattering near this frequency. The second is that with decreasing $k\lambda_{D-}$ the condition $\lambda \gg \lambda_{D-}$ becomes satisfied, and accordingly the correlation with the ions grows important. Therefore, in agreement with the qualitative discussion given above, the main profile constricts to that determined by the Doppler shift of the average ion velocity.

3. The peak near the electron plasma frequency shows a halfwidth corresponding to the Landau damping. Within the frame of the test particle treatment, this result is not surprising, since we only consider collective effects. Actually, however, the halfwidth of this peak should be expected essentially larger since in the range $k\lambda_{D-} \ll 1$ collisional phenomena contribute more than Landau damping (see p. 124).

4. For very small values of $k\lambda_{D-}$ the central peak again develops an offside maximum. This is due to collective ion phenomena similar to those responsible for the peak near ω_{p-}.

5. The offside maximum moves toward the wings and narrows if the electron temperature rises beyond the ion temperature. This can be understood by remembering that collective phenomena will be excited in an ion plasma for phase velocities $\omega_f/k = (\Theta_-/m_+)^{1/2}$.

6. If there is relative drift motion of electrons and ions then, the line profile is asymmetric, and one of the peaks shows a rapid increase. Again this phenomenon is due to the excitation of collective ion waves. As discussed in Chapter III, the relative electron–ion drift can even produce an instability.

4.4. Application of the Fluctuation–Dissipation Theorem

As we have seen, the incorporation of particle correlation effects in the light-scattering theory is not yet fully satisfactory. We also understood on p. 306 that the scattering phenomenon depends basically on the fluctuation spectrum of the system.

It consequently suggests itself to relate the scattering theory to the calculations of the conductivity via the fluctuation–dissipation theorem (see Chapter I, Section 5)

$$\Theta \frac{k^2}{\omega^2} \operatorname{Re} \left(\frac{\sigma(k, \omega)}{\varepsilon(k, \omega)} \right) = \lim \frac{(2\pi)^4 e^2}{TV} \langle | \tilde{\tilde{n}}(\mathbf{k}, \omega)|^2 \rangle_0 \tag{4.56}$$

However, such an attempt appears useful only if one applies theories of conductivity that are more refined with respect to the ion dynamics than those presented here.

5. Radiation Transport

5.1. The Transport Equation

Having calculated the macroscopic coefficients governing the behavior of electromagnetic waves in a fully ionized system, we turn to study the propagation of electromagnetic energy in a system which can be characterized by a complex dielectric constant $\varepsilon_\omega(\mathbf{r})$ which may exhibit a weak space dependence. More precisely, we require that the basic condition of geometrical optics

$$\lambda/L \ll 1 \tag{5.1}$$

hold, where L is the characteristic length for the variation of $\varepsilon_\omega(\mathbf{r})$.

From Maxwell's equations for continuous media

$$\begin{aligned} \nabla \cdot \tilde{\mathbf{H}}_\omega &= 0; & \nabla \cdot (\varepsilon_\omega \tilde{\mathbf{E}}_\omega) &= 0 \\ \nabla \times \tilde{\mathbf{H}}_\omega &= -i\omega\varepsilon_0\varepsilon_\omega \tilde{\mathbf{E}}_\omega\,; & \nabla \times \tilde{\mathbf{E}}_\omega &= i\omega\mu_0 \tilde{\mathbf{H}}_\omega \end{aligned} \tag{5.2}$$

we find the wave equation

$$\Delta \tilde{\mathbf{E}}_\omega + (\omega^2/c^2)\, \varepsilon_\omega(\mathbf{r})\, \tilde{\mathbf{E}}_\omega + \nabla(\nabla\varepsilon_\omega/\varepsilon_\omega \cdot \tilde{\mathbf{E}}_\omega) = 0 \tag{5.3}$$

Within the approximation of geometrical optics, we may (locally) consider a plane wave with the direction $\mathbf{s}$ and drop the third term in (5.3), applying (5.1), and find

$$\frac{\partial^2 \tilde{\mathbf{E}}_\omega}{\partial s^2} + \frac{\omega^2}{c^2}\, \varepsilon_\omega \tilde{\mathbf{E}}_\omega = 0 \tag{5.4}$$

From the WKB method, we find the approximate solution

$$\tilde{\mathbf{E}}_\omega(s) = \mathbf{C}_\omega \exp\left(i(\omega/c) \int \varepsilon_\omega^{1/2}\, ds\right) \tag{5.5}$$

and correspondingly

$$\tilde{\mathbf{H}}_\omega(s) = \mathbf{D}_\omega \exp\left(i(\omega/c) \int \varepsilon_\omega^{1/2}\, ds\right) \tag{5.6}$$

which yields, for the modulus of the Poynting vector,

$$S = (2\pi)^{-1} \int \tilde{E}_{\omega_1} \tilde{H}^*_{\omega_2} \exp\left[-i(\omega_1 - \omega_2)t\right] d\omega_1\, d\omega_2 \tag{5.7}$$

Time averaging over a period large in comparison to all oscillation periods contributing to (5.7), gives

$$\langle S \rangle = \pi^{-1} \operatorname{Re} \left\{ \iint_{\omega_1, \omega_2 > 0} \tilde{E}_{\omega_1} \tilde{H}^*_{\omega_2} \exp\left[-i(\omega_1 - \omega_2)t\right] d\omega_1\, d\omega_2 \right\} \tag{5.8}$$

Substituting (5.5) and (5.6), we find

$$\begin{aligned} \langle S \rangle = \pi^{-1} \operatorname{Re} \Bigg\{ \iint_{\omega_1, \omega_2 > 0} & C_{\omega_1} D^*_{\omega_2} \exp\left(-\tfrac{1}{2} \int (\kappa_{\omega_1} + \kappa_{\omega_2})\, ds\right) \\ & \times \exp\left[i\left(c^{-1} \int (\omega_1 n_{\omega_1} - \omega_2 n_{\omega_2})\, ds - (\omega_1 - \omega_2)t\right)\right] d\omega_1\, d\omega_2 \Bigg\} \end{aligned} \tag{5.9}$$

where we have introduced the refractive index n_ω and the spatial absorption coefficient κ_ω by

$$n_\omega = \mathrm{Re}(\varepsilon_\omega^{1/2}); \qquad \kappa_\omega = (2\omega/c)\,\mathrm{Im}(\varepsilon_\omega^{1/2}) \approx (\mu_0/\varepsilon_0)^{1/2}\,[\sigma(\omega)/n_\omega] \tag{5.10}$$

Now we study a wave packet with all frequencies ω_1, ω_2 close to ω. Then we can use $\kappa_{\omega_1} + \kappa_{\omega_2} \approx 2\kappa_\omega$, and

$$\frac{n_{\omega_1}\omega_1}{c} - \frac{n_{\omega_2}\omega_2}{c} = (\omega_1 - \omega_2)\frac{d}{d\omega}\left(\frac{n_\omega \omega}{c}\right) = \frac{\omega_1 - \omega_2}{u_g(\omega)} \tag{5.11}$$

where $u_g(\omega)$ is the local group velocity at ω. So we arrive at

$$\langle S\rangle_\omega = \pi^{-1} \exp\left[-\int \kappa_\omega\, ds\right] \mathrm{Re} \int_{\omega_1,\omega_2>0} C_{\omega_1} D^*_{\omega_2} \times \exp\left[i(\omega_1 - \omega_2)\left(\int \frac{ds}{u_g(\omega)} - t\right)\right] d\omega_1\, d\omega_2 \tag{5.12}$$

Thus $\langle S\rangle_\omega$ satisfies the differential equation

$$\partial\langle S\rangle_\omega/\partial s = -\kappa_\omega \langle S\rangle_\omega - [1/u_g(\omega)]\,\partial\langle S\rangle_\omega/\partial t \tag{5.13}$$

and describes the energy transport by wave packets of given direction.

Superposition of such wave packets with directions close to **s** results in a wave pencil of small angular extension $\Delta\Omega$. The radiation intensity is defined by

$$I_\omega = \langle S\rangle_\omega/\Delta\Omega\,\Delta\omega \tag{5.14}$$

Using Snellius' law in the form

$$n_\omega^{\;2}\,\Delta\Omega = \text{constant} \tag{5.15}$$

and using (5.13) and (5.14), we find the **radiation transport equation**†

$$\frac{1}{u_g}\frac{\partial}{\partial t}\frac{I_\omega}{n_\omega^{\;2}} + \frac{\partial}{\partial s}\frac{I_\omega}{n_\omega^{\;2}} = -\kappa_\omega \frac{I_\omega}{n_\omega^{\;2}} \tag{5.16}$$

Obviously (5.16) accounts on the right-hand side for absorption but not for emission and scattering. This is not surprising since the Maxwell

† We draw attention to the fact that the approximation of geometrical optics used in our derivation is actually not sufficient to decide whether n_ω^{-2} is to be placed in or outside the differential operator $\partial/\partial s$. A more accurate derivation has to account for the term with $\tilde{\mathbf{E}}_\omega \cdot \nabla\varepsilon_\omega$ in (5.3). This requires to distinguish the polarization directions and to apply the WKB method to a higher order in λ/L.

equations (5.2) neither account for thermal emission nor for fluctuations. Including emission, we have instead of (5.16)

$$\frac{1}{u_g}\frac{\partial}{\partial t}\frac{I_\omega}{n_\omega^2} + \frac{\partial}{\partial s}\frac{I_\omega}{n_\omega^2} = \frac{j_\omega - \kappa_\omega I_\omega}{n_\omega^2} \tag{5.17}$$

where j_ω is the emission coefficient (defined as the power generated per unit volume, time, frequency at ω and unit solid angle, flowing in the direction $\mathbf{s}$ of the ray). Here we can substitute the emission coefficients calculated in Chapter V. Close to equilibrium we may apply Kirchhoff's law

$$j_\omega = n_\omega^2 \kappa_\omega B_\omega \tag{5.18}$$

where B_ω is the blackbody radiation given by Planck's formula

$$B_\omega = \frac{2\hbar\omega^3}{(2\pi)^3 c^2}\frac{1}{e^{\hbar\omega/\Theta} - 1} \approx \frac{2\omega^2}{(2\pi)^3 c^2}\,\Theta \tag{5.19}$$

(Equation 5.19 accounts for both directions of polarization.)

If we include scattering, too, we can—in analogy to the Boltzmann equation—represent these scattering processes on the right-hand side of (5.15) by an additional "collision term" in the form

$$(1/n_\omega^2)(\delta I_\omega/\delta s)_{\mathrm{sc}} = (1/n_\omega^2)\int (I_{\omega'}(\Omega') - I_\omega(\Omega))\,\sigma_{\omega'}(\omega, \Omega' - \Omega)\,d\omega'\,d\Omega' \tag{5.20}$$

where $\sigma_{\omega'}(\omega, \Omega' - \Omega)$ is the differential scattering cross section as defined in (4.27).

5.2. Some Remarks on the Boundary Conditions

In addition to the transfer equation discussed in the preceding we need boundary conditions. They present a rather involved problem. Let us assume that the plasma is restricted to a volume V. Then the radiation distribution outside of this volume V provides the boundary conditions for the solution of the transport equation.

Usually the plasma develops a sheath close to its boundary. In this sheath the plasma variables experience variations so strong that the application of the transfer equation is precluded. This sheath comprises our boundary problem which—in its general form—can not be solved yet.

There are, however, two cases which are tractable and of principal interest: The case of a discontinuous surface and the case of perpendicular incidence. If the surface sheath can be represented by a discontinuity, then trivially our problem is solved by Fresnel's formula for the reflection of radiation. If we have perpendicular incidence and a finite sheath ($0 < x < a$), describable by a complex dielectric constant, equation (5.3) reads

$$d^2E/dx^2 + k^2(x)E = 0; \qquad k^2 = (\omega^2/c^2)\,\varepsilon(x) \tag{5.21}$$

We further assume $k(x) \equiv k_0$ for $x \leqslant 0$ and $k(x) \equiv k_1$ for $x \geqslant a$ which provides the solutions

$$\begin{aligned} E &= E_0(e^{ik_0x} + \rho e^{-ik_0x}), \qquad & x \leqslant 0 \\ E &= E_0\tau e^{ik_1x}, & x \geqslant a \end{aligned} \tag{5.22}$$

where ρ and τ are the coefficients of reflection and transmission of the amplitudes (not of the intensities!).

To solve (5.21) in the interval $[0, a]$ we make the usual phase ansatz in the slightly modified form (Geffcken, 1941)

$$E(x) = A \exp\left[i \int (k(x) - \phi(x))\,dx\right] \tag{5.23}$$

resulting in the Riccati differential equation

$$\phi' + 2ik\phi = k' + i\phi^2 \tag{5.24}$$

We expect that $\phi(x)$ will prove to be a small quantity in most cases, since with $\phi(x) \equiv 0$ the ansatz (5.23) transforms into the WKB solution. Differentiating (5.22) and (5.23) logarithmically and equating the results at $x = a$ yields the initial condition

$$\phi(a) = 0 \tag{5.25}$$

and at $x = 0$ the reflection coefficient

$$\rho = \phi(0)/[2k_0 - \phi(0)] \tag{5.26}$$

So our problem is solved if we can determine $\phi(0)$ from (5.24) and (5.25).

In general an explicit solution of the Riccati differential equation is not possible, and we therefore construct the equivalent integral equation which can be solved by iteration.

Using the method of the variation of the constant and (5.25), we find

$$\phi(x) = \exp\left[-2i \int_0^x k(\xi)\,d\xi\right] \int_a^x \exp\left[2i \int_0^z k(\xi)\,d\xi\right] (k'(z) + i\phi^2(z))\,dz \tag{5.27}$$

For small values of ϕ, we may consider linear contributions only. Then we have

$$\phi(0) = -\int_0^a \exp\left[2i \int_0^z k(\xi)\, d\xi\right] k'(z)\, dz \tag{5.28}$$

and the reflection coefficient is simply given by $\rho = \phi(0)/2k_0$. Since EE^* describes the intensity of our wave, the usual reflection coefficient of the intensities is given by $R = \rho\rho^*$. This yields the final result

$$\begin{aligned} R &= \tfrac{1}{4}\left| (1/k_0) \int_0^a \exp\left[2i \int_0^x k(\xi)\, d\xi\right] k'(x)\, dx \right|^2 \\ &= \tfrac{1}{4}\left| \varepsilon^{-1/2}(0) \int_0^a \exp\left[2i(\omega/c) \int_0^x \varepsilon^{1/2}(\xi)\, d\xi\right] (d\varepsilon^{1/2}/dx)\, dx \right|^2 \end{aligned} \tag{5.29}$$

In the limit $a \to 0$, we recognize the linearized Fresnel formula

$$\lim_{a\to 0} R = \left| \frac{\varepsilon^{1/2}(a) - \varepsilon^{1/2}(0)}{2\varepsilon^{1/2}(0)} \right|^2 \tag{5.30}$$

In (5.29), we may interpret

$$\bar{\rho}(x) := -\tfrac{1}{2}\varepsilon^{-1/2}(0)\, d\varepsilon^{1/2}(x)/dx \tag{5.31}$$

as a "differential reflection coefficient" corresponding to the Fresnel formula. This differential coefficient is multiplied by the factor

$$\exp\left[2i(\omega/c) \int_0^x \varepsilon^{1/2}(\xi)\, d\xi\right] \tag{5.32}$$

which accounts for absorption (via Im $\varepsilon^{1/2}$) and interference phenomena in our sheath.

Of course, our linear approach cannot account for the effect of reflection in the sheath on the reflection coefficient R: All multiple reflection phenomena require the application of (5.27) in full.

In general, a numerical evaluation of the Riccati differential equation is more convenient. Using the simplified relations

$$\operatorname{Re} \varepsilon \approx 1 - \frac{\omega_{\mathrm{p-}}^2}{\omega^2}; \qquad \operatorname{Im} \varepsilon \approx \frac{1}{8}\left(\frac{2}{\pi}\right)^{1/2} \frac{\omega_{\mathrm{p-}}^3}{\omega^3} \frac{\ln \Lambda}{\Lambda} \approx \frac{1}{\Lambda} \frac{\omega_{\mathrm{p-}}^3}{\omega^3} \tag{5.33}$$

we have performed these calculations for a sheath with a density variation

$$n_-(x) = \tfrac{1}{2}n_1[1 - \cos(\pi x/a)] \tag{5.34}$$

In Figures VI.8 and VI.9, we give the reflection coefficient for $\omega_{\mathrm{p}1}/\omega = 0.2$, and respectively, 0.8; and $\Lambda_1 = 10$ and, respectively, ∞.

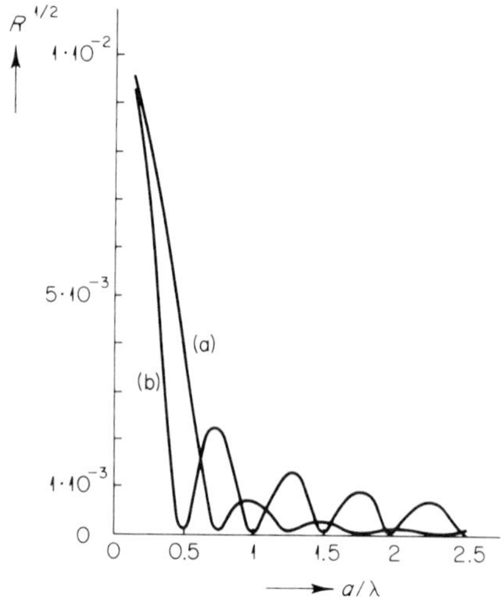

FIGURE VI.8. Reflection coefficient R as a function of the plasma sheath thickness a measured in wavelengths λ: $\omega_{p1} = 0.2\omega$, $\Lambda_1 = \infty$. (a) Density variation according to (5.34); and (b) linear density variation.

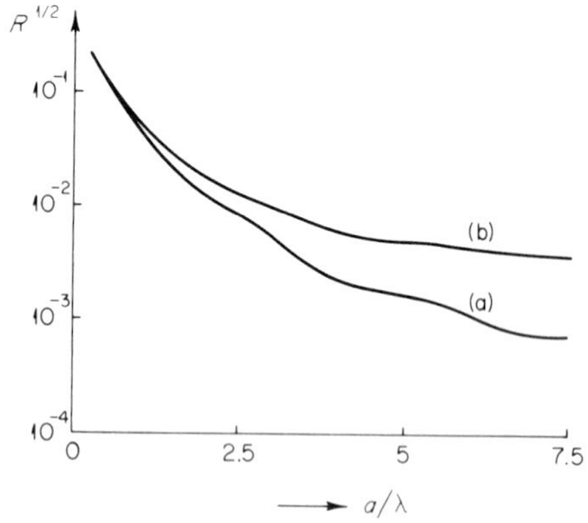

FIGURE VI.9. Reflection coefficient R as a function of the plasma sheath thickness a measured in wavelengths λ: $\omega_{p1} = 0.8\omega$. (a) $\Lambda_1 = \infty$; (b) $\Lambda_1 = 10$.

REFERENCES AND SUPPLEMENTARY READING

GENERAL REFERENCES

Bekefi, G. (1966). "Radiation Processes in Plasmas." Wiley, New York.

Dawson, J. (1964). Lecture Notes. Princeton Univ., Princeton, New Jersey.

Dawson, J. (1968). "Advances in Plasma Physics," Vol. I (A. Simon and W. B. Thompson, eds.). Wiley (Interscience), New York.

SECTION 1

Dupree, T. H. (1963). *Phys. Fluids* **6**, 1714.
Dupree, T. H. (1964). *Phys. Fluids* **7**, 923.

SECTION 2

Birmingham, T., Dawson, J., and Oberman, C. (1965). *Phys. Fluids* **8**, 297.
Bunemann, O. (1959). *Phys. Rev.* **115**, 503.
Fried, B. D. (1959). *Phys. Fluids* **2**, 337
Furth, H. P. (1963). *Phys. Fluids* **6**, 48.
Kahn, F. D. (1962). *J. Fluid Mech.* **14**, 321.
Weibel, E. S. (1959). *Phys. Rev. Lett.* **2**, 83.

SECTION 3

Bernstein, I. B., and Trehan, S. K. (1961). *Nucl. Fusion* **1**, 3.
Dawson, J., and Oberman, C. (1962). *Phys. Fluids* **5**, 517.
Dawson, J., and Oberman, C. (1963). *Phys. Fluids* **6**, 394.
Margenau, H. (1958). *Phys. Rev.* **109**, 6.

SECTION 4

Bernstein, I. B., Trehan, S. K., and Weenink, M. P. H. (1964). *Nucl. Fusion* **4**, 61.
Ramsden, S. A. (1970). *In* "Physics of Hot Plasmas" (B. J. Rye and J. C. Taylor, eds.). Oliver and Boyd, Edinburgh.
Rosenbluth, M. N., and Rostoker, N. (1962). *Phys. Fluids* **5**, 776.
Rostoker, N., and Rosenbluth, M. N. (1960). *Phys. Fluids* **3**, 1.

SECTION 5

Geffcken, W. (1941). *Ann. Phys. (Leipzig)* **40**, 835.
Kofink, W. (1947). *Ann. Phys. (Leipzig)* **1**, 119.
Rawer, K. (1939). *Ann. Phys. (Leipzig)* **35**, 385.

Appendix Characteristic Quantities of the Plasma

There are a number of characteristic quantities which occur frequently in the theory of the plasma. Knowledge of the dependence of these parameters on the data of the system is of interest not only for numerical evaluation but also for theoretical approximation procedures. We therefore list in the following some of these characteristic data, their mutual relationships, and numerical values.

In view of the applications it seems advised to present the characteristic quantities in a detailed form. To this end we study the characteristic times and characteristic lengths separately for the interaction with each component of the system and for a specified velocity of the test particle.

The system is neutral and consists of two components in equilibrium at a given temperature Θ with a density $n_+ = n_- = n$. Their individuals are singly charged and characterized by their masses m_μ. Index-free quantities refer to the particle kind under consideration.

Clearly, the characteristic quantities for the interaction of the test particle with a component must be expressible in terms of n, Θ alone. That means their interdependence can be described by one suitably chosen parameter, e.g., Λ.

Characteristic Frequencies and Lengths

In the correlation-free treatment of the Coulomb system the only characteristic frequency is the **plasma oscillation frequency**:

$$\omega_p = (4\pi ne^2/m)^{1/2} \tag{1}$$

The corresponding characteristic length for the correlationless plasma is the **Debye length**:

$$\lambda_{\mathrm{D}} = (\Theta/4\pi ne^2)^{1/2} = v_{\mathrm{th}\,\mu}/3^{1/2}\omega_{\mathrm{p}\mu} \tag{2}$$

If weak correlations are taken into account then as a further characteristic frequency the **collision frequency**

$$\nu_{\mathrm{c}} = \frac{3}{2^{1/2}}\,\omega_{\mathrm{p}}\,(\ln \Lambda)/\Lambda \tag{3}$$

occurs.

The characteristic length corresponding to this collision frequency is the **mean-free path**

$$\lambda_{\mathrm{c}} = (2\Theta/m)^{1/2}/\nu_{\mathrm{c}} \tag{4}$$

Apart from this mean-free path, the **classical interaction radius**

$$r_w = e^2/\Theta \tag{5}$$

and the **average particle distance**

$$r_0 = (4\pi n/3)^{-1/3} \tag{6}$$

are useful quantities for the description of a plasma with correlations.

The collision frequency and the mean-free path given above characterize the relaxation phenomena caused by correlation effects in a rather crude way. More precisely, one has to distinguish an infinite number of relaxation times depending on the physical processes under consideration. We shall give in the following a general formulation of these relaxation times for arbitrary cylinder symmetrical combinations of velocity moments. In the evaluation we select three specific quantities which have been generally used in the literature (Montgomery and Tidman, 1964):

the **slowing down frequency**

$$\nu_{\mathrm{s}}(u) = -\frac{1}{u}\frac{\partial}{\partial t}u_{\parallel} \tag{7}$$

the **deflection frequency**

$$\nu_{\mathrm{d}}(u) = \frac{1}{u^2}\frac{\partial}{\partial t}u_{\perp}{}^2 \tag{8}$$

and the **energy exchange frequency**

$$\nu_{\epsilon}(u) = \frac{1}{u^2 - u_0{}^2}\frac{\partial}{\partial t}u^2 \tag{9}$$

where $\mathbf{u}$ designates the individual velocity of the test particle and u_0 the velocity which yields $\partial u^2/\partial t = 0$.

In addition to the quantities for a test particle with a given velocity we also present the average times and lengths defined by

$$\langle \tau_i \rangle = \int 4\pi u^2 \, du \, \bar{f}_M(u) \frac{1}{\nu_i(u)} \tag{10}$$

$$\langle \lambda_i \rangle = \int 4\pi u^2 \, du \, \bar{f}_M(u) \frac{u}{\nu_i(u)} \tag{11}$$

where $\bar{f}_M(u)$ is the Maxwellian distribution. We prefer here to give the average times since the average frequencies tend to infinity for $m/m_\mu \to 0$. This effect is—due to the divergence of the Coulomb cross section—caused by the contribution of very low velocities.

General Formalism

We expand the velocity distribution $\bar{f}(\mathbf{v})$ of the test particle in terms of spherical harmonics

$$\bar{f}(\mathbf{v}) = \sum_l {}^{(l)}f(v) \, P_l(\cos\theta), \qquad \cos\theta = (\mathbf{v} \cdot \mathbf{u})/vu \tag{12}$$

and consider an arbitrary function $s(\mathbf{v})$ also expanded in the form

$$s(\mathbf{v}) = \sum_l s_l(v) \, P_l(\cos\theta) \tag{13}$$

Then the time variation of the velocity average of this function due to collisions with particles of kind μ is given by

$$\left(\frac{\partial\langle s\rangle}{\partial t}\right)_\mu = 4\pi \sum_l \frac{1}{2l+1} \int_0^\infty v^2 s_l(v) \, I({}^{(l)}f \mid \bar{f}_\mu) \, dv \tag{14}$$

In our case of a cylinder symmetrical disturbance the Fokker–Planck collision term has the form (Rosenbluth, Mc Donald, and Judd 1957)

$$I({}^{(l)}f \mid \bar{f}_\mu) = l(l+1)\, \alpha_\mu \, {}^{(l)}f + \frac{m}{m_\mu v^2} \frac{d}{dv} (\beta_\mu \, {}^{(l)}f) + \frac{1}{v^2} \frac{d}{dv} \left(\gamma_\mu \frac{d\, {}^{(l)}f}{dv}\right) \tag{15}$$

where the quantities α_μ, β_μ, and γ_μ are given by

$$\alpha_\mu = \frac{8\pi^2 e^4}{3m^2 v^3} \ln \Lambda \left\{ \frac{1}{v^2} \int_0^v v_1^4 \, \bar{f}_\mu(v_1) \, dv_1 \right.$$
$$\left. - 2v \int_v^\infty v_1 \bar{f}_\mu(v_1) \, dv_1 - 3 \int_0^v v_1^2 \, \bar{f}_\mu(v_1) \, dv_1 \right\} \tag{16}$$

$$\beta_\mu = \frac{4\pi e^2}{m} \ln \Lambda \int_0^v v_1^2 \bar{f}_\mu(v_1)\, dv_1 \tag{17}$$

$$\gamma_\mu = \frac{1}{3}\left\{\frac{4\pi e^2}{m}\right\}^2 \ln \Lambda \left\{\frac{1}{v}\int_0^v v_1^2 \bar{f}_\mu(v_1)\, dv_1 + v^2 \int_v^\infty v_1 \bar{f}_\mu(v_1)\, dv_1\right\} \tag{18}$$

Integration by parts yields from (14) using (15)

$$\left(\frac{\partial\langle s\rangle}{\partial t}\right)_\mu = 4\pi \sum_l \frac{1}{2l+1}\left\{l(l+1)\int_0^\infty \alpha_\mu v^2\, {}^{(l)}f\, s_l\, dv - \frac{m}{m_\mu}\int_0^\infty \beta_\mu \frac{ds_l}{dv}\, {}^{(l)}f\, dv \right.$$
$$\left. + \int_0^\infty \frac{d}{dv}\left(\gamma_\mu \frac{ds}{dv}\right) {}^{(l)}f\, dv\right\} \tag{19}$$

To study the test particle problem we assume the distribution

$$\bar{f}(\mathbf{v}) = \delta(\mathbf{v} - \mathbf{u}) \tag{20}$$

with the expansion coefficients

$${}^{(l)}f(v) = \frac{2l+1}{4\pi v^2}\, \delta(v - u) \tag{21}$$

which yields from (19) the results

$$\left(\frac{\partial}{\partial t}\langle s(\mathbf{u})\rangle\right)_\mu = \sum_l \left[l(l+1)\, \alpha_\mu\, s_l(u) - \frac{m}{m_\mu}\frac{\beta_\mu}{u^2}\frac{ds_l}{du} + \frac{1}{u^2}\frac{d}{du}\left(\gamma_\mu \frac{ds_l}{du}\right)\right] \tag{22}$$

or with the definitions

$$\begin{aligned}
\nu_{1\mu}(u) &= -2\alpha_\mu(u) = \frac{\nu_c}{y^3}\left[\operatorname{erf}(x)\left(1 - \frac{1}{2x^2}\right) + \frac{1}{\pi^{1/2}x}\, e^{-x^2}\right] \\
\nu_{2\mu}(u) &= \frac{\beta_\mu(u)}{u^3} = \frac{\nu_c}{y^3}\left[\operatorname{erf}(x) - \frac{2x}{\pi^{1/2}}\, e^{-x^2}\right] \\
\nu_{3\mu}(u) &= -\frac{1}{u^3}\frac{d\gamma_\mu}{du} = \frac{\nu_c}{y^3}\left[\frac{\operatorname{erf}(x)}{2x^2} - \frac{2}{\pi^{1/2}}\, e^{-x^2}\left(x + \frac{1}{2x}\right)\right] \\
\nu_{4\mu}(u) &= \frac{\gamma_\mu(u)}{u^4} = \frac{\nu_c}{y^3}\frac{1}{2x}\left[\operatorname{erf}(x) - \frac{2x}{\pi^{1/2}}\, e^{-x^2}\right]
\end{aligned} \tag{23}$$

with $y^2 = mu^2/2\Theta$ and $x^2 = m_\mu y^2/m$, we find

$$\left(\frac{\partial}{\partial t}\langle s(u)\rangle\right)_\mu = -\sum_l \left[\frac{l(l+1)}{2}\nu_{1\mu} + \frac{m}{m_\mu}\nu_{2\mu}\, u\frac{d}{du} + \nu_{3\mu}\, u\frac{d}{du} - \nu_{4\mu}\, u^2\frac{d^2}{du^2}\right] s_l(u) \tag{24}$$

From this equation we recognize that all relaxation frequencies may be composed of four characteristic frequencies, where the frequency $\nu_{1\mu}$ essentially describes the effect of scattering, $\nu_{2\mu}$ the effect of the change in modulus, and $\nu_{3\mu}$, $\nu_{4\mu}$ account for corrections arising from the motion of the collision partners.

According to the definitions (7), (8), (9) the relaxation frequencies for slowing down, deflection, and energy exchange can be written in the form

$$\begin{aligned} \nu_{s\mu} &= \nu_{1\mu} + \frac{m}{m_\mu}\nu_{2\mu} + \nu_{3\mu} \\ \nu_{d\mu} &= 2\nu_{1\mu} \\ \nu_{\epsilon\mu} &= \frac{2u^2}{u^2 - u_0^2}\left(\frac{m}{m_\mu}\nu_{2\mu} + \nu_{3\mu} - \nu_{4\mu}\right) \end{aligned} \tag{25}$$

Results

The most detailed information is contained in Figures A.1–A.4. Figures A.1 and A.2 present the characteristic relaxation frequencies and lengths for a test electron in an electron component or a test ion in an ion component. Figures A.3 and A.4 provide the corresponding quantities for electrons scattered in a proton component. The corresponding results for the relaxation of an ion in an electron component have not been given since these effects are negligible in comparison to the ion–ion results.

The dependencies exhibited in Figures A.1–A.4 are involved in the range $u \lesssim v_{\text{th}\mu}$. For velocities $u > v_{\text{th}\mu}$, however, they can be readily interpreted on the basis of the u^{-4} variation of the Coulomb cross section.

It may be expected that the relaxation frequencies for the simultaneous interaction with two components can be found from the addition of the results in Figures A.1 and A.2 or A.3 and A.4, respectively. This is correct for slowing down and deflection effects. As far as the energy relaxation is concerned, one has to observe that the quantity u_0 in the definition (9) is not the average thermal velocity and in particular depending on the mass ratio m/m_μ. This, in principle, excludes the addition of the results for the energy exchange. On the other hand the energy exchange time given in Figure A.1 is sufficient since the corresponding contribution of the ion component is negligible due to the high mass ratio anyhow.

The dependence of the velocity u_0 on the mass ratio m/m_μ is demonstrated in Figure A.6. It is one of the noteworthy results that the velocity u_0 for which no average energy exchange occurs is always smaller than the average thermal velocity from which it deviates remarkably even in the case $m = m_\mu$.

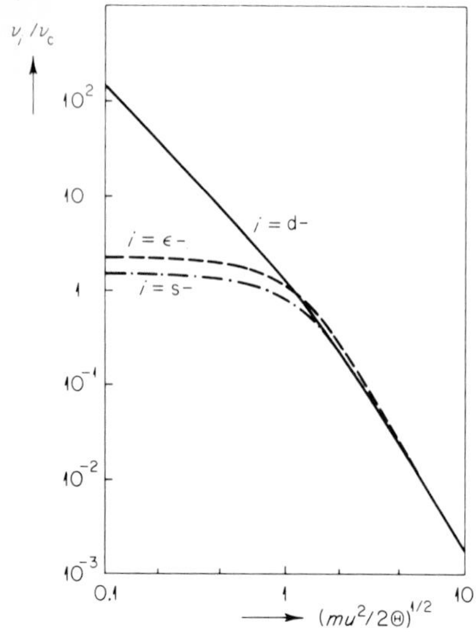

FIGURE A.1. Slowing down, deflection, and energy exchange frequencies for electrons of velocity u colliding with thermal electrons of temperature Θ.

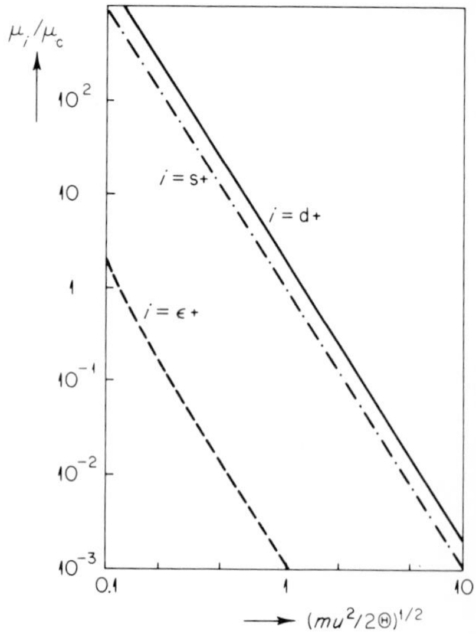

FIGURE A.2. Slowing down, deflection, and energy exchange frequencies for electrons of velocity u colliding with thermal H ions of temperature Θ.

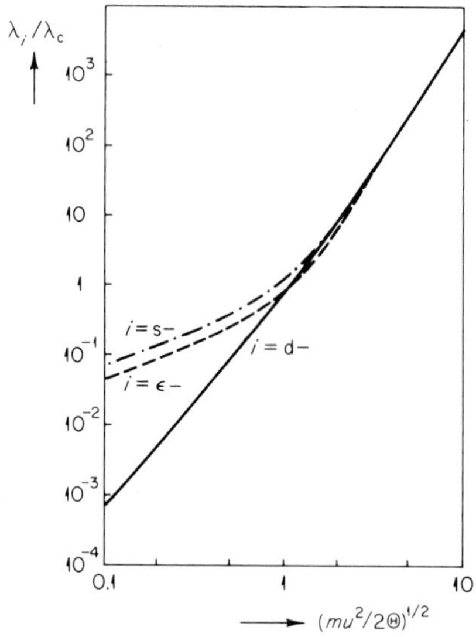

FIGURE A.3. Slowing down, deflection, and energy exchange lengths for electrons of velocity u colliding with thermal electrons of temperature Θ.

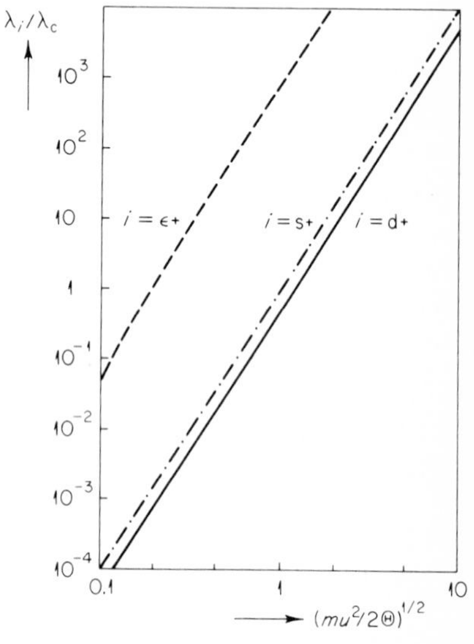

FIGURE A.4. Slowing down, deflection, and energy exchange lengths for electrons of velocity u colliding with thermal H ions of temperature Θ.

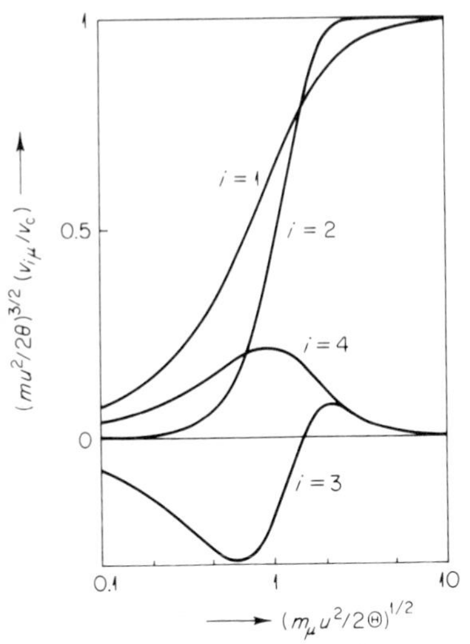

FIGURE A.5. Dependence of the frequencies $\nu_{1\mu}, \ldots, \nu_{4\mu}$ according to (A.23) and (A.24) on the test particle velocity u.

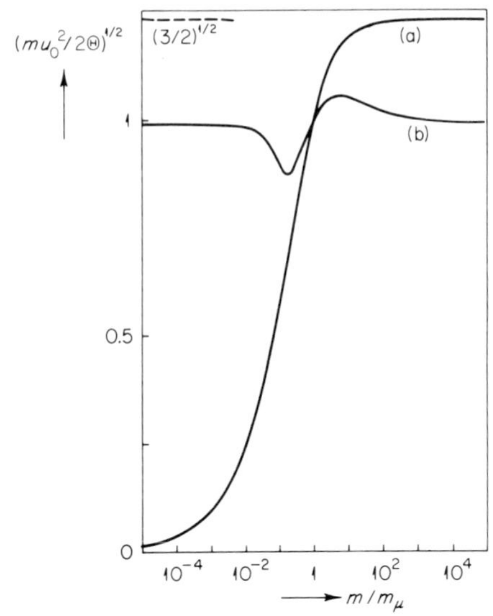

FIGURE A.6. Dependence of the test particle velocity u_0 yielding $\partial u^2/\partial t = 0$ on the mass ratio m/m_μ (a) collisions with partners of mass m only; (b) collisions with equal number of partners of masses m and m_μ.

To facilitate the calculation of relaxation frequencies for a physical phenomenon different from those presented in Figures A.1–A.4 we show in Figure A.5 the four frequencies $\nu_{1\mu}, \ldots, \nu_{4\mu}$ as a function of the test particle velocity and the mass ratio.

TABLE A1

RELATIONS BETWEEN CHARACTERISTIC LENGTHS

		r_w	r_0	λ_D	λ_c
r_w	=	1	$3^{1/3}\Lambda^{-2/3}$	$3\Lambda^{-1}$	$\frac{9}{2}\frac{\ln\Lambda}{\Lambda^2}$
r_0	=	$\frac{1}{3^{1/3}}\Lambda^{2/3}$	1	$3^{2/3}\Lambda^{-1/3}$	$\frac{3^{5/3}}{2}\frac{\ln\Lambda}{\Lambda^{4/3}}$
λ_D	=	$\frac{1}{3}\Lambda$	$\frac{1}{3^{2/3}}\Lambda^{1/3}$	1	$\frac{3}{2}\frac{\ln\Lambda}{\Lambda}$
λ_c	=	$\frac{2}{9}\frac{\Lambda^2}{\ln\Lambda}$	$\frac{2}{3^{5/3}}\frac{\Lambda^{4/3}}{\ln\Lambda}$	$\frac{2}{3}\frac{\Lambda}{\ln\Lambda}$	1

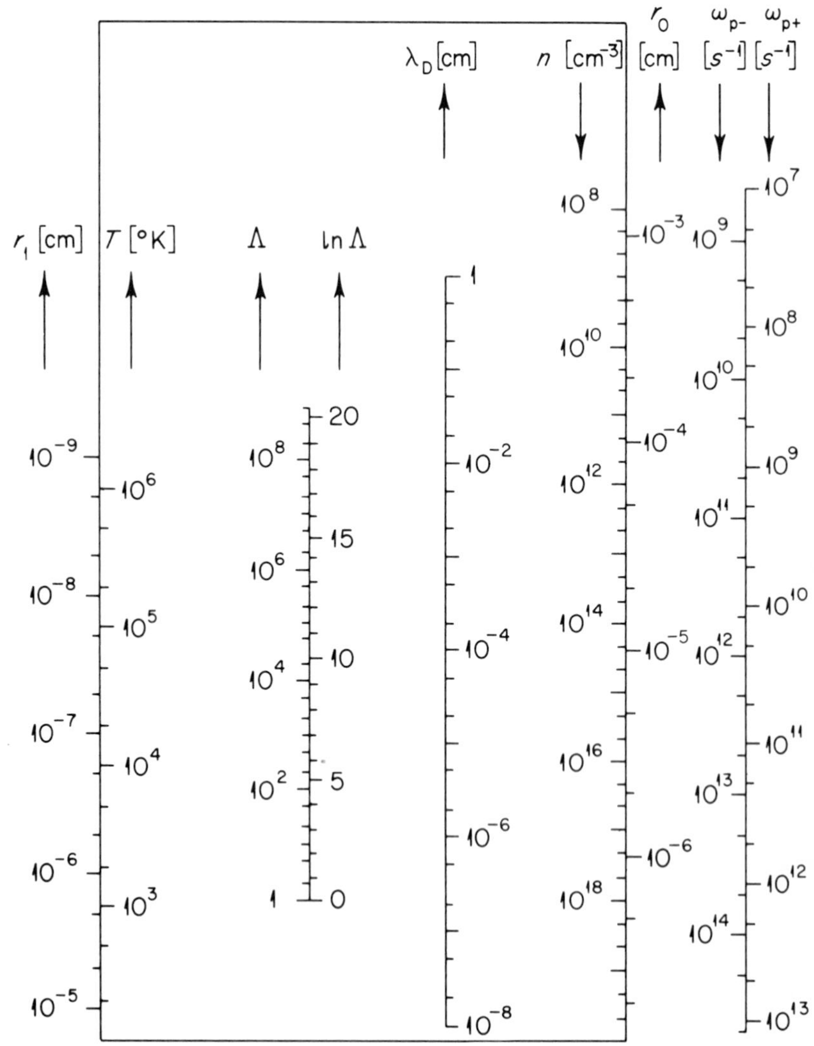

FIGURE A.7. The above figure demonstrates the magnitude of some of the characteristic data. Within the frame we have a nomogram for the calculation of λ_D, Λ, $\ln \Lambda$ and consequently λ_c. The scales outside the frame relate directly r_w to $T = \Theta/\kappa_B$ resp. r_0, ω_{p-} and ω_{p+} to n.

TABLE A2

AVERAGE DATA FOR ELECTRON COLLISIONS

Collision partners		$i = \mathrm{s}$	$i = \mathrm{d}$	$i = \epsilon$
Electrons	$\langle\lambda_i\rangle/\lambda_c$	2.45	2.31	1.84
	$\langle\tau_i\rangle\ \nu_c$	1.70	1.45	1.24
Protons	$\langle\lambda_i\rangle/\lambda_c$	3.75	1.88	$3.44 \cdot 10^3$
	$\langle\tau_i\rangle\ \nu_c$	2.26	1.13	$2.07 \cdot 10^3$
Electrons and protons	$\langle\lambda_i\rangle/\lambda_c$	1.43	1.03	1.84
	$\langle\tau_i\rangle\ \nu_c$	0.91	0.63	1.24

The average quantities are given in Table A2 and in the nomographs of Figure A.7.

REFERENCES

Montgomery D. C. and Tidman, D. A. (1964). "Plasma Kinetic Theory," Mc Graw-Hill, New York.

Rosenbluth, M. N., McDonald, W. M. and Judd, D. L. (1957). *Phys. Rev.* **107**, 1.

Author Index

Subject Index

E

S

T

V